The Ethics of "Geoengineering" the Global Climate

In the face of limited time and escalating impacts, some scientists and politicians are talking about attempting "grand technological interventions" into the Earth's basic physical and biological systems ("geoengineering") to combat global warming. Early ideas include spraying particles into the stratosphere to block some incoming sunlight, or "enhancing" natural biological systems to withdraw carbon dioxide from the atmosphere at a higher rate. Such technologies are highly speculative and scientific development of them has barely begun.

Nevertheless, it is widely recognized that geoengineering raises critical questions about who will control planetary interventions, and what responsibilities they will have. Central to these questions are issues of justice and political legitimacy. For instance, while some claim that climate risks are so severe that geoengineering must be attempted, others insist that the current global order is so unjust that interventions are highly likely to be illegitimate and exacerbate injustice. Such concerns are rarely discussed in the policy arena in any depth, or with academic rigor. Hence, this book gathers contributions from leading voices and rising stars in political philosophy to respond. It is essential reading for anyone puzzled about how geoengineering might promote or thwart the ends of justice in a dramatically changing world.

The chapters in this book were originally published in the journals *Ethics, Policy & Environment* and *Critical Review of International Social and Political Philosophy.*

Stephen M. Gardiner is Professor of Philosophy at the University of Washington, Seattle, and is author of *A Perfect Moral Storm: The Ethical Challenge of Climate Change* and *Debating Climate Ethics*, as well as many articles on climate justice and the ethics of geoengineering.

Catriona McKinnon is Professor of Political Theory at the University of Exeter, author of *Climate Change and Future Justice* and numerous articles on climate ethics and justice.

Augustin Fragnière is a trained philosopher and environmental scientist, who has published on climate ethics, geoengineering and sustainability theory.

The Ethics of "Geoengineering" the Global Climate

Justice, Legitimacy and Governance

Edited by
Stephen M. Gardiner, Catriona McKinnon and Augustin Fragnière

Routledge
Taylor & Francis Group
LONDON AND NEW YORK

First published 2021
by Routledge
2 Park Square, Milton Park, Abingdon, Oxon, OX14 4RN

and by Routledge
52 Vanderbilt Avenue, New York, NY 10017

Routledge is an imprint of the Taylor & Francis Group, an informa business

British Library Cataloguing in Publication Data
A catalogue record for this book is available from the British Library

ISBN13: 978-0-367-50154-9

Typeset in Myriad Pro
by Newgen Publishing UK

Publisher's Note
The publisher accepts responsibility for any inconsistencies that may have arisen during the conversion of this book from journal articles to book chapters, namely the inclusion of journal terminology.

Contents

Citation Information

The chapters in this book were originally published in different issues of *Ethics, Policy & Environment* and *Critical Review of International Social and Political Philosophy*. When citing this material, please use the original page numbering or DOI for each article, as follows:

Chapter 1

The Tollgate Principles for the Governance of Geoengineering: Moving Beyond the Oxford Principles to an Ethically More Robust Approach
Stephen M. Gardiner and Augustin Fragnière
Ethics, Policy & Environment, volume 21, issue 2 (2018), pp. 143–174

Chapter 2

Climate Change, Climate Engineering, and the 'Global Poor': What Does Justice Require?
Marion Hourdequin
Ethics, Policy & Environment, volume 21, issue 3 (2018), pp. 270–288

Chapter 3

Indigeneity in Geoengineering Discourses: Some Considerations
Kyle Powys Whyte
Ethics, Policy & Environment, volume 21, issue 3 (2018), pp. 289–307

Chapter 4

Recognitional Justice, Climate Engineering, and the Care Approach
Christopher Preston and Wylie Carr
Ethics, Policy & Environment, volume 21, issue 3 (2018), pp. 308–323

Chapter 5

Institutional Legitimacy and Geoengineering Governance
Daniel Edward Callies
Ethics, Policy & Environment, volume 21, issue 3 (2018), pp. 324–340

Chapter 6

Legitimacy and Non-Domination in Solar Radiation Management Research
Patrick Taylor Smith
Ethics, Policy & Environment, volume 21, issue 3 (2018), pp. 341–361

Chapter 7

Toward Legitimate Governance of Solar Geoengineering Research: A Role for Sub-State Actors
Sikina Jinnah, Simon Nicholson and Jane Flegal
Ethics, Policy & Environment, volume 21, issue 3 (2018), pp. 362–381

Chapter 8

Fighting risk with risk: solar radiation management, regulatory drift, and minimal justice
Jonathan Wolff
Critical Review of International Social and Political Philosophy, DOI: 10.1080/13698230.2020.1694214

Chapter 9

The Panglossian politics of the geoclique
Catriona McKinnon
Critical Review of International Social and Political Philosophy, DOI: 10.1080/13698230.2020.1694216

Chapter 10

Democratic authority to geoengineer
Holly Lawford-Smith
Critical Review of International Social and Political Philosophy, DOI: 10.1080/13698230.2020.1694219

Chapter 11

A mission-driven research program on solar geoengineering could promote justice and legitimacy
David R. Morrow
Critical Review of International Social and Political Philosophy, DOI: 10.1080/13698230.2020.1694220

Chapter 12

Geoengineering the climate and ethical challenges: what we can learn from moral emotions and art
Sabine Roeser, Behnam Taebi and Neelke Doorn
Critical Review of International Social and Political Philosophy, DOI: 10.1080/13698230.2020.1694225

Notes on Contributors

Wylie Carr, US Fish and Wildlife Service, Atlanta, GA, USA.

Daniel Edward Callies, Institute for Practical Ethics, University of California, San Diego, CA, USA.

Neelke Doorn, Ethics and Philosophy of Technology Section, Department of Values, Technology and Innovation, Faculty of Technology, Policy and Management, Delft University of Technology, Delft, Netherlands.

Jane Flegal, Environment Program, William and Flora Hewlett Foundation, Menlo Park, CA, USA.

Augustin Fragnière, Institute of Geography and Sustainability, University of Lausanne, Lausanne, Switzerland.

Stephen M. Gardiner, Program on Ethics, Department of Philosophy, University of Washington, Seattle, WA, USA.

Marion Hourdequin, Department of Philosophy, Colorado College, Colorado Springs, CO, USA.

Sikina Jinnah, Departments of Politics and Environmental Studies, UC Santa Cruz, Santa Cruz, CA, USA.

Catriona McKinnon, Department of Politics, University of Exeter, Exeter, UK.

Holly Lawford-Smith, School of Historical and Philosophical Studies, University of Melbourne, Melbourne, Australia.

David R. Morrow, Forum for Climate Engineering Assessment, American University, Washington, DC, USA and Institute for Philosophy and Public Policy, George Mason University, Fairfax, VA, USA.

Simon Nicholson, School of International Service, American University, Washington, DC, USA.

Christopher Preston, University of Montana, Missoula, MT, USA.

Sabine Roeser, Ethics and Philosophy of Technology Section, Department of Values, Technology and Innovation, Faculty of Technology, Policy and Management, Delft University of Technology, Delft, Netherlands.

Patrick Taylor Smith, Department of Philosophy, University of Twente, Enschede, Netherlands.

Behnam Taebi, Ethics and Philosophy of Technology Section, Department of Values, Technology and Innovation, Faculty of Technology, Policy and Management, Delft University of Technology, Delft, Netherlands.

Kyle Powys Whyte, Department of Philosophy, Department of Community Sustainability, Michigan State University, East Lansing, MI, USA.

Jonathan Wolff, Blavatnik School of Government, University of Oxford, Oxford, UK.

Introduction: Geoengineering, Political Legitimacy and Justice

Stephen M. Gardiner, Catriona McKinnon and Augustin Fragnière

Geoengineering is commonly defined as "the deliberate large-scale manipulation of the planetary environment to counteract anthropogenic climate change" (The Royal Society 2009). Technologies which might be deployed to attempt geoengineering are either speculative or only in the very early stages of development. Those currently being considered fall into two main camps. Carbon Dioxide Removal (CDR) techniques aim to decrease climate change by *withdrawing* significant amounts of carbon dioxide, the main anthropogenic greenhouse gas, from the atmosphere. Proposed methods include direct air capture, ocean fertilization, enhanced weathering and large-scale afforestation. Solar Radiation Management (SRM) technologies endeavour to *offset* the warming effects of rising greenhouse concentrations by decreasing the amount of energy the Earth's surface receives from the Sun, usually through increasing the planet's albedo (i.e. its reflectivity). In the early literature, the techniques attracting most attention are spraying sulfate particles in the stratosphere (stratospheric sulfate injection or "SSI") and marine cloud brightening ("MCB"). Other possibilities include making deserts, oceans and roofs more reflective.

It is sensible to ask why we should pay any attention to these speculative technologies. On first encountering the idea of geoengineering, many people dismiss it as a bizarre distraction from the urgent business of mitigation and adaptation. Still, in our view, there are at least three significant developments that suggest that we (philosophers, political theorists and the general public) should be taking the emerging debate about geoengineering seriously. First, early moves toward geoengineering are already afoot and have some momentum. In particular, limited geoengineering research is already happening in a few places around the world, and the idea has started to be taken seriously by some policymakers. Second, the push for geoengineering is happening in part because it is becoming widely accepted that the current international approach to climate policy is inadequate for meeting the stated goal of "avoiding dangerous anthropogenic interference" in the climate system by staying below 2°C, or preferably 1.5°C (e.g. Rogelj et al 2019). As a result, geoengineering is being promoted by some as "Plan B". Finally, at least some kinds of geoengineering are quickly becoming *implicitly* accepted in some policy communities, and without much wider discussion. For example, most of the Intergovernmental Panel on Climate Change (IPCC) scenarios for staying below 2°C *already assume* the rollout of carbon dioxide removal technologies on a substantial scale. Given this, there is reason for concern about the trajectory of other forms of geoengineering, such as SRM and specifically stratospheric sulfate injection. We comment on each of these developments below.

Already Afoot?

Despite their speculative nature, interest in geoengineering technologies is growing rapidly in science and policy circles. For several decades, pursuit of such technologies was largely dismissed as irrelevant, and indeed counterproductive, to serious progress on climate change. However, some now advocate for geoengineering techniques as possible "tools in the toolkit" for future climate action. This "partial mainstreaming" began in 2006 when Paul Crutzen, a prominent atmospheric chemist, argued that serious research was needed because of ongoing political inertia on reducing global emissions (Crutzen 2006). Crutzen's focus was on stratospheric sulfate injection. He stressed that SSI was "by far not the best solution" (212) and that "the very best would be if emissions of the greenhouse gases could be reduced so much that the stratospheric sulfur release experiment would not need to take place" (217). Still, he also insisted that "currently, this looks like a pious wish" (217).

Crutzen's intervention was controversial in the science community. Nevertheless, in the intervening decade or so, discussion of geoengineering has developed at a considerable pace. In particular, a number of reports from scientific academies (e.g. Royal Society 2009; National Research Council 2015) and policy institutes (e.g. Bipartisan Policy Center's Task Force on Climate Remediation 2011; Chhetri *et al.* 2018) have emerged.[1] Notably, such reports typically remain cautious (even sceptical) about climate engineering, emphasizing the priority of mainstream mitigation strategies, but also suggest that additional research is needed. Meanwhile, a significant and growing academic literature has developed on geoengineering science, policy, ethics and governance. Notably, the zeitgeist – the general intellectual climate within the area – with respect to SRM research seems to be that the most that is justified right now is self-regulation, perhaps using guiding principles and codes of conduct, such as the Oxford Principles (The Royal Society 2009; Rayner *et al.* 2013; Hubert *et al.* 2016). Relatedly, many reports on governance to date advocate an "allowed zone" for research (Cicerone 2006; Morrow 2017). Still, such views are not universal (cf. Gardiner 2011a, 2013; Hamilton 2013; Smith 2019).

Beyond the academic community, geoengineering is also starting to catch the attention of climate diplomats and legislators, especially in the US. For example, the head of the IPCC, Hoesung Lee, has said that the Panel should be examining geoengineering – including its governance – very seriously (Goldberg 2016). Moreover, in 2016 the US Senate appropriations committee passed a spending bill that included funding to support geoengineering research (Nuccitelli 2016) and allocations have recently been made to the National Oceanic and Atmospheric Administration (NOAA) for research on SRM specifically (Fialka 2020). These developments sit alongside reports of concerns expressed by major figures in the SRM research community that research is "under provided", and that governance is needed to push it forward.

1 In the interests of full disclosure, McKinnon was one of the authors of the Forum for Climate Engineering Assessment report (Chhetri *et al.* 2018), while Gardiner participated in the Bipartisan Policy Center group but resigned before their report was released (cf. Romm 2011).

A Context of Failure

Soon we may come to think of the present time as a relatively comfortable transition period, during which discussions of geoengineering were no longer dismissed as completely fanciful or taboo, but nevertheless remained relatively contained and largely academic. However, that period may now be drawing to a close. Mainly this is because over the last decade the failures of the international response to climate change have turned from deeply disturbing to outright alarming. Global emissions are now up by more than forty percent since the early 1990s, when the international community committed itself to avoiding "dangerous anthropogenic interference" with the climate systems. In addition, recent political developments mean that the prospects for meeting mainstream climate targets currently look bleak. In particular, the scientific community has stated recently that at this point a *transformative political effort* is required to meet the aspirational international goal of limiting global temperature rise to the 1.5°C agreed in Paris (2015), which many believe is necessary to prevent severe harms, especially to vulnerable populations (IPCC 2018). Using the frame of the carbon budget, to have a 50/50 chance of staying below 1.5°C, we have less than 12 years left of current emissions. Moreover, even if current commitments under the Paris Agreement were met – which looks highly optimistic at this point – they would fall substantially short of fulfilling this aim.

One upshot of the ongoing failure of international climate policy is that geoengineering technologies are increasingly being promoted as a necessity in the fight against climate change, often as central to a new "Plan B". Moreover, this kind of argument is being taken increasingly seriously at the highest levels (Fialka 2020; for robust criticism of this framing, see Fragnière and Gardiner 2016).

Geoengineering in IPCC Scenarios

Strikingly, most of the pathways forward considered by the IPCC *already assume* significant use of one kind of geoengineering – carbon dioxide removal – even though these technologies remain largely speculative, especially for deployment at scale. The IPCC states: "All pathways that limit global warming to 1.5°C with limited or no overshoot project the use of carbon dioxide removal (CDR) on the order of 100–1000 GtCO2 over the 21st century". Moreover, they add that even this may not be enough: "reversing warming after an overshoot of 0.2°C or larger during this century would require upscaling and deployment of CDR at rates and volumes that might not be achievable given considerable implementation challenges". Given this, some geoengineering advocates argue that we have already entered the realm where scientists and policymakers should push hard to develop the necessary technology for SRM as well as CDR, including by conducting field experiments. For example, at least two outdoor experiments are currently in preparation, by groups at Harvard University (for SSI) and the University of Washington (for MCB).

Given these three developments, we think it is urgent that normative reflection be undertaken now to provide us with guidelines and policy recommendations. It is widely held that the ethical and governance challenges of geoengineering are at least as daunting as the technical challenges, and perhaps more so (Royal Society 2009; Gardiner 2011b; Keith 2013; McKinnon 2019). Reasons for this include that, although geoengineering technologies might provide some benefits in the fight against climate change, they may also bring about significant harm, result in serious disparities and create new risks. In addition, such technologies pose difficult questions about who should control them, what principles should govern their choices and what responsibilities those engaged in geoengineering would have to humanity and the rest of nature.

The academic literature provides several proposals of general principles for geoengineering governance (e.g. cf. Jamieson 1996; Rayner *et al.* 2013; Gardiner and Fragnière 2018). The aim of this special issue is to move forward the normative discussion of *political legitimacy* and *justice* specifically. Our starting assumption is that it would be unwise to proceed with particular technologies and develop governance systems without at the same time addressing hard questions about justice and legitimacy. One reason for this is that clarifying the central normative questions is key to understanding obstacles to, parameters for, and constraints on research, policy and governance. In particular, *it matters for policy if some kinds of geoengineering turn out to be easier from the point of view of justice and legitimacy than others*. Suppose, for instance, that reasonably legitimate forms of stratospheric sulfate injection turn out to be much more difficult to achieve than comparable forms of carbon dioxide removal (e.g. because SSI encourages concentrations of power that make political legitimacy harder). Or suppose that some ways of doing SRM (e.g. marine cloud brightening) are easier to calibrate and so have better prospects for justice than SSI. In our view, such normative lessons would be important for geoengineering research, policy and governance. Hence, any actual pursuit of a turn to geoengineering ought not proceed without tackling the normative questions head on.

This book draws on papers initially presented at two conferences. The first – *Geoengineering: Political Legitimacy and Justice* – took place at the University of Washington in Seattle, US, in November 2017, and was sponsored by the National Science Foundation (grant number 1549983). The second – *Geoengineering, Justice, and Legitimacy* – took place at the Leverhulme Programme on Climate Justice at the University of Reading, UK, in September 2018, and was supported both by the National Science Foundation (grant number 1549983) and the Leverhulme Trust (grant number DS-2014-002). These conferences brought together scholars from different disciplines (philosophy, political science, law, atmospheric sciences) to consider how a just and legitimate governance system for geoengineering technologies should be designed. The chapters in this book were originally published in the journals *Ethics, Policy & the Environment* and *Critical Review of International Social and Political Philosophy*.

Outline of chapters

In Chapter 1, Stephen Gardiner and Augustin Fragnière propose a new set of governance principles for geoengineering, based on the paradigm case of stratospheric sulfate injection. Gardiner and Fragnière's principles – the Tollgate Principles – emerge through an engagement with an early, well-known and influential proposal for the governance of geoengineering research, the Oxford Principles. With a close focus on the values of justice and legitimacy, Gardiner and Fragnière argue that the Oxford Principles are too narrow, overly procedural and lack sufficiently robust ethical content. By contrast, their own Tollgate Principles aim to be much more realistic about the context of geoengineering, to be much less complacent about its objectives and to sharpen the ethical focus of conversations about how to govern research and development as it progresses.

The next three chapters focus on the need to take a broad approach to justice in assessing geoengineering policies. In Chapter 2, Marion Hourdequin assesses an argument offered by the prominent geoengineering scientist David Keith and his group at Harvard that we have a moral duty to the world's poor to pursue research and development of SRM, and especially SSI (Horton and Keith 2016). Finding this approach wanting, Hourdequin argues that a consequentialist and distributive conception of justice is too narrow a perspective on which to ground such a duty. Instead, she proposes a multidimensional account of climate justice that encompasses distributive, procedural and recognitional considerations. Such an approach is necessary in part because, she argues, "the current concentration of power in relation to SRM research and development is untenable" and encourages "paternalism, parochialism, and expertise imperialism". By contrast, a multidimensional approach to climate justice does a better job of taking seriously the interests and the participation of "the global poor" in geoengineering decisions.

Chapters 3 and 4 echo this theme. In Chapter 3, Kyle Powys Whyte provides insights into problems of multidimensional justice by drawing on the historical experiences of indigenous peoples. Whyte shows how intricately tied questions of climate and geoengineering justice are with other forms of domination and vulnerabilities, especially for those who have been confronting radical environmental change for a very long time because of forced displacements.

In Chapter 4, Christopher Preston and Wylie Carr develop a care-based approach to assessing the justice of geoengineering schemes. Contrary to the dominant distributive paradigm, their approach focuses on recognitional justice, and elements such as context, relationships, power and vulnerability. They illustrate their method with case studies conducted in Kenya, the Solomon Islands and the North American Arctic.

The next three chapters deal with the regulation of geoengineering research and deployment, in particular in the case of SRM, either through the lens of legitimacy or by reference to risks of domination. In Chapter 5, Daniel Callies notes that even though most people involved in the geoengineering debate agree that legitimate governance is necessary if we want to move forward with research and development, there is still little discussion among political philosophers about what would constitute legitimacy in this case. To fill this gap, Callies proposes a set of normative principles for the legitimacy of geoengineering that emerges from a particular conception of legitimacy developed out of Allen Buchanan's influential approach to institutional legitimacy.

In Chapter 6, Patrick Taylor Smith investigates the risk of domination entailed by SRM research. Building on republican theories of justice as non-domination, he argues that even indoor SRM research can alter power relations between countries and that a governance regime should not rely only on western scientists constraining themselves. Smith also maintains that transparency and information-sharing is not enough to avoid potential domination of developed countries over developing countries. Strikingly, he concludes that one way to respond to this worry would be for the developed world to facilitate geoengineering capacity-building *in developing countries*.

Chapter 7 investigates questions of legitimacy and governance with respect to SRM research in the context of proposed outdoor experiments in the United States. Sikina Jinnah, Simon Nicholson and Jane Flegal argue that public engagement is a necessary condition for any legitimate governance regime, but that it is not sufficient. Building on orchestration theories of governance, they contend that a few crucial US states have an important role to play in geoengineering governance, and propose the creation of state-level advisory commissions to oversee SRM research.

Chapters 8 and 9 engage with the ethics of risk as it applies in the context of geoengineering, and how risk can be navigated in ways that serve justice. In Chapter 8, Jonathan Wolff focuses on the common idea that there are "risk-risk tradeoffs" involved in making decisions about research and any possible deployment of SSI. Specifically, it is often observed that geoengineering is typically proposed as a way to minimize the risks of severe climate impacts; but SSI technologies create risks of their own. Wolff's chapter reflects on the different structures of risky situations, and claims that decision-making about geoengineering is a case of moral hazard (albeit not pure moral hazard). He argues that, knowing this, decision-makers should impose regulation on the adoption of new technologies such as geoengineering. This creates a new problem of regulatory capture and shows that the regulators require regulation. Wolff proposes a new principle for this domain: the Negative Minimum Equity Principle. This principle serves the ends of justice by protecting people likely to be made worst off by policy decisions on geoengineering.

In Chapter 9, Catriona McKinnon addresses a specific new risk created by SRM: that of "termination shock". If SRM were to be ceased abruptly then temperatures would very quickly increase beyond pre-deployment levels with catastrophic consequences. McKinnon's chapter locks horns with recent arguments made by two geoengineering experts that the risk of termination shock can be minimized by a handful of relatively simple policy measures. Addressing each of these measures, she argues that they reveal a heroic optimism about the prospects for long-term political cooperation and trust between states and private companies. She argues that given the large uncertainties about the impacts of geoengineering, we are ethically compelled to put worst-case scenarios front and centre in policy-oriented thinking about research, development and deployment.

Chapters 10, 11 and 12 explore, in different ways, the extent to which societal control should be exercised over geoengineering research and possible deployment, and what ethically adequate societal control would look like.

In Chapter 10, Holly Lawford-Smith engages with how people think about geoengineering in different scenarios. She argues that democratic authorization is necessary for the legitimacy of a unilateral deployment of SRM: as a matter of principle, when people are in fundamental and reasonable disagreement over a decision to be taken, the will of the demos should decide the outcome. Ideally, democratic authorization involves asking all people affected by the decision for their view. Lawford-Smith argues that this ideal can be approximated by representative sampling, and she reports on the outcomes of a sampling exercise in the Australian context. The results raise serious concerns about the legitimacy of unilateral deployment.

Chapter 11 – by David Morrow – addresses the risks created by research into solar radiation management geoengineering as it is proceeding under the status quo. At present, research efforts are not coordinated or overseen by a single body and are driven by the curiosity of the individual researchers involved. He argues that this state of affairs makes it less likely that SRM research could serve the ends of justice and legitimacy. What is needed instead is a well-designed, mission-driven SRM research programme. He envisages the programme as being internationally representative, as providing a place for societal deliberation and public engagement, and as aiming to limit the international domination of vulnerable states that SRM research could facilitate.

Finally, in Chapter 12, Sabine Roeser, Behnam Taebi and Neelke Doorn explore the ways in which encounters with works of art could enrich ethical deliberation about geoengineering research and deployment. They describe how the moral emotions that art often stimulates enrich ethical thinking about technologies such as geoengineering. The role of moral emotions, and the ways in which art can enable exploration of these emotions, is overlooked in the foresight practices commonly used to stimulate thinking about the ways in which geoengineering could develop in the future.

To conclude, taken together, the chapters in this book demonstrate the need for ongoing engagement from experts in ethics and political philosophy with the geoengineering debate, and the value of such engagement. Crucially, many of the concerns at the heart of both climate change and geoengineering are ethical. Existing discussions in science, policy and the public already implicitly reflect that fact. It is time for such discussions to show this explicitly, at greater depth, and with genuine engagement with relevant areas of expertise. This book makes many of the necessary first steps. It is therefore essential reading for anyone puzzled or worried about how geoengineering might legitimately promote or illegitimately thwart the ends of justice in a world suffering climate change.

References

Bipartisan Policy Center's Task Force on Climate Remediation (2011). *Geoengineering: A National Strategic Plan for Research on the Potential Effectiveness, Feasibility, and Consequences of Climate Remediation Technologies*. Bipartisan Policy Center, Washington D.C.

Chhetri, N., Chong, D., Conca, K., Falk, R., Gillespie, A., Gupta, A., Jinnah, S., Kashwan, P., Lahsen, M., Light, A., McKinnon, C., Thiele, L.P., Valdivia, W., Wapner, P., Morrow, D., Turkaly, C. and Nicholson, S. (2018). *Governing Solar Radiation Management*. Washington, D.C.: Forum for Climate Engineering Assessment.

Cicerone, R.J. (2006). Geoengineering: Encouraging research and overseeing implementation. *Climatic Change*, 77 (3–4), 221–226.

Crutzen, P. (2006). Albedo enhancement by stratospheric sulphur injections: A contribution to resolve a policy dilemma? *Climatic Change* 77: 211–219.

Fialka, J. (2020). NOAA gets go ahead to study controversial climate plan B. *Scientific American*, January 23.

Fragnière, A. and Gardiner, S.M. (2016). Why geoengineering is not 'plan B'. In C.J. Preston (ed.) *Climate Justice and Geoengineering: Ethics and Policy in the Atmospheric Anthropocene*. London: Rowman & Littlefield International, 15–32.

Gardiner, S.M. and Fragnière, A. (2018). The Tollgate Principles for the governance of geoengineering: Moving beyond the Oxford principles to an ethically more robust approach. *Ethics, Policy & Environment*, 21 (2), 143–174 (also in this volume).

Gardiner, S.M. (2011a). Some early ethics of geoengineering: A commentary on the values of the Royal Society report. *Environmental Values*, 20, 163–188.

Gardiner, S.M. (2011b). *A Perfect Moral Storm: The Ethical Tragedy of Climate Change*. New York: Oxford University Press.

Gardiner, S.M. (2013). The desperation argument for geoengineering. *PS: Political Science & Politics*, 46 (1), 28–33.

Goldberg, S. (2016). UN climate science chief: It's not too late to avoid dangerous temperature rise. *The Guardian*, May 2016. https://www.theguardian.com/environment/2016/may/11/un-climate-change-hoesung-lee-global-warming-interview

Hamilton, C. (2013). *Earthmasters*. New Haven, CT: Yale University Press.

Horton, J. and Keith, D. (2016). Solar geoengineering and obligations to the global poor. In C.J. Preston (ed.) *Climate Justice and Geoengineering: Ethics and Policy in the Atmospheric Anthropocene*. London: Rowman & Littlefield International, 79–92.

Hubert, A.-M., Kruger, T. and Rayner, S. (2016). Geoengineering: code of conduct for geoengineering. *Nature*, 537 (7621), 488–488.

IPCC (Intergovernmental Panel on Climate Change) (2018). *Global Warming of 1.5°C: Summary for Policymakers,* https://www.ipcc.ch/site/assets/uploads/sites/2/2018/07/SR15_SPM_High_Res.pdf

Jamieson, D. 1996. Ethics and Intentional Climate Change. *Climatic Change*, 33 (3), 323–336.

Keith, D.W. (2013). *A Case for Climate Engineering*. Boston, MA: MIT Press.

McKinnon, C. (2019). The Panglossian politics of the geoclique. *Critical Review of International Social and Political Philosophy* (also in this volume).

Morrow, D. (2017). *International Governance of Climate Engineering: A Survey of Reports on Climate Engineering, 2009-2015.* http://ceassessment.org/wp-content/uploads/2017/06/Morrow-WPS001.pdf

National Research Council (2015). *Climate Intervention: Reflecting Sunlight to Cool Earth*. Washington D.C.: The National Academies Press.

Nuccitelli, D. (2016). Scientists debate experimenting with climate hacking to prevent catastrophe. *The Guardian*, June 2016. https://www.theguardian.com/environment/climate-consensus-97-per-cent/2016/jun/01/scientists-debate-experimenting-with-climate-hacking-to-prevent-catastrophe

Rayner, S., Heyward, C., Kruger, T., Pidgeon, N., Redgwell, C. and Savulescu, J. (2013). The Oxford principles. *Climatic Change*, 121 (3), 499–512.

Rogelj, J., Forster, P.M., Kriegler, E., Smith, C.J. and Séférian, R. (2019). Estimating and tracking the remaining carbon budget for stringent climate targets. *Nature*, 571 (7765), 335–342.

Romm, J. (2011). Dysfunctional, lopsided geoengineering panel tries to launch greenwashing euphemism 'climate remediation'. *Think Progress*, October 6.

Royal Society (2009). *Geoengineering the Climate: Science, Governance and Uncertainty*. London: Royal Society.

Smith, P.T. (2019). Legitimacy and non-domination in solar radiation management research. *Ethics, Policy & Environment* (also in this volume).

The Tollgate Principles for the Governance of Geoengineering: Moving Beyond the Oxford Principles to an Ethically More Robust Approach

Stephen M. Gardiner and Augustin Fragnière

ABSTRACT
This article offers a constructive critique of the Oxford Principles for the governance of geoengineering and proposes an alternative set of principles, the Tollgate Principles, based on that critique. Our main concern is that, despite their many merits, the Oxford Principles remain largely instrumental and dominated by procedural considerations; therefore, they fail to lay the groundwork sufficiently for the more substantive ethical debate that is needed. The article aims to address this gap by making explicit many of the important ethical questions lurking in the background, especially around values such as justice, respect and legitimacy.

It is widely accepted in the scientific community that climate change poses a severe threat to current and future generations, as well as to the rest of nature (IPCC, 2014). Nevertheless, the countries of the world are not currently on track to meet their stated goal of avoiding dangerous anthropogenic interference with the climate system (UNFCCC, 1992), understood in terms of the internationally agreed targets of limiting average global temperature rise to 2 degrees Celsius and pursuing efforts to achieve 1.5 degrees (Climate Action Tracker, 2015; UNEP, 2017; UNFCCC, 2015). Moreover, the political prospects for further robust action remain questionable. Consequently, many scientists are concerned that in practice deliberate large-scale technological interventions into the climate system ('geoengineering') are already, or may soon become, unavoidable if the 1.5 or 2 degrees targets are to be met (Anderson & Peters, 2016; Bawden, 2016; EASAC, 2018; Kriegler et al., 2018; Shepherd, 2016). At the same time, it is generally recognized that a drive toward geoengineering would have serious social implications, that 'ethical considerations are central to decision-making in this field', and that 'analysis of ethical and social issues associated with research and deployment' should be a central research priority (Shepherd et al., 2009, pp. 39, 53).

One of the earliest interventions comes in the form of the ground-breaking Oxford Principles (Rayner et al., 2013; Rayner, Redgwell, Savulescu, Pidgeon, & Kruger, 2009), which remain influential.[1] In this paper, we build on the Oxford proposal, focusing on its ethical dimensions and in particular the ethical adequacy of its framing of geoengineering.[2] First, we offer a detailed constructive critique of the Oxford Principles. Second, we propose an alternative set of principles based both on that critique and also on some standard work in practical ethics. We name these 'the Tollgate Principles', in part after the village pub in which the guidelines were originally developed, but also because in our view respecting the principles is 'the price that must be paid' by any attempt to frame and introduce an ethically defensible geoengineering policy.[3] One upshot of the Tollgate Principles is that geoengineering becomes a much more ethically demanding enterprise than is often suggested. This has implications for how geoengineering policy is likely to evolve and especially for the prospects for 'ethical geoengineering'.

1. The Oxford Principles

One approach to generating principles is broadly 'bottom-up'. It proceeds by identifying the ethically salient features of geoengineering based on existing reports, experience from related cases, and so on. Another approach is broadly 'top-down'. It confronts the issue of geoengineering from the perspective of foundational or mid-level theory (e.g. in moral philosophy, international political theory, global justice, etc.) and seeks to apply such theory directly to geoengineering. In our view, both approaches have a role to play and ideally will ultimately become integrated. As a way to push the debate forward, we employ the bottom-up approach. In doing so, we also identify some of the issues relevant to a top-down strategy.

Our starting point is the influential Oxford Principles, first put forward by a small group of distinguished academics at Oxford University in 2009. These principles were given qualified endorsements by the UK House of Commons report (House of Commons Science and Technology Committee, 2010) and the Asilomar report (ASOC, 2010), and their spirit and content were subsequently fleshed out in an article in *Climatic Change* (Rayner et al., 2013). The principles have played a pioneering role in the geoengineering debate, and we have a great deal of respect for the authors' contribution. Our hope is to continue their necessary and important work by enriching the ethical discussion and preparing the ground for a wider, and possibly top-down, debate. Our background concern is that, despite the explicit intention to foster the debate about the 'overarching societal values' that should govern geoengineering policy (Rayner et al., 2013, p. 503), the original Oxford Principles are largely instrumental and dominated by procedural considerations. As a result, they do not sufficiently lay the groundwork for the more substantive ethical debate that is needed, especially around values such as justice, respect and legitimacy.[4]

The Oxford authors summarize their principles as follows:

(OP1) Geoengineering to be regulated as a public good.

(OP2) Public participation in geoengineering decision-making.

(OP3) Disclosure of geoengineering research and open publication of results.

(OP4) Independent assessment of impacts.

(OP5) Governance before deployment.[5]

Before assessing these principles directly, we offer some quick clarifications about our approach. First, the Oxford group follows the Royal Society in defining geoengineering as: 'the deliberate large-scale manipulation of the planetary environment to counteract anthropogenic climate change' (Shepherd et al., 2009, p. 1). While some would reject this definition (e.g. as too permissive and overly vague) and others are skeptical of the term 'geoengineering' itself (e.g. Heyward, 2013; Jamieson, 2013), we aim to sidestep definitional discussions in this paper by assuming that we are discussing the paradigm case of stratospheric sulfate injection (SSI). The extent to which other interventions share the features that make all or some of the Tollgate Principles appropriate, and the question of whether these deserve the label 'geoengineering' are topics for another occasion (cf. Gardiner, 2016).[6]

Second, in their original form, each Oxford principle was accompanied by a brief text (Rayner et al., 2009), and in the later article, each is supplemented by a longer comment (Rayner et al., 2013). While it is not entirely clear whether the supplements are intended to define the principles, draw out implications, or something else, we shall assess each principle in conjunction with its accompanying remarks.

Third, one background question concerns the scope of the Oxford principles, and in particular whether they are intended to guide research and deployment, research alone, or even just early, small-scale research (e.g. excluding large-scale field trials). We believe that these tasks are not entirely separable, since governance of research (even near-term research) cannot help but be influenced by the wider aspiration of potential deployment and the norms that would govern such deployment. Nevertheless, in this paper, we will not take a stand on the interpretive issue. Instead, we assume that the ultimate aim of developing principles is to frame geoengineering – from early research through deployment – in ways that facilitate successful governance. Seen in this light, the question is how far the Oxford Principles assist in this project, however they were initially intended.

Fourth, the Oxford principles are often criticized as being too high-level or abstract to be useful (e.g. Nature Editorial, 2012). This is a common concern about governance principles in all areas. However, we agree with the Oxford authors that abstraction in this context need not be a problem and may be an advantage (Rayner et al., 2013). For instance, offering high-level principles often allows one to avoid prejudging more specific issues prematurely, to identify such issues, and to facilitate appropriately formed, justified and authoritative judgments about them. These are important elements of successful governance. Hence, our Tollgate principles will also be high level.

Finally, in any case, our primary intention is to influence the framing, tone and direction of geoengineering governance, rather than focus on the specific designation or wording of particular principles. Again, our aim is to enrich the ethical dimensions of the conversation and prepare the ground for a wider, and possibly top-down, discussion. Although we believe that the Tollgate principles are useful, we do not see them as the final word, but rather as another step on an ongoing journey.

1.1. Regulating a Public Good

The first Oxford principle states:

> **Oxford Principle 1 (OP1): Geoengineering to be regulated as a public good**
>
> While the involvement of the private sector in the delivery of a geoengineering technique should not be prohibited, and may indeed be encouraged to ensure that deployment of a suitable technique can be effected in a timely and efficient manner, regulation of such

> techniques should be undertaken in the public interest by the appropriate bodies at the state and/or international levels. (Rayner et al., 2009)

Unfortunately, this principle provides a problematic framing of geoengineering.[7] As John Virgoe put it in his testimony to the UK House of Commons, 'once you peer below the surface of the public good, it becomes quite hard to define it and you get into some difficult ethical territory' (House of Commons, 2010, Ev 12). On our view, this is due to an awkward ambiguity. The phrase 'public good' has informal, colloquial uses, but also a number of closely related technical meanings in economics and international relations, often with specific, but potentially conflicting, policy connotations.[8] This threatens to make framing geoengineering as a 'global public good' seriously misleading, especially in the public sphere. Consequently, we will advocate for a more transparent, and explicitly ethical, approach which emphasizes central values, such as justice and political legitimacy.

Let us begin with three uses of 'public good' and 'global public good'. First, one minimal colloquial understanding is that of something that is not (or not primarily) a private concern, but a public one that should be governed or regulated as such. This sense of 'public good' is suggested by the 2009 text accompanying OP1 (above).

Second, in economics and public policy, the common technical conception of 'public good' defines a pure public good as a good that is both nonrival and nonexcludable. A good is *nonrival* if and only if one person's consumption of the good does not limit or inhibit another person's consumption. A good is *nonexcludable* if and only if, once it is available to some, others cannot be prevented from consuming it. A standard example is the good provided by a lighthouse. It is nonrival: one sailor's being able to see the rocks does not limit or inhibit others being able to do the same thing, and *vice versa*. It is also nonexcludable: once the lighthouse is illuminating the rocks for some sailors, others cannot be prevented from seeing them too. These features of pure public goods are also emphasized in the geoengineering context, including by the Oxford authors (Rayner 2011, p. 11).

Third, a further colloquial conception of 'public good' envisions something that is 'good for everyone' (i.e. universally beneficial), or at least benefits everyone affected. This claim is also often invoked in more technical discussions in international relations and economics, in that it motivates both the standard examples (such as the lighthouse) and the basic interest in public goods more generally. Indeed, importantly, for some the universal benefit claim becomes a matter of definition, such that it is included alongside the claims of nonrivalness and nonexcludability in the technical meaning of 'global public good' or 'public good'.[9] Most notably, in its seminal work the United Nations Development Program defines a global public good as 'a public good with benefits that are strongly universal in terms of countries ... people ... and generations'.[10] Similarly, Scott Barrett, who thinks of geoengineering as a global public good, includes universal benefit in his definition, stating 'when provision succeeds, global public goods make people everywhere better off' and 'global public goods are thus universally to be desired' (2007b, p. 1). Notably, in their 2013 description of the first Oxford Principle and its notion of public good, the Oxford authors also write: 'the global climate must be managed jointly, *for the benefit of all*, and with appropriate consideration for future generations' (Rayner et al., 2013, p. 505).

Universal benefit conceptions of 'global public good' and 'public good' are highly relevant to geoengineering, since versions of the claim that geoengineering in general, and SSI in particular, are 'good for everyone' are popular among some scientific advocates of pursuing research. For instance, Ken Caldeira claims: '... for most reasonable climate change metrics, if any party acted in their own self-interest [in implementing SSI] *every party would be better off* than if no party had acted' (Caldeira, 2012; emphasis added). Similarly, others claim that in their model simulations '*all regions benefit* by deployment of solar geoengineering at the level of any other regions preference' (Ricke, Moreno-Cruz, & Caldeira, 2013, p. 5; emphasis added).

Now, there is a lively philosophical debate about how the terms 'global public good' and 'public good' are usually understood in international relations, economics and public policy, how they should be defined, whether some definitions are deflationary – including perhaps radically so – and what this means for calling geoengineering, and especially SSI, a global public good (Gardiner, 2013b, 2014b; Morrow, 2014). Here, however, we shall set such concerns aside and instead focus on the ethical significance of the various public good framings.

A central concern is that framing geoengineering as a global public good risks painting too rosy a picture of the governance challenge.[11] First, the universal benefit conception is most naturally read as making an overt appeal to a fundamental ethical consideration, the promotion of human welfare, and so as making ethics central. However, paradoxically, framing geoengineering as universally beneficial often has the effect of marginalizing ethical concerns. Probably this is because it is initially tempting to assume that universal improvements in welfare are so attractive as to render further ethical discussion idle. In particular, it may be thought that, if a public good is universally beneficial, no one has good reason to object to its supply.[12] Since no one suffers (indeed all benefit), what could the objection be? Moreover, even if an objection could be found, wouldn't it be overridden by the weight of the universal benefit?

Second, conceptions of 'global public good' that appeal to the technical features of nonrivalness and nonexcludability encourage a similarly optimistic view of the prospects for, and ethics of, provision. For instance, in the lighthouse case, the supply of the lighthouse can be achieved by a small group of sailors motivated to secure their own safety. In providing the light for themselves, they also procure it for all sailors who travel in that area. Moreover, the fact that the others do not contribute (but 'free ride') need not undermine provision. The action and motivation of the small group suffices to supply the relevant good to all. In addition, if one assumes that the intervention is universally beneficial, providing the public good appears not only ethically unobjectionable, but also laudable. Finally, given this, if for some reason no group emerges to provide the lighthouse, there are strong reasons for government intervention to fill the gap. Since (by hypothesis) provision is in everyone's interest, such intervention should not be controversial. Moreover, this conclusion is reflected by experience, given the tradition of *public provision* for many public goods (Barrett, 2007b, p. 3; Bodansky, 2012, p. 20).

Third, the lighthouse paradigm seems highly relevant to many forms of geoengineering. For one thing, in reducing incoming radiation (solar radiation management [SRM]) or lowering the atmospheric concentration of carbon dioxide (carbon dioxide removal [CDR]) for itself, an actor simultaneously reduces those things for others, even if they do not cooperate. If such action is universally beneficial, this seems not only unobjectionable, but also ethically laudable.

For another thing, if the intervention is one that can be provided by an individual actor or a small group following its own self-interest (a 'single-best effort public good'), then it appears either that such actors should be permitted and encouraged to intervene, or else (if none emerge) that governments (individually or collectively) should fill the gap themselves. This is highly relevant to geoengineering since SSI in particular is often described as a 'single best effort' public good (e.g. Barrett, 2007a).[13] This encourages the thought that widespread cooperation in the *supply* of the public good may not be needed.

Putting these ideas together, the rosy picture is that if geoengineering (understood as SSI or something relevantly similar) is a global public good, then: (1) it *benefits everyone*; (2) it is (therefore) ethically unobjectionable; (3) some may be motivated to *supply it for all*; (4) failing that, there are strong grounds for (and a corresponding tradition of) *public provision* and (5) unilateral or small group provision is ethically both permissible and laudable, even without wider cooperation.

Unfortunately, this rosy picture is seriously misleading. One sign of this comes from the minimal conception of a public good as something that is a public concern that should be governed or regulated as such. While this seems correct for paradigm cases of geoengineering, it is important to emphasize that this is mainly because the deep ethical issues (e.g. justice, political legitimacy) raised by SSI imply that geoengineering should not be treated merely as a private concern, and in particular ought not be left to the mechanisms of an unfettered economic market, as if it were a paradigm example of a private good. In short, the idea that SSI is a global public good in the minimal sense is motivated by the thought that ethical concerns ought to be central to governance. But why might this be the case?

a. Universal Benefit

Let us begin with the universal benefit claim. Traditional public goods such as the lighthouse provide unambiguous benefits for all those affected. In such cases, the claim of universal benefit is meant to be descriptive. However, descriptive claims of universal benefit are likely to be false for geoengineering in general, and for specific paradigm cases of geoengineering, such as SSI.

On the one hand, the scope is much too wide. There is nothing about injecting sulfates into the stratosphere *as such* – that is, just any old injection, done in any way, and in any amount – that guarantees that everyone will benefit from that injection. Thus, the universal benefit claim is false for SSI considered as such and could only be made for specific implementations of SSI.

On the other hand, even if one restricts the scope to specific implementations, the descriptive claim is so strong as to be implausible. The very idea that a particular geoengineering intervention could benefit *absolutely everyone* affected by it is an extremely demanding one, given that the effects of climate engineering are, and are intended to be, global and also to span multiple generations, possibly over thousands of years. Indeed, the whole idea of truly universally beneficial geoengineering might be thought so descriptively demanding as to be fanciful. Moreover, despite what some advocates maintain, there are clear reasons for believing that some will suffer. For instance, early work suggests that SSI introduces new global risks that threaten severe impacts for some populations, such as a disruption of temperature and precipitation patterns, a net decrease in precipitations globally, and a risk of termination shock (National Research Council, 2015). Even if there might be net benefits to many, the universal benefit claim seems likely to be overblown.[14]

Things are not much better if one assesses the universal benefit claim as a normative standard rather than an empirical claim. At one point, the Oxford authors suggest a normative interpretation when they write: 'the global climate *must* be managed jointly, *for the benefit of all*, and with appropriate consideration for future generations. In short, geoengineering *must* be regulated *so as* to promote the general good' (Rayner et al., 2013, p. 505; emphases added). This passage appears to imply that the first Oxford principle expresses an *ethical requirement*: that the kinds of geoengineering that *ought* to be considered – that is, that should be the subject of scientific research and public policy – are those that benefit all those affected (as well as having the other characteristics of genuine public goods).

Unfortunately, universal benefit seems not to be the most reasonable standard from an ethical point of view. On the one hand, to require that a purely physical and technological intervention satisfies the universal benefit requirement seems *unduly demanding*. Not only does insisting on net benefits to all those affected (future generations and possibly nonhuman nature included) impose a very strong and perhaps empirically unsatisfiable requirement, but it seems highly plausible that some geoengineering interventions would be morally justifiable even if they allow for some harm. For example, schemes that protect basic human rights at some net economic costs to the more affluent might nevertheless be ethically defensible.

On the other hand, paradoxically, the universal benefit requirement also seems in some ways *excessively permissive*. Specifically, even if universal benefits were possible, this would not suffice to justify implementation, since there may be other grounds for opposing geoengineering (e.g. political legitimacy, procedural and distributive justice, relationship to nature, etc.). Indeed, doing something that is good for someone's welfare is often impermissible, as, for instance, when it violates other rights that they have.[15] For example, Americans might claim that China, Russia or Iran have no right to implement a form of SSI without their consent, even if such geoengineering would clearly benefit the United States.

In light of such concerns, making universal benefit a strict requirement for the permissibility of geoengineering appears hasty and unreasonable, at least at this early stage.[16] Moreover, the concerns suggest that much more needs to be said about what ethical requirements geoengineering interventions should meet. This pushes us in the direction that we are suggesting: full-blown moral and political philosophy.[17]

b. *Rivalness*

Let us turn now to the more technical claims associated with public goods, beginning with nonrivalness: that one person's consumption neither limits nor inhibits another person's consumption. This is often seen as the most central technical characteristic of a public good.

At first glance, paradigm cases of geoengineering such as SSI do appear to be nonrival in one sense. Once a specific intervention is provided, all 'consume' the effects of that intervention without prejudice to others doing the same. Considered as an overall bundle, the effects of a specific geoengineering intervention are not 'used up' when one party experiences them. They are not a scarce resource. For example, temperature reductions in Haiti do not 'use up' reductions in Bangladesh; droughts in India do not 'use up' floods in Northern Europe.

Still, it is worth noticing that this is a peculiar sense of 'nonrival', and (whatever its technical credentials) it is not clear how normatively relevant it is, especially in the case of geoengineering. To see this, consider another sense in which geoengineering is rival. There are many possible geoengineering policies. Consequently, those contemplating the implementation of geoengineering are likely to face choices between many particular kinds and levels of intervention. This creates rivalry that is normatively relevant. Suppose, for example, that Russia prefers SSI that caps global temperature rise at +2.4°C, and Tuvalu prefers a cap of +1.5 degrees. Both cannot be satisfied at the same time. Hence, the different interventions are rival in that *the selection of one scheme precludes the implementation of others.*[18] Consequently, groups favoring interventions that are not chosen may see their interests marginalized.

This seems the more important kind of rivalry, both politically and ethically. In particular, the former sense of rivalry – effects in one area do not 'use up' effects in another – loses its intuitive importance when the effects are no longer universally beneficial, but cover important losses as well as gains. For instance, it seems deeply misleading to characterize a global cap at 2.4°C as 'nonrival' just because the benefits of a longer growing season in Siberia do not 'use up' or preclude the loss of Tuvalu through sea level rise. Notably, such cases raise central issues of justice, political legitimacy, and other values. To obscure this behind the rhetoric of 'nonrivalness' is a serious moral failing.

We conclude that, even if geoengineering is nonrival in a narrow technical sense, it is not clear that this has particular normative significance. Instead, ethical concerns, especially of justice, are central and unavoidable.

c. Excludability

Let us turn now to the question of whether geoengineering is *nonexcludable*. Technically, for some good *x* to be nonexcludable means that 'once *x* is available to some, others cannot be prevented from consuming *x*'. On the one hand, from the point of view of the provider of the good, there is a clear reason for thinking that geoengineering is nonexcludable. Since it is – by definition – a global intervention, none are shielded from the effects of a specific geoengineering scheme, and so cannot in some sense be 'prevented from consuming it'. This is important since, in the absence of proper regulation, the providers of at least some SSI schemes might choose interventions that disproportionately benefit them, and disproportionately harm others. Interestingly, this might be construed as a more morally relevant form of exclusion, as it gives providers a level of control over geoengineering outcomes denied to those who merely consume the effects. As has already been suggested, this raises numerous issues of justice.

On the other hand, the issue can also be approached from the point of view of the nonproviders. Nonproviders cannot exclude themselves from the effects of a geoengineering scheme: *there is no opting out*. Consequently, nonproviders are vulnerable to the decisions of providers, and this raises serious ethical questions, including those of justice, domination, rights and responsibility. For one thing, for most interventions there are risks and costs, as well as benefits, to be distributed, and their imposition on some people rather than others requires ethical justification. For another, nonexcludability raises the worrying possibility of hostile interventions, such as predatory geoengineering (e.g. aiming to damage one's political rivals) or parochial geoengineering (e.g. discriminating against future generations). Once again, geoengineering requires regulation not because of the (probably fanciful) universal benefits it could provide, but because of its potential for harm, injustice and other ethical infractions.

d. The Relevant Public

Let us close with two more general concerns about the framing of the first Oxford Principle. First, there is the question of how to understand the relevant 'public' and so the scope of the related public good. From an ethical point of view, the relevant public is global, intergenerational and ecological, so that the first principle should refer to the global, intergenerational and ecological good. In our view, this point should be emphasized in the principle itself. Still, it is not clear that the conventional framings even have it in mind. For instance, in its original version, the first Oxford Principle used only the phrase 'public good'. This might be understood in various ways, including as referring only to the national good, or the good of the current generation, or the short-term economic good. In the later version, the Oxford authors make it clear in their comments on the first Oxford principle that they believe that considerations of global and intergenerational justice must be taken into account (Rayner et al., 2013, p. 524). However, in our view, these considerations are so central that they should appear in the principle itself.

e. Injustice

The second concern is that framing the first principle in terms of regulating a public good implicitly marginalizes other morally relevant understandings of the climate problem. For example, there is a huge difference between framing geoengineering as the *supply of a universal benefit*, and framing it as *a rectification of injustice*. Note, for instance, the 'supply of universal benefit' framing tends to be perceived as exclusively forward-looking. By contrast, in many cases, climate engineering is best understood as a *response* to the infliction of some risk or damage. Plausibly, this involves an important backwards-looking component, and one which strongly encourages questions of responsibility, including of who should do what, and how it should be done.

Given the above, what should we conclude? Notice that we have not argued that it is impossible that some geoengineering scheme could be found that is a genuine global, intergenerational and ecological public good including in the technical sense of being nonrival and nonexcludable, nor have we claimed that such a scheme would necessarily be ethically undesirable if it could be found. Nevertheless, the existence of such a scheme seems empirically unlikely in the real world, and (much more importantly) the focus on the technical senses of 'global public good' and 'public good' threatens to be seriously misleading. Instead, the real bite of the first Oxford Principle is in its claim that geoengineering should be subject to public oversight and on behalf of the public interest. This is a common, but nontechnical understanding of 'public good', and is justified largely by the importance of ethical issues such as justice, political legitimacy and the human relationship to nature. In light of this, we suggest that a better framing principle would be:

> **1st Tollgate Principle (Framing): Geoengineering should be administered by or on behalf of the global, intergenerational and ecological public, in light of their interests and other ethically relevant norms.**
>
> Geoengineering raises complex issues of global, intergenerational and ecological ethics – including issues of justice, legitimacy, domination, rights, responsibility and the human relationship to nature. It should be framed as such and administered by or on behalf of the global, intergenerational and ecological public in light of relevant norms. This has implications for deployment, governance and research.

The first Tollgate Principle helps to bring into focus an important, but neglected question (that OP1 does not): what are the primary objectives of geoengineering governance?[19] Moreover, while it allows for the possibility of a pure public good solution, it also includes much else. As well as being much more realistic, it puts the emphasis in the right place.

1.2. Public Participation

The second Oxford principle is labeled 'public participation in geoengineering decision-making'. This suggests some concern for legitimacy and justice. However, closer examination reveals that the principle is much more narrowly construed than the surface language implies, and so is likely to mislead. The principle states:

> **Oxford Principle 2 (OP2): Public participation in geoengineering decision-making.**
>
> Wherever possible, those conducting geoengineering research should be required to notify, consult, and ideally obtain the prior informed consent of, those affected by the research activities. The identity of affected parties will be dependent on the specific technique which is being researched – for example, a technique which captures carbon dioxide from the air and geologically sequesters it within the territory of a single state will likely require consultation and agreement only at the national or local level, while a technique which involves changing the albedo of the planet by injecting aerosols into the stratosphere will likely require global agreement.[20]

Despite its name, this principle risks sharply limiting the role of the public in geoengineering decision-making in at least five ways. The first limitation is that it appears that the principle applies only to research and *only to one kind of research*, studies that actually affect people. It does not advocate for public participation in decisions either about research more generally, or about deployment. For example, the principle does not suggest that the public should be consulted about (a) whether to engage in an aggressive, dedicated research program on geoengineering (such as a Manhattan or Apollo-style project), or (b) whether to pursue emergency-oriented research (such as atmospheric SRM) or longer-term projects (such as some CDR) or both, or (c) whether to push forward with a global governance scheme for geoengineering, and so on. Instead, the sole area for consultation seems to be 'field testing' that has tangible implications for particular people or groups. This restriction is unmotivated and puzzling.[21]

The second way in which the principle is limited is that it applies *only to those affected* by research. In their discussion of Principle 2, the authors of the Oxford Principles acknowledge that further discussion is needed about precisely what 'affected' means in this context. However, the text suggests that they lean toward a narrow interpretation that includes only those direct and enduring *material* effects (e.g. Rayner et al., 2013, p. 505). This suggests that those who are not materially affected are excluded from decision-making both in these cases and (given the first point) from geoengineering decision-making in general.

This approach encourages three presumptive objections. First, given that this is the only Oxford principle concerning public participation, it implies that *those who are not directly and materially affected have no say at all*, on either field testing or geoengineering research more generally. This assumption seems deeply objectionable, yet no reason is offered for it. This exclusion of the 'unaffected' is surprising if (as the first Tollgate principle states) geoengineering involves major issues of legitimacy, justice and ecological values on a global and intergenerational scale. Is this not their (i.e. your) planet as much as anyone else's?

The second objection is the flip-side of this. The narrow scope of the principles suggests a further covert assumption: that *some* (e.g. researchers, the organizations that sponsor them) *have the right to do whatever they like to the planet*, so long as either they do not materially affect other people currently alive, or else reach an agreement with those directly and materially affected that they can proceed. This assumption is not just striking but also deeply contentious. Intentionally messing with the basic physical structure of the planet arguably means much more, morally and politically, than is captured just in terms of its direct material effects on specific people. This worry is especially relevant in cases of large-scale experiments and in a context when many of those affected, such as future generations and other species, have no voice.

The third objection is that the principle brings on serious practical issues with a strong ethical edge. For one thing, since we are talking about *prior* consent, it raises questions about how we identify before doing the experiment who will be materially affected and who will not, and (more importantly perhaps) who gets to make that call. For another thing, given the complexity of the climate system, it may be difficult to identify the victims even after the experiment, since (for example) this requires being able to pin down what would have happened otherwise. (For more on both issues, see below.) In addition, notice that there is no *ex post* principle – for example, for dealing with a process for compensating people who we didn't know in advance would be affected.[22]

Returning to the main narrative, the third way in which the second Oxford Principle sharply limits the role of the public in geoengineering decision-making is that the requirement is to 'notify, consult, and ideally obtain ... prior informed consent'. However, given that geoengineering is a global, intergenerational and ecological issue, *many of those affected are not available* to be notified or consulted or informed, and are not in a position to consent. Moreover, given that the principle mandates that the requirement holds only 'when possible', the default assumption seems to be that those who cannot be consulted will not have their concerns taken into account.

We think this implication should be resisted. Specifically, there might be other ways to take the interest of future generations and nonhuman beings into account, such as appropriate representative institutions (Gardiner, 2014a; Gonzalez-Ricoy & Gosseries, 2017). In mentioning prior consent only, the Oxford Principles risk brushing them aside (see discussion under point 5 below).

The first three limitations suggest that the title of the second principle is deeply misleading. The notion that it advocates for robust 'public participation' is illusory. Moreover, the limits suggested do nothing to address, and may even encourage, the threat of a pronounced bias in geoengineering policy, globally, intergenerationally and across species. Since the second principle is one of only two political principles in the Oxford set, this is a serious matter. However, there are also issues about the requirement itself. These emerge in the fourth and fifth limitations.

The fourth limitation concerns the second Oxford principle's *description of the parties in charge of applying it*. According to that description, the requirement applies to 'those conducting geoengineering research'. To begin with, this is most naturally read as the research scientists themselves. However, the Oxford Principles are supposed to advise governments and other sociopolitical entities on how to govern geoengineering. This raises a problem. Obviously, scientists wanting to carry out field trials have some ethical responsibilities. Still, it would be bizarre to claim that it is *solely* the responsibility of these specific scientists to ensure that adequate public participation occurs, and that governments and other bodies should simply step aside except to enforce that requirement. Such a move would seem a serious abdication of responsibility.

Given this, it seems wise to reject the natural reading, and instead to understand 'those conducting geoengineering research' as the social and political bodies in authority. However, here we must emphasize that one of the central issues of geoengineering ethics is that it is not clear that we currently have such bodies, or that existing authorities are either equipped or authorized to take on the task (e.g. Gardiner, 2014a). Since the second Oxford principle obscures this issue, it is urgently in need of clarification and reframing. In our view, such a reframing would emphasize the need to develop governance institutions alongside research and not just before deployment (see our critique of Principle 5).

The fifth limitation concerns the *notion of consent*. The first thing to notice is that prior consent is just one way of securing political legitimacy. Even though it is a widely held principle, there are circumstances where it fails to deliver the expected normative benefits (for example, as already mentioned, with respect to future generations and the rest of nature). Therefore, what we should be after with this second principle is not consent per se, but a wider conception of political legitimacy (within which consent might play a significant role).

The second thing to notice is that the participation principle is silent on the grounds for consent. It offers no guidance on what is *relevant* for consent in this case.[23] The reason for this is likely that the Oxford authors argue in their 2013 comment on OP2 that the understanding of consent, and public participation in general, varies greatly across political cultures, so that the measures necessary to insure public participation should not be specified at the level of principles (Rayner et al., 2013, p. 506).

Still, this may be unnecessarily pessimistic. Moreover, whatever the form that consent actually takes, the *relevant grounds* are important both to current advocates of geoengineering and their critics. For example, on the one hand, advocates will not be happy with refusals to consent – effectively vetoes – that rest on spurious, deeply self-interested or ideological reasons. On the other hand, critics will claim that ethically serious consent will occur only when important values have been respected, including legitimacy of decision-making, compensation provisions, ecological values and so on.[24] Again, the notion of consent seems only to scratch the surface of the problem and cannot be adopted as a principle without more substantive discussion.

These concerns raise a background worry about the Oxford approach. Namely, on too many issues that really matter, it remains silent, shifting the responsibility elsewhere. For instance, importantly, the second principle might be accused of *placing undue burdens on research subjects*. When asked for consent, are they really expected to determine by themselves when and how geoengineering is acceptable? *Is it fair to make them the sole custodians of this responsibility?* Isn't this an abdication of responsibility by social and political institutions and the wider public?

If the second Oxford Principle is too narrow, what should we say instead? The first Tollgate principle states that geoengineering should be administered by or on behalf of the global, intergenerational and ecological public, in light of their interests and ethically relevant norms. So, a new second principle ought to deal with the initial implications of this in terms of how decisions should be made and by whom.

The first point to be made is that, although the *moral subjects* of geoengineering are those affected by it across the globe, time and species, the *agents* properly speaking are (a) a particular generation of humans that chooses to initiate a geoengineering scheme, and then (b) the successive generations charged with managing the scheme (and its consequences) over time. This is important because most schemes are expected to last at least many decades, and probably several centuries.

The second point is that, given the first, it is far from clear that a simple informed consent model, drawn from medical contexts and based on the consent of those currently alive who are directly and materially affected, is at all appropriate (Gardiner, 2013a). Arguably, the kind of 'public participation' needed involves much more substantial moral and political norms, including those of appropriately global and intergenerational procedural justice (Gardiner, 2014a, 2017).

The third point is that these norms should include political legitimacy (not forgetting inclusion and diversity), justice, ecological values and much else. Hence, institutions are needed which are capable of administering geoengineering and subject to appropriate checks and balances.

Given these points, we suggest two principles:

> **2nd Tollgate Principle (Authorization): Geoengineering decision-making (e.g. authorizing research programs, large-scale field trials, deployment) should be done by bodies acting on behalf of (e.g. representing) the global, intergenerational and ecological public, with appropriate authority and in accordance with suitably strong ethical norms, including of justice and political legitimacy.**[25]
>
> Ethical geoengineering would be the task of a sequence of generations (or their agents) operating on behalf of a wider global, intergenerational and ecological public. Institutions would be needed that are ethically authorized to carry out, and capable of managing, such a task (a) in light of and in accordance with appropriate norms of global, intergenerational and ecological ethics, including those pertaining to political legitimacy, justice, the human relationship to nature and especially the perspectives of the most vulnerable groups, while

(b) ensuring reasonable relationships with other institutions (e.g. through suitable checks and balances).

Since it is also true that people materially affected by field tests might have a special say about the where, what and how of those tests, we propose supplementing the authorization principle with a more specific consultation principle modeled on the second Oxford principle:

3rd Tollgate Principle (Consultation): Decisions about geoengineering research activities should be made only after proper notification and consultation of those materially affected and their appropriate representatives, and after due consideration of their self-declared interests and values.

Consultation can be achieved through a variety of deliberative procedures. However, a priority should be placed on methods that stress full information and autonomy, and are well-placed to provide genuinely representative feedback that is neither superficial nor easily manipulated. The process should be especially sensitive to respecting the self-determination and self-understanding of affected groups, taking particular care in eliciting responses from historically marginalized or oppressed populations (cf. Whyte, 2013, 2016). Indeed, there is a strong presumption that such groups should play a central part in deliberation, including by participating in leadership roles.

At this point, it starts to become clear that the first three Tollgate principles are much broader than the first two Oxford Principles and likely to be more demanding. Possibly, they are so demanding as to cast doubt on the prospect of highly ethical geoengineering being achievable in the context of our current geopolitical environment. In our view, this is not an objection to the new principles. Indeed, it is an advantage that they highlight what is really at stake. In addition, if ultimately our approach casts doubt on whether geoengineering can satisfy high ethical standards, this at least makes it clear that we should also investigate whether there are lower, perhaps very minimal, ethical standards that specific proposals might more easily satisfy, whether these too are realistically achievable, and what that means for how we understand the risks of geoengineering (Gardiner & Fragnière, 2017). Unfortunately, the Oxford Principles obscure these important questions.

1.3. Full Disclosure

The third Oxford principle presses for full disclosure of geoengineering research and open publication of results, including negative results:

Oxford Principle 3 (OP3): Disclosure of geoengineering research and open publication of results.

There should be complete disclosure of research plans and open publication of results in order to facilitate better understanding of the risks and to reassure the public as to the integrity of the process. It is essential that the results of all research, including negative results, be made publicly available.

The House of Commons accepts this 'requirement' and proposes that it 'should be unqualified' (House of Commons 2010, p. 32).

The first thing to notice is that the third Oxford principle raises important questions, especially when understood in a completely unqualified way. Full disclosure is not endorsed in all settings and might also create risks. For example, one of the main concerns in geoengineering policy is that of a rogue actor (or actors) implementing a scheme by themselves and for their own purposes.[26] The usual assumption is perhaps that openness should at least ensure that the type of geoengineering attempted by a rogue actor would be well-informed and so more likely to avoid the worst consequences of geoengineering. However, this thought assumes that 'worst consequences' will mean the same thing to the rogue as to the researchers. This need not be the case. For example, suppose the mainstream geoengineering community investigates technique X but then abandons it because X inevitably involves negative consequences for some region or another. Full disclosure of this might reveal to rogue actor A that X is good for it but bad for its main rival B. In that case, the openness of the research may facilitate A's choice of X. The mainstream reason to reject X is in fact not a reason against it for A and may even be a reason in its favor. Consequently, there are at least two notable risks of full disclosure: facilitating rogue geoengineering and facilitating geoengineering with problematic consequences.

The Oxford authors discuss a similar case involving articles discussing a mutation of the avian flu virus (H5N1), and point to the decisions of the US National Science Advisory Board for Biosecurity to rule in favor of open publication. They conclude that security concerns do not always trump the benefits of full disclosure. Specifically, they say that, as in the bird flu case, closed geoengineering has risks of its own, and there are benefits on the other side. The third Oxford principle emphasizes two: facilitating a better understanding of the risks and reassuring the public of the integrity of the process. These emphases suggest that the goals of the disclosure principle are both *epistemic* and *ethical*. Full disclosure brings all the available knowledge into the public arena to inform and be scrutinized by researchers and the general public alike. This can be valuable in numerous ways. For instance, disclosure can help to ensure that analysis is rigorous and that technical mistakes are not made; it plays a role in ensuring that important possibilities are not overlooked,[27] and it can help to create a wider community of inquiry, and especially one that is more diverse (intellectually, nationally and socially) than a closed approach would produce.[28]

The thought, then, is that full disclosure can facilitate a high quality scientific and public discourse. However, we should notice that concealed information is not the only threat to this discourse, and scientific rigor not its only object. Full disclosure is intended to promote inclusion and diversity, and thereby trust and accountability, as well as reliability.

Once the nonepistemic reasons for openness are pointed out, this has further implications. If the concern is with inclusion, trust and accountability as well as reliability, this suggests that matters such as appropriate authorization of research projects, checks and balances, and so on, become important. If the question is how to create and facilitate a reliable, inclusive, trustworthy and accountable research process, unconditional openness is *only one answer*, and only a partial one at that. This issue is pressing if climate change involves a serious risk of moral corruption, especially when it comes to intergenerational issues (Gardiner, 2011a). For example, openness about scientific research will not directly aid the global poor or future generations in securing inclusion, trust and accountability, since they are not around to press their claims. Others must do this for them, and more thought is needed about how this is to be done. Again, the relevant Oxford Principle seems both insufficient to the task and to obscure the central issues.

This discussion suggests an expanded version of the third principle:

4th Tollgate Principle (Trust): Geoengineering policy should be organized so as to facilitate reliability, trust and accountability across nations, generations and species.

Geoengineering policy should be organized so as to facilitate reliability, trust and accountability. This suggests high-levels of openness and disclosure in publication of research plans and results, and in geoengineering decision-making more generally. It also suggests that geoengineering policy should accept demanding norms of inclusion, across nations and other demographic groups, at all levels, including the basic design of policy and institutions. In addition, if any material is protected, it should be subject to a review process involving stakeholders or their representatives to maximize trust. Special attention should be paid to the concerns of those groups affected by geoengineering decisions who cannot be effectively represented, such as future generations and nonhuman nature.

1.4. Independent Assessment

These concerns resurface when we come to the fourth Oxford principle. This calls for independent assessment of impacts of geoengineering research, encompassing both environmental and socioeconomic impacts.

Oxford Principle 4: Independent assessment of impacts

An assessment of the impacts of geoengineering research should be conducted by a body independent of those undertaking the research; where techniques are likely to have transboundary impact, such assessment should be carried out through the appropriate regional and/or international bodies. Assessments should address both the environmental and socio-economic impacts of research, including mitigating the risks of lock-in to particular technologies or vested interests.

The central concern of this principle, that is ensuring the integrity and reliability of the research and development process, seems to fall under the 4th Tollgate principle just discussed, albeit here with a specific focus on environmental and socio-economic impacts. Nevertheless, there are some more specific issues worth identifying.

The first is what counts as a geoengineering impact. This involves contentious questions, since what counts as geoengineering, what counts as geoengineering-specific research, what counts as a geoengineering 'experiment' and what would count as relevant impacts are all matters of dispute. The last issue is particularly important. It echoes one that already arises in the climate change discussion. In adaptation policy, there is an issue about how to separate out global warming impacts from other impacts. For example, how do we decide how many of the deaths and diseases inflicted during a particular hurricane are the results of global warming when (say) global warming only increases the frequency of such extreme events, and other factors, such as background poverty, poor infrastructure, and so on, are such that the hurricane has worse effects than might have occurred elsewhere? As Dale Jamieson memorably remarks: no one's death certificate will ever read 'climate change' (Jamieson, 2005).

In general, to attribute impacts to global warming or to geoengineering involves making decisions on what constitutes the natural and socioeconomic baselines against which they will be judged, and also what normative baselines of reasonable claims should be presupposed. On the face of it, these matters are even more difficult in the geoengineering case, and likely to be severely contested. One reason for this is that the relevant baseline for assessment of impacts applied by advocates of SSI is often the severe climate change avoided, rather than 'normal' conditions. Yet catastrophic baselines tend to obscure the underlying ethical issues (Gardiner, 2013a).[29]

The second issue is what counts as 'independent' assessment, and who is to do it. Clearly, it cannot be the original researchers. But can it be researchers from the same discipline, nation or international research community? Or do those affected by geoengineering research have the right to independent assessment, and also the authority to assess according to their own values and then demand compensation on their terms for negative impacts? And who will decide that? In short, the fourth Oxford Principle raises major conceptual, epistemic and political issues, and values are at the heart of them.

In addition to these 'internal' questions, there are wider issues. Why is the review to which research subjects are entitled limited to the impacts and their assessment? On the one hand, one might have expected a review of the objectives, methods and assumed constraints on the geoengineering research as well. On the other hand, 'assessment' seems too limited. As already suggested, affected countries, for example, might reasonably ask for a procedural role in the generation of research plans and the design of geoengineering institutions before accepting any transboundary impacts.

Some of these issues are mentioned by the Oxford authors, but without specific answers being provided. In their view, the role of the principles is to provide general rules that have to be 'interpreted and implemented in different ways' according to the specific technology under consideration (Rayner et al., 2013, p. 504). Although we agree that it is unreasonable to expect robust answers to these questions from a single article, we nonetheless think that the principles should reflect such complexities and not obscure them. We shall return to this issue in a moment. First, let us turn to the fifth principle.

1.5. Governance

The fifth Oxford principle makes a claim about the necessity of governance. It states:

> **Oxford Principle 5: Governance before deployment**
>
> Any decisions with respect to deployment should only be taken with robust governance structures already in place, using existing rules and institutions wherever possible.

This principle is straightforwardly endorsed by the House of Commons report, which also advocates for further research domestically and internationally. Nevertheless, there remain serious concerns.

First, while the idea that deployment requires robust structures of governance seems correct, the principle is striking for what it does not say: namely, that field-testing and perhaps some other forms of research require governance too. The previous discussion suggests ample grounds for this (e.g. inclusion, trust, accountability, values in design).

Second, the narrowness of the Oxford governance principle is all the more surprising since the distinction between deployment and research is not secure when it comes to field testing (e.g. Parson & Ernst, 2013, p. 326). In particular, some have argued that there is no real difference between the more important kinds of field testing and actually deploying geoengineering (Robock, Bunzl, Kravitz, & Stenchikov, 2010). This is acknowledged by the Oxford authors in the accompanying text. Still, it is surprising that the Oxford principles themselves are silent on this issue, and that they insist on treating the two topics so differently. In particular, whereas deployment is to require 'robust' governance, field testing is treated as if a sharply limited approach will do: only those who can be shown to be materially affected in advance have a voice, and only insofar as they are entitled to independent assessment of the impacts and to be 'notified, consulted' and 'ideally' provide informed consent.

Third, a drift away from the fifth Oxford principle is already suggested by our previous discussion of the second (participation) and fourth (independent assessment) principles, since it is not clear that these can do without some substantial level of governance. For example, there is the issue of who decides when transboundary effects are likely (second principle) and who is responsible for getting the independent assessment done (fourth principle). Since this cannot be the original researchers – otherwise it would not be independent – these principles seem to presuppose a social project that already requires a significant level of governance.

Fourth, arguably there are reasons for thinking that the Oxford authors should have taken a much stronger stance, along these lines:

> No deployment or field testing before governance, and strong priority to the governance issue, including perhaps only limited scientific and technical investment in geoengineering before it is clearer whether ethical governance is feasible and what the constraints are.

One such reason is that (as the Royal Society says) the ethical and governance obstacles might be more difficult to overcome than the scientific and technical ones. Another reason is that we might want to avoid the conjurer's trick scenario, where the research is undertaken by an exclusive group according to their own values, objectives, constraints and conception of the problem (e.g. conditions of deployment), and then put forward for action at the last minute, under pressure of a quickly emerging crisis.

Fifth, and more generally, we have to be cautious about the word 'governance', as it may be interpreted in more or less extensive ways (Gardiner, 2011b). In our view, the question goes beyond the mere monitoring and control of geoengineering technologies, and is instead one of moral and political *justification*. Indeed, the core question is whether and how the decisions made about geoengineering can be defended, especially to those seriously affected by them. Thus, a distinct term such as 'ethical accountability' may be required.

Rather than focusing too much on the sheer architecture of institutional arrangements, the most urgent task for geoengineering policy is arguably to work out the content of the normative criteria, such as political legitimacy and justice, that governance institutions should meet. The devising of a new institutional arrangement that is suitable to the governance of geoengineering has to proceed from these normative criteria, not the other way around. However, those that are especially vulnerable to climate change and geoengineering (the global poor, future generations and nonhuman nature) are largely voiceless and difficult to represent in a purely procedural way. This makes the need for ethical accountability at the same time morally necessary and theoretically challenging.

In light of this discussion, we propose an alternative to the fifth Oxford principle:

5th Tollgate Principle (Ethical Accountability): Robust governance systems (including of authority, legitimacy, justification and management) are increasingly needed[30] and ethically necessary[31] at each stage from advanced research[32] to deployment.

Robust structures of consultation, administration and justification are ethically necessary before deployment, but a significant level should also be in place prior to commencing field testing (domestic and international), and some governance is desirable even before that, as advanced research programs emerge. These structures should be flexible in light of new information and socio-political-ecological realities, and should aim to avoid lock-in and the entrenchment of vested interests. (Jamieson, 1996; Lin, 2016; McKinnon, 2018)

2. Jamieson's Principles

At this point, we might note a more general way in which the Oxford principles, on which the first five Tollgate principles are based, seem lacking: namely, none of them speak directly to the *substance* of a justifiable geoengineering program.[33] This is a surprising omission.[34] Fortunately, we do get some guidance from elsewhere. First, we will discuss principles for geoengineering offered more than a decade earlier in a classic article by Dale Jamieson, but curiously ignored by the Oxford principles and most other reports. Jamieson proposes that geoengineering should be *technically feasible, reliably predictable, socioeconomically preferable* and *respect well-founded ethical norms*, such as democratic decision-making, a prohibition on irreversible changes, and respect for nature (Jamieson, 1996). We shall build on and add to this approach. Second, we will consider some more general resources in practical ethics. Given constraints of space, we will only gesture at the relevant issues very briefly, to give a sense of the direction we believe should be pursued. We defer a deeper discussion to another occasion.

2.1. *Technical Feasibility*

Jamieson's first principle requires that geoengineering be 'technically feasible'. In one way, this requirement looks trivial. Suppose climate engineering is defined as 'intentional intervention in the climate system to address global climate change'. If the question is: 'Is it possible for humans to intervene in the climate system and make a difference?', the answer seems highly likely to be 'yes'. The phenomenon of anthropogenic climate change already implies that humans are capable of having influence unintentionally, and so do have the power to have effects. The real question is thus not whether large-scale intervention is technically feasible, but whether *desirable* forms of large-scale intervention are technically feasible.

Jamieson's other three principles speak to this. Hence, the first must be understood in terms of them, as demanding that desirable forms of geoengineering must be technically feasible. However, this is not to say that the first principle lacks bite. In particular, some believe that ethically reasonable forms of geoengineering are not technically feasible. For example, some claim that we will not be able to produce reliable and socioeconomically preferable techniques in the time available. If this is so, then even if ethical geoengineering is possible in principle, in practice it is not, and so pursuit of it is likely to be wasteful of (scientific and political) time and energy. Given the scarce (scientific and political) resources available to address climate change, this would be a significant concern. Consequently, something like Jamieson's first principle has a role to play even as the focus turns to the other three. We therefore propose:

6th Tollgate Principle (Technical Availability): For a geoengineering technique to be policy-relevant, *ethically defensible forms* of it must be technically feasible on the relevant timeframe.

Since geoengineering is defined with respect to specific policy goals (counteracting anthropogenic climate change), relevant techniques must be technically feasible ways of both addressing these goals, and satisfying appropriate parameters and constraints (internal and external). Both the goals themselves and the associated parameters and constraints are heavily ethical (cf. Gardiner & Weisbach, 2016, chapters 2–3). Therefore, the central concern with technical feasibility is with the (potential) availability of ethically defensible forms of geoengineering. These include techniques that may plausibly be deployed to counteract climate change on a timeframe relevant to meeting core ethical objectives, such as moderating threats to human and nonhuman welfare, or protecting basic human rights. That being said, ethically defensible forms may also include ethically laudable, decent, and perhaps even 'tolerably indecent' approaches, which raises issues of its own. (Gardiner & Fragnière, 2017)[35]

2.2. Predictability

Jamieson's second principle is that acceptable geoengineering should be reliably predictable. While this condition may initially seem uncontroversial, in the geoengineering context it is sometimes violated. For instance, some emergency arguments for SSI may implicitly set aside reliable prediction (e.g. framing SSI as a 'Hail Mary' pass undertaken in the *hope* that it will push the climate in a better direction, without side effects that are too bad). We agree with Jamieson that a reasonable level of evidence is required (e.g. mere hope does not suffice), and in particular that serious assurance is needed that intervention will not make matters significantly worse. While we also acknowledge that what counts as a reasonable level of evidence might vary with the circumstances, including the nature of the threat and the available alternatives, we would still insist that interventions meet scientific standards appropriate to the circumstances. Given this, we propose:

7th Tollgate Principle (Predictability): For a geoengineering technique to be policy-relevant, ethically defensible forms of it must be reasonably predictable on the relevant timeframe and in relation to the threat being addressed.

The policy relevance of particular geoengineering techniques depends in part on whether ethically defensible forms are (or may plausibly become) reasonably predictable, including on a timescale relevant to useful deployment and with suitable assurances that deployment would not involve additional threats or levels of risk comparable to or more severe than those the intervention is intended to alleviate (cf. Gardiner, 2006; Hartzell, 2012; Shue, 2010). The assurances should be grounded in comprehensive scientific appraisals that meet standards of evidence appropriate to the proposed action and situation. What that would mean under specific scenarios is a matter for further deliberation in accordance with the other Tollgate Principles.

2.3. Preferability

Jamieson's third principle – that geoengineering be socioeconomically preferable to its alternatives – gets closer to the heart of the issue. It raises the question of how we should assess the desirability of a given scheme. First, the principle raises the point that it is not really the climate indicators (e.g. global temperature, precipitation patterns) that matter, ultimately. What matters is the effects on, and treatment of, ethically significant subjects in terms of indicators such as welfare, human rights, justice, political entitlements and so on. In light of this, we should be careful about adopting too narrow an account of the indicators. For instance, one adjustment we would make is to frame the indicators more widely to include ecological values, such as the interests of nonhuman animals and ecosystems.

Second, Jamieson's principle suggests that the partial truth in the universal benefit principle – and so latent in the public good framing – is the idea that the kind of geoengineering we are most interested in is one that treats all those affected well, or at least decently, in terms of ethical parameters, and especially the most basic parameters, such as their fundamental rights and interests. Instead of 'universal benefit' then, what we are looking for, or aspiring to (if we are considering geoengineering at all), is a geoengineering policy that protects those affected better than any alternative, and protects them at a fundamental level. The relevant alternatives here are presumably allowing severe climate change to occur unchecked, or any other feasible policy interventions that do not include some geoengineering.

However, third, and most importantly of all, this need not mean that the *technical intervention alone* provides the right kind of protection *all by itself*. Instead, the requirement should apply to the overall global, social, economic and political system. This system should interface with the intervention in an integrated way, so as to provide the best protection available (Gardiner, 2013a, p. 513). In other words, it is the *overall social world* in which geoengineering occurs that should satisfy the requirement, and this can be done even if the intervention standing alone does not. In particular, the global social, economic and political system can intervene directly or indirectly to protect those affected by both climate change and geoengineering from negative impacts (e.g. through aid, migration programs, etc.). Offhand, this seems at least as feasible as attempting to provide protection exclusively through technical means, and perhaps much more so. Yet it involves a very different vision of what dealing with climate change, including in part through geoengineering, actually involves.[36]

In our view, these points suggest:

8th Tollgate Principle (Protection): Climate policies that include geoengineering schemes should be socially and ecologically preferable[37] to other available climate policies, and focus on protecting basic ethical interests and concerns (e.g. human rights, capabilities, fundamental ecological values).

Geoengineering interventions should be understood as parts of more general climate policies, and such policies should be assessed comparatively. For example, we should avoid framing the relevant question as one of which kind of action (e.g. geoengineering, mitigation, adaptation) or technique (e.g. SSI, CDR, building sea walls) is to be preferred. Instead, the attention should be on overall policies, where these are likely to include several kinds of action and techniques, and on the social system as a whole, including whether fundamental social change is warranted to address the climate threat, either instead of or in conjunction with any geoengineering interventions. Policies should be evaluated on broad ethical grounds, reflecting social values (including political and economic values) as well as ecological values, and with a central focus on protecting basic interests and concerns.

2.4. Respecting Ethical Norms

Jamieson's fourth principle requires that geoengineering interventions *respect well-founded ethical norms*, such as democratic decision-making, a prohibition on irreversible changes, and respect for nature.

a. General Norms

We begin by proposing a more general approach, initially grounded in W. D. Ross' pluralistic approach to ethical values (Ross, 2002 [1930]). Ross provides a list of principles that he thinks of as '*prima facie* duties'. These are often thought to provide a 'default' list of generally morally salient concerns for ethical deliberation (e.g. Timmons, 2002, p. 208), and this is the sense in which we invoke them here.[38] One advantage of this approach is that the relevance of such concerns is a matter of wide and overlapping consensus across different normative theories and ethical traditions. Another is that they provide a foundation on which developments of 'top-down' approaches to the ethics of geoengineering and its governance can occur, not least because they anchor discussions within and between contrasting theories and traditions.

Since their initial appearance, Ross' duties have been interpreted, updated and supplemented by others. A contemporary list of principles of ethical salience might read (see Resnik, 1998):

- **Non-maleficence**: Do not harm yourself or other people.
- **Reparation**: Repair harms you have done.
- **Justice**: Treat people fairly: treat equals equally, unequals unequally.[39]
- **Autonomy**: Allow rational individuals to make free and informed choices.
- **Gratitude**: Appropriately acknowledge benefits that others have given you.
- **Beneficence**: Help yourself and other people.
- **Efficiency**: Maximize the ratio of benefits to harms for all people.
- **Fidelity**: Keep your promises and agreements.
- **Honesty**: Do not lie, defraud, deceive or mislead.
- **Self-improvement**: Strive to make yourself ethically better.

In our view, most of these principles are relevant to geoengineering,[40] and an ethically defensible approach to geoengineering would address them, showing how and why they should be incorporated in institutions and policies. We defer this for another occasion. Our main purpose here is merely to signal the claim, and point out that conventional proposals –

such as the Oxford principles – seem to ignore most (if not all) of the Rossian concerns. We thus propose:

9th Tollgate Principle (Respecting General Ethical Norms): Geoengineering policy should respect general ethical norms that are well-founded and salient to global environmental policy (e.g. justice, autonomy, beneficence).

Ethical responses to climate change, including those involving geoengineering techniques, must be appropriately responsive to (i.e. fully take into account) general and well-founded ethical norms, many of which underlie the basic concerns, motivation, and parameters of climate policy in the first place. Such norms might initially be generated from common principles of ethical salience (e.g. Ross' *prima facie* duties), but can also provide a grounding for more 'top-down' discussions within and between different normative theories and ethical traditions.

b. Ecological Norms

In addition to the (revised) Rossian list, other matters become ethically salient in geoengineering and similar environmental cases that are worthy of inclusion. In particular, the list makes no mention of relevant environmental or ecological norms, yet these also seem of central importance to geoengineering governance. Consider two major areas.

First, one set of norms concerns the parameters and constraints of environmental governance. Jamieson's prohibition on irreversible changes is one example. We are not confident in this specific constraint (e.g. given the rapid pace of climate change, it is not clear what 'irreversible' means or why it is a particularly salient concern). Nevertheless, we believe that the category remains important. As better illustrative examples, we propose:

- **Sustainability**: Interventions should be compatible with and promote the long-term survival and flourishing of human and nonhuman forms of life.

- **Precaution**: Where there are threats of serious or irreversible harm, a lack of full scientific certainty should not be used as a reason for postponing suitably effective measures.[41]

The second area is humanity's relationship to nature. No mention is made of this in the Rossian list; yet it is a major concern in the ethics of geoengineering. One sign that this is a problem is that (presumably) there would be major ethical objections to a geoengineering intervention that protected current and future people, but only at the price of eradicating all other species and all natural places from the planet.[42] Less drastically, concerns about mass extinction and the destruction of distinctive ecosystems are central to many peoples' worries about climate change, and therefore also geoengineering. As a result, some – such as Jamieson – suggest that an appeal to respect for nature is crucial to spurring the motivation necessary for addressing climate change (Jamieson, 2010, 2014), and he includes this in his list of well-founded ethical norms.

In light of such concerns, we suggest adding two further principles of ethical salience to our list:

- **Respect for nature**: Respect natural organisms, ecological communities and wild places.

- **Ecological accommodation**: Seek to live within, among, and together with the rest of nature.

These principles require further specification. Still, arguably, this is no less so than for other principles, including the Oxford principles. More importantly, the lack of precision is not so important in this context. The role of the principles is merely to pick out ethical matters that should be taken into account. Working out what they mean in detail, how they should be incorporated into decision-making, and specifically how they can be integrated with other salient issues is a matter for subsequent discussion. Yet, importantly, none of this implies that the principles of ethical salience do nothing. On the contrary, they cast doubt on certain policies and help to remove blindspots. For instance, in some cases, there are clear violations of their spirit, as when some call for the wholescale domination and annexation of nature for the sake of human aggrandizement.

We conclude by proposing one further Tollgate Principle:

10th Tollgate Principle (Respecting Ecological Norms): Geoengineering policy should respect well-founded ecological norms, including norms of environmental ethics and governance (e.g. sustainability, precaution, respect for nature, ecological accommodation).

3. Conclusion

Our discussion of the Oxford principles has yielded the following set of conclusions and alternative proposals. First, we should reject the public goods framing of geoengineering because it is overly simplistic, obscures the choices to be made within and between geoengineering options, thereby conceals the ethical and especially justice dimensions of the issue, and is also misleading about the important question of public and private provision. Instead, we should replace the first Oxford Principle with:

1st Tollgate Principle (Framing): Geoengineering should be administered by or on behalf of the global, intergenerational and ecological public, in light of their interests and other ethically relevant norms.

Second, we should broaden the second Oxford principle concerning public participation, since its actual provisions are too narrow and more properly understood as requiring only consultation for a narrow range of research subjects of field testing. Instead, we should endorse:

2nd Tollgate Principle (Authorization): Geoengineering decision-making (e.g. authorizing research programs, large-scale field trials, deployment) should be done by bodies acting on behalf of (e.g. representing) the global, intergenerational and ecological public, with appropriate authority and in accordance with suitably strong ethical norms (e.g. justice, political legitimacy).

3rd Tollgate Principle (Consultation): Decisions about geoengineering research activities should be made only after proper notification and consultation of those

materially affected and their appropriate representatives, and after due consideration of their self-declared interests and values.

Third, we should subsume both the full disclosure and independent assessment principles under a wider principle highlighting trust and accountability. For one thing, we should enlarge the scope of concern to include not just disclosure of research plans and results and assessment of impacts, but also participation in design, inclusion and diversity of scientists, scientific discipline, nationality and demographic groups (including national, subnational, regional, etc.). For another, we should acknowledge that what counts as geoengineering research, deployment and the very notion of impacts needs to be understood in socio-economic and ethical terms as well as scientific. Hence, the domain will necessarily involve a range of value assumptions that should be deliberated on and endorsed within a wider process. For those reasons, it is better to replace these principles with:

4th Tollgate Principle (Trust): Geoengineering policy should be organized so as to facilitate reliability, trust and accountability across nations, generations and species.

Fourth, we should broaden the requirement of the fifth principle that robust governance structures be in place before deployment. On the one hand, serious structures are required prior to full deployment and in fact are presupposed by the Oxford principles concerning field testing. One important reason for this is that for at least some techniques, there is not a sharp separation between field testing and deployment for SRM sulfate injection. On the other hand, governance is too narrow a term for what is at stake, and the 'wait and see' approach presupposes that reflection on these issues will not bring on serious constraints on what kinds of geoengineering are feasible. Hence, ideally we need a coevolution of consultation, administration and justification with scientific research. We would not want to see a large waste of research effort on socio-political-ecological unfeasible approaches, just as not of socio-political-ecological effort on scientifically unfeasible approaches.

5th Tollgate Principle (Ethical Accountability): Robust governance systems (including of authority, legitimacy, justification and management) are increasingly needed and ethically necessary at each stage from advanced research to deployment.

Finally, we need substantive principles speaking to the very core of the issue, namely questions of ethics and justice. Inspired by Dale Jamieson's early work, the aim of these principles is to make sure that geoengineering policy is not biased in favor of the powerful of the current generation and not decided on the basis of shallow criteria such as economic efficiency alone:

6th Tollgate Principle (Technical Availability): For a geoengineering technique to be policy-relevant,*ethically defensible forms* of it must be technically feasible on the relevant timeframe.

7th Tollgate Principle (Predictability): For a geoengineering technique to be policy-relevant, ethically defensible forms of it must be reasonably predictable on the relevant timeframe and in relation to the threat being addressed.

8th Tollgate Principle (Protection): Climate policies that include geoengineering schemes should be socially and ecologically preferable to other available climate

policies and focus on protecting basic ethical interests and concerns (e.g. human rights, capabilities, fundamental ecological values).

9th Tollgate Principle (Respecting General Ethical Norms): Geoengineering policy should respect general ethical norms that are well-founded and salient to global environmental policy (e.g. autonomy, justice).

10th Tollgate Principle (Respecting Ecological Norms): Geoengineering policy should respect well-founded ecological norms, including norms of environmental ethics and governance (e.g. sustainability, precaution, respect for nature, ecological accommodation).

These revisions have implications. The first is that the Tollgate principles are more demanding than the Oxford principles at all levels between research and deployment. The second is that their objectives seem threatened by current sociopolitical realities. In particular, really fulfilling the objectives would require incorporating the concerns of and reasonable protections for the global population, future generations and nature. This is daunting task for which current institutions appear poorly prepared. One advantage of the Tollgate Principles is that they help to draw attention to the multiple dimensions in which this may be so, and so encourage us to wrestle with these difficult questions in a way that previous attempts do not.

Notes

1. Since the first version of the Oxford Principles, a large number of reports have emerged, many of which propose their own principles (e.g. Shepherd et al., 2009; ASOC, 2010; Bipartisan Policy Center, 2011; Bodle & Oberthür, 2014; Schäfer, Lawrence, Stelzer, Born, & Low, 2015; National Research Council, 2015; for a useful synthesis, see Morrow, 2017). We focus on the Oxford Principles because of their historical role, continued relevance, strong overlap with many subsequent proposals, and ongoing influence, including in philosophical circles (e.g. Moellendorf, 2014). Additional relevant proposals have emerged while this article was under review (e.g. Hubert, 2017; UNESCO, 2017) and include some developments that accord with our Tollgate approach. We defer discussion of those proposals for another occasion.
2. We intend 'ethical' in a broad sense, to include the concerns of moral and political philosophy, as well as the normative dimensions of international relations, politics, law and economics. As philosophers, our concerns are primarily with underlying normative ideals and concepts. We are not, for instance, focused on assessing the legal dimensions of potential geoengineering governance or its relationship to existing international law, though our principles are relevant to the ethical foundations and implications of such law. We also do not attempt to provide details of implementation (including lower-level principles of implementation), but rather offer general guidance for that task.
3. Ironically, the Tollgate was subsequently bought out by a 'fabulously wealthy', 'pioneering champion of luxury eco-chic' and transformed from a community pub into a gathering place for the social, economic and political elite ('the poshest pub in Britain'). See Moir (2013).
4. Notably, in their introduction the Oxford authors compare their proposal to the Belmont Principles governing medical research (2013, p. 504). Those principles – *respect for persons, beneficence* and *justice* – are very substantive. However, in the Oxford Principles themselves, the only substantive requirement lies implicitly in the public good framing of the first principle, which we criticize below. Since it is clear that paradigm forms of geoengineering

pose acute issues of respect, beneficence and justice (and other values), we strive to include these considerations in our Tollgate principles.

5. Similar principles of public participation, transparency and independent assessment have been endorsed by virtually every subsequent report (ASOC, 2010; Bipartisan Policy Center, 2011; Bodle & Oberthür, 2014; National Research Council, 2015; Schäfer et al., 2015; Shepherd et al., 2009). They constitute what might be called the core procedural principles. However, as we will see, their interpretation is far from straightforward.
6. Notably, those who reject the term 'geoengineering' often propose instead separating 'solar radiation management' (SRM) techniques from 'carbon dioxide removal' (CDR) techniques. However, arguably, this simply recreates the problem of diversity at a lower level. For instance, some SRM techniques (e.g. painting roofs white) have few of the ethics- and policy-relevant features of SSI and more in common with some CDR techniques (e.g. reforestation). Similarly, the paradigm case of CDR – direct air capture on a truly massive scale – bears many similarities with SSI (Gardiner 2011, pp. 344–5).
7. This section draws heavily on Gardiner (2013b, 2014b).
8. Notably, several uses feature in presentations of the first Oxford Principle by the Oxford authors (see below).
9. This is so for both 'public good' and 'global public good'. For examples, see Gardiner (2013b, 2014b).
10. This definition is repeatedly emphasized by the editors and motivates their identification of a minimum (Pareto-style) threshold for impure global public goods (Kaul, Grunberg, & Stern, 1999; p. 510, 2–3, 11, 16; Kaul, Condeicao, Le Goulven, & Mendoza, 2003; p. 605; cf. World Bank 2001, p. 129). Nevertheless, it is worth noting that there is a fair amount of internal movement on the definitional issue, especially over time. Notably, (i) in the 1999 foundational work, the UNDP (a) includes universal benefit in the definition of 'global public good' even while (b) defining 'public good' itself narrowly, in terms of nonrivalry and nonexcludability alone (Kaul et al., 1999, p. 511); whereas (ii) in later work (a) they opt for a more neutral *definition* of 'global public good' that does not include the universal benefit claim, even though (b) the universal benefit claim continues to play an important role in motivating the interest in global public goods and their relevance for public policy.
11. This is not at all to say that this is what the Oxford authors intend. The point here is that the initial framing of geoengineering as a public good encourages the rosy picture, particularly when it constitutes the sole substantive principle. So, there is reason to favor a more transparent principle, particularly in the public sphere.
12. For example, perhaps Barrett's inference from 'global public goods make people everywhere better off' to 'Global public goods are *thus* universally to be desired' (2007b, p. 1; emphasis added).
13. Note that we do not assume that other forms of geoengineering, such as CDR, are also 'single-best effort' public goods.
14. CDR technologies may also face significant drawbacks. For example, the currently most favored technology, known as BECCS, might result in 'land grabs', public health issues and problems of carbon storage security (Shrader-Frechette, 2015).
15. Cf. Gardiner's example of your breaking into my house in order to clean it (Gardiner, 2013b).
16. In the 2013 version, the Oxford Principles authors mention two weaker interpretations of 'for the benefit of all'. First, the Pareto principle requires that at least some benefit and no one be made worse off. Second, the Kaldor-Hicks criterion holds that some can be made worse off if those benefitting from the scheme can compensate them in principle. However, these principles run into similar objections as the universal benefit requirement. For instance, the Pareto requirement seems too strong: for example, why is it impermissible for very rich people to endure small costs if it saves the rest? Similarly, the Kaldor-Hicks requirement seems much too weak. As compensation need not actually be paid, there is no meaningful sense in which everyone benefits: some are simply sacrificed to benefit others.
17. Perhaps one could resurrect a universal benefit criterion by adopting a very expansive conception of universal benefit that includes other ethical considerations beyond welfare. In

their later work, the Oxford authors may be suggesting something like this when they say 'specifying exactly what counts as "the benefit of all" requires consideration of global and intergenerational justice' (Rayner et al., 2013). Nevertheless, this is almost all that they say, other than offering a couple of examples that highlight concepts (i.e. Pareto, Kaldor-Hicks) more familiar from welfarist approaches than theories of justice. Moreover, it is unclear how an expansive conception fits with other claims about public goods they appear to endorse (e.g. what would it mean to say that the expansive kind of global public good is nonrival and nonexcludable?). More importantly, the expansive conception would also push us toward full-blown moral and political philosophy, as it raises many of the same ethical questions. Finally, presupposing a nonstandard and expansive account conceals more than it reveals. It is therefore deeply misleading in the interdisciplinary policy context of geoengineering. Consequently, we favor a more transparent approach that makes clear the centrality and multiplicity of the ethical concerns (see below).

18. Similarly, different levels of carbon dioxide removal (e.g. 50 or 150 ppm) cannot all be achieved simultaneously.
19. (Morrow, 2017), p. 10. For instance, Morrow highlights that while existing reports 'generally agree that a primary objective ... is to manage the physical and political risks of [geoengineering] research and potential deployment ... that agreement masks an important disagreement about the specific risk-risk trade-off involved in researching [geoengineering] as opposed to prohibiting or foregoing research' (Morrow, 2017, p. 10).
20. The principles of public participation, transparency and independent assessment are endorsed by virtually every report so far (ASOC, 2010; Bipartisan Policy Center, 2011; Bodle & Oberthür, 2014; National Research Council, 2015; Schäfer et al., 2015; Shepherd et al., 2009). They constitute what might be called the core procedural principles. However, as we will see shortly, their interpretation is far from straightforward.
21. See also Blomfield (2015) and Smith (2017).
22. Compensation is mentioned with respect to deployment in the commentary of Principle 5. However, note that compensation after the fact for a harm from a project is very different from consultation in advance about whether the project should proceed. See below.
23. For different ways of understanding what would count as consent to geoengineering, see Wong (2015).
24. Note again the problem that only those directly affected will have any say.
25. Ultimately, principles of geoengineering governance, including our principles and the Oxford Principles, also require such authorization. In our view, the presenting of such principles (whether by academics, policy groups or organizations) is primarily a bottom-up process to promote discussion prior to, and as an aid to, eventual authorization. It would be ethically unreasonable, as well as politically unrealistic, to see such principles as 'top-down' impositions on the global order.
26. In practice, the relevant 'rogues' may be restricted to actors with the implicit backing from major powers (e.g. Gardiner, 2013a).
27. As well as technical possibilities, research choices are often strongly influenced by the nonscientific values of researchers or funding institutions. These values intervene at many levels, including in the basic design of the research program, and (as we are seeing) in the principles and institutions that govern it. Given this, it is very important that the process is scrutinized at every level, and from a variety of perspectives. This is well known in philosophy of science (e.g. medical research).
28. As well as assisting in the first two ways, this may play an important role in the development of the subject, and of wider scientific and social ties between participating institutions, disciplines and societies. For example, openness is assumed to facilitate and encourage global participation in geoengineering research. The hope is that this assists in making geoengineering research a genuinely global public project at least among researchers. Among other things, this may aid in disrupting attempts to create parochial

or predatory programs, and so undermine the prospects for a geoengineering arms race.

29. David Morrow indicates to us that this suggests another way in which the Oxford approach is limited. The baseline concerns naturally arise for deployment and large-scale, long-term field trials, so perhaps the fourth Oxford principle should be restricted to very low-impact and near-term research.
30. That is, as the stakes get higher.
31. That is, to satisfy basic normative requirements, such as of justice and political legitimacy.
32. Advanced, mission-driven, geoengineering research programs can benefit from governance even prior to field tests for various reasons, including to ensure that projects are directed toward appropriate ends and in light of the other Tollgate Principles (see also Blomfield, 2015; Smith, 2017). This is partly for the good for such programs to ensure that their research remains relevant to their mission.
33. Tellingly, most reports on geoengineering are similarly limited. For instance, questions of justice are absent or only mentioned in passing. The closest to a robust treatment of the question is the recent EuTRACE report which devotes five pages to issues of justice. However, even in this case, the only substantive element in the proposed principles of governance is a principle of harm minimization (Schäfer et al., 2015).
34. However, recall that this may partly be due to the framing effects of the public goods principle, which may implicitly assume that, because geoengineering is a public good, there will be no objections and that in principle geoengineering is an unambiguously good thing.
35. Some (especially advocates of geoengineering) worry that asking for ethical defensibility sets the bar too high; others (especially skeptics) are concerned that it may be compatible even with truly lamentable forms of intervention (cf. Gardiner & Fragnière, 2017). Still, much of this debate is really about what the ethical criteria actually are, and what the alternatives might be. One advantage of the Tollgate framework is that it highlights and thereby facilitates that debate.
36. One reason this approach is important is the tendency of some framings of geoengineering (e.g. as 'Plan B') implicitly to isolate geoengineering techniques from comparisons with other policy options (Fragnière & Gardiner, 2016).
37. Ideally, the chosen policy would be superior on each dimension; however, the principle does not require this: it maintains only that these are the most central, ethically relevant dimensions.
38. Ross maintains (i) that when only one duty is relevant, it holds sway, but when more than one are present, the concerns have to be weighed against each other. We shall not go this far and instead regard the relevant duties more weakly, as (ii) principles of ethical salience that capture concerns that should be taken into account when they arise, but which may not be decisive, even if they arise alone. However, Ross also (iii) does not order his principles according to any kind of priority rule, and (iv) does not assert that his list is complete, but leaves it open-ended. We follow him in these claims.
39. Note that for Ross, justice is 'a distribution of happiness between (…) people in proportion to merit' (Ross 2002 [1930], p. 26). However, most contemporary conceptions of justice are considerably broader, and it is the broader notions we have in mind here.
40. For example, Jamieson's norm of democratic decision-making can be seen as grounded in autonomy and justice.
41. This characterization is based a common, minimal definition (e.g. UNESCO, 2017; UNFCCC, 1992). Our views on precaution are more complex and potentially more demanding (e.g. Gardiner, 2006; Hartzell, 2012; Shue, 2010); however, we set such issues aside here.
42. Cf. Gardiner's 'Dome World' scenario (Gardiner, 2011a).

Acknowledgments

For comments, we thank Alex Lenferna and two external referees (one of whom identified himself as David Morrow). The views expressed remain our own.

Disclosure statement

No potential conflict of interest was reported by the authors.

Funding

The research was supported by the Smith School of Enterprise and Environment at Oxford University (in the form of a visiting fellowship for Gardiner in 2012, when the first draft was written); the Swiss National Science Foundation [Fragnière, Grant P300P1_161110]; and the National Science Foundation [Grant 1549983].

References

Anderson, K., & Peters, G. (2016). The trouble with negative emissions. *Science, 354*(6309), 182–183.

ASOC. (2010). *The asilomar conference recommendations on principles for research into climate engineering techniques.*

Barrett, S. (2007a). The incredible economics of geoengineering. *Environmental and Resource Economics, 39*(1), 45–54.

Barrett, S. (2007b). *Why cooperate? The incentive to supply global public goods.* New York: Oxford University Press.

Bawden, T. (2016, January 8). COP21: Paris deal far too weak to prevent devastating climate change, academics warn. *The Independent.* Retrieved from http://www.independent.co.uk/environment/climate-change/cop21-paris-deal-far-too-weak-to-prevent-devastating-climate-change-academics-warn-a6803096.html

Bipartisan Policy Center. (2011). *Geoengineering: A national strategic plan for research on the potential effectiveness, feasibility, and consequences of climate remediation technologies.* Washington, DC: Author.

Blomfield, M. (2015). Geoengineering in a climate of uncertainty. In J. Moss (Ed.), *Climate change and justice* (pp. 39–58). Cambridge, UK: Cambridge University Press.

Bodansky, D. (2012). What's in a concept? Global public goods, international law, and legitimacy. *European Journal of International Law, 23*(3), 651–668.

Bodle, R., & Oberthür, S. (2014). *Options and proposals for the international governance of geoengineering.* Berlin, Germany: Federal Environmental Agency of Germany.

Caldeira, K. (2012, April 12). Ethics of geoengineering (Food). *Geoengineering Google Group.*

Climate Action Tracker. (2015, December 8). 2.7C is not enough - we can get lower. Retrieved from http://climateactiontracker.org/assets/publications/briefing_papers/CAT_Temp_Update_COP21.pdf

EASAC. (2018). *Negative emission technologies: What role in meeting Paris agreement targets?* (No. 35). Halle, Germany: Author.

Fragnière, A., & Gardiner, S. M. (2016). Why geoengineering is not "Plan B.". In C. J. Preston (Ed.), *Climate justice and geoengineering: Ethics and policy in the atmospheric anthropocene* (pp. 15–32). London: Rowman and Littlefield.

Gardiner, S. M. (2006). A core precautionary principle. *Journal of Political Philosophy, 14*(1), 33–60.

Gardiner, S. M. (2011a). *A perfect moral storm: The ethical tragedy of climate change.* New York: Oxford University Press.

Gardiner, S. M. (2011b). Some early ethics of geoengineering the climate: A commentary on the values of the Royal Society report. *Environmental Values, 20*(2), 163–188.

Gardiner, S. M. (2013a). The desperation argument for geoengineering. *PS: Political Science & Politics, 46*(1), 28–33.
Gardiner, S. M. (2013b). Why geoengineering is not a "global public good", and why it is ethically misleading to frame it as one. *Climatic Change, 121*(3), 513–525.
Gardiner, S. M. (2014a). A call for a global constitutional convention focused on future generations. *Ethics & International Affairs, 28*(3), 299–315.
Gardiner, S. M. (2014b). Why "global public good" is a treacherous term, especially for geoengineering. *Climatic Change, 123*(2), 101–106.
Gardiner, S. M. (2016). Geoengineering: Ethical questions for deliberate climate manipulators. In S. M. Gardiner & A. Thompson (Eds.), *The Oxford handbook on environmental ethics*. Oxford: Oxford University Press.
Gardiner, S. M. (2017). The threat of intergenerational extortion: On the temptation to become the climate mafia while masquerading as an intergenerational Robin Hood'. *Canadian Journal of Philosophy*, *47*(2–3), 368–394.
Gardiner, S. M., & Fragnière, A. (2017, April). If a climate catastrophe is possible, is everything permitted? Paper presented at the Western Political Science Association, Vancouver, BC.
Gardiner, S. M., & Weisbach, D. A. (2016). *Debating climate ethics*. New York: Oxford University Press.
Gonzalez-Ricoy, I., & Gosseries, A. (Eds.). (2017). *Institutions for future generations*. Oxford: Oxford University Press.
Hartzell, L. (2012). Precaution and solar radiation management. *Ethics, Policy and Environment, 15*(2), 158–171.
Heyward, C. (2013). Situating and abandoning geoengineering: A typology of five responses to dangerous climate change. *PS: Political Science & Politics, 46*(1), 23–27.
House of Commons Science and Technology Committee. (2010). *The regulation of geoengineering*.
Hubert, A.-M. (2017). Code of conduct for responsible geoengineering research. Retrieved from http://www.ucalgary.ca/grgproject/files/grgproject/revised-code-of-conduct-for-geoengineering-research-2017-hubert.pdf
IPCC. (2014). K. Pachauri & L. A. Meyer, Core Writing Team, R (Eds.). *Climate change 2014: Synthesis report*. Geneva, Switzerland: IPCC.
Jamieson, D. (1996). Ethics and intentional climate change. *Climatic Change, 33*(3), 323–336.
Jamieson, D. (2005). Adaptation, mitigation, and justice. In W. Sinnott-Armstrong & R. B. Howarth (Eds.), *Perspectives on climate change* (pp. 221–253). Oxford, UK: Elsevier.
Jamieson, D. (2010). Climate change, responsibility, and justice. *Science and Engineering Ethics, 16*(3), 431–445.
Jamieson, D. (2013). Some whats, whys and worries of geoengineering. *Climatic Change, 121*(3), 527–537.
Jamieson, D. (2014). *Reason in a dark time: Why the struggle against climate change failed – And what it means for our future*. New York: Oxford University Press.
Kaul, I., Condeicao, P., Le Goulven, K., & Mendoza, R. U. (eds). (2003). *Providing global public goods*. Oxford: Oxford University Press.
Kaul, I., Grunberg, I., & Stern, M. A. (eds). (1999). *Global public goods*. Oxford: Oxford University Press.
Kriegler, E., Luderer, G., Bauer, N., Baumstark, L., Fujimori, S., Popp, A., … van Vuuren, D. P. (2018). Pathways limiting warming to 1.5°C: A tale of turning around in no time? *Philosophical Transactions of the Royal Society A: Mathematical, Physical and Engineering Sciences, 376*(2119), 20160457.
Lin, A. C. (2016). The missing pieces of geoengineering research governance. *Minnesota Law Review, 100*, 2509–2576.
McKinnon, C. (2018). Sleepwalking into lock-in? Avoiding wrongs to future people in the governance of solar radiation management research. *Environmental Politics, 118*(17), 1–19.
Moellendorf, D. (2014). *The moral challenge of dangerous climate change: Values, poverty, and policy*. New York: Cambridge University Press.

Moir, J. (2013, September 8). The poshest pub in Britain. *Daily Mail*. Retrieved from: http://www.dailymail.co.uk/debate/article-2415665/The-Wild-Rabbit-The-poshest-pub-Britain-owned-tycoons-wife-stuffed-celebs.html

Morrow, D. R. (2014). Why geoengineering is a public good, even if it is bad. *Climatic Change, 123*, 95–100.

Morrow, D. R. (2017). International governance of climate engineering: A survey of reports on climate engineering, 2009–2015. *FCEA Working Paper Series*, 1–41.

National Research Council. (2015). *Climate intervention: Reflecting sunlight to cool earth*. Washington D.C.: The National Academies Press.

Nature Editorial. (2012). A charter for geoengineering [editorial]. *Nature, 485*:415.

Parson, E. A., & Ernst, L. N. (2013). International governance of climate engineering. *Theoretical Inquiries in Law, 14*(1), 307–338.

Rayner, S. (2011). The challenges of geoengineering governance. In *Pilot workshop on governing geoengineering in the 21st century: Asian perspectives* (pp. 10–11). Singapore: RSIS Centre for Non-Traditional Security (NTS) studies.

Rayner, S., Heyward, C., Kruger, T., Pidgeon, N., Redgwell, C., & Savulescu, J. (2013). The Oxford principles. *Climatic Change, 121*(3), 499–512.

Rayner, S., Redgwell, C., Savulescu, J., Pidgeon, N., & Kruger, T. (2009). *Memorandum on draft principles for the conduct of geoengineering research*. London, UK: House of Commons Science and Technology Committee.

Resnik, D. B. (1998). *The ethics of science: An introduction*. London: Routledge.

Ricke, K., Moreno-Cruz, J., & Caldeira, K. (2013). Strategic incentives for geoengineering coalitions to exclude broad participation. *Environmental Research Letters, 8*, 014021.

Robock, A., Bunzl, M., Kravitz, B., & Stenchikov, G. L. (2010, January 29). A test for geoengineering? *Science, 324*, 530–531.

Ross, D. (2002 [1930]). *The right and the good*. New York: Oxford University Press.

Schäfer, S., Lawrence, M., Stelzer, H., Born, W., & Low, S. (2015). *The european transdisciplinary assessment of climate engineering (EuTRACE)*.

Shepherd, J. (2016, February 17). What does the Paris agreement mean for geoengineering? Retrieved from http://blogs.royalsociety.org/in-verba/2016/02/17/what-does-the-paris-agreement-mean-for-geoengineering/

Shepherd, J., Caldeira, K., Cox, P., Haigh, J., Keith, D., Launder, B., Watson, A. (2009). *Geoengineering the climate*: Science, governance and uncertainty (pp. 1–98). London, UK: The Royal Society.

Shrader-Frechette, K. (2015). Biomass incineration: Scientifically and ethically indefensible. In A. Maltais & C. McKinnon (Eds.), *The ethics of climate governance* (pp. 155–172). London: Rowman & Littlefield International.

Shue, H. (2010). Deadly delays. In S. M. Gardiner, S. Caney, D. Jamieson, & H. Shue (Eds.), *Climate ethics: Essential readings*. Oxford: Oxford University Press.

Smith, P. T. (2017, November). *Legitimacy and domination in SRM research. Paper presented at conference on geoengineering*. Seattle: legitimacy and justice at the University of Washington.

Timmons, M. (2002). *Moral theory: An introduction*. Lantham, MD: Rowman and Littlefield.

UNEP. (2017). *The emissions gap report 2017*. Nairobi: Author.

UNESCO. (2017). Declaration of ethical principles in relation to climate change.

UNFCCC. (1992). *United Nations framework convention on climate change*. New York: United Nations.

UNFCCC. (2015). Paris agreement.

Whyte, K. P. (2013). Justice forward: Tribes, climate adaptation and responsibility. *Climatic Change, 120*(3), 517–530.

Whyte, K. P. (2016). Indigenous experience, environmental justice and settler colonialism. In B. Bannon (Ed.), *Nature and experience: Phenomenology and the environment* (pp. 157–174). London: Rowman and Littlefield.

Wong, P.-H. (2015). Consenting to geoengineering. *Philosophy & Technology, 29*(2), 173–188.

World Bank. (2001). *Global development finance*. Washington, DC: Author.

Climate Change, Climate Engineering, and the 'Global Poor': What Does Justice Require?

Marion Hourdequin

ABSTRACT
In recent work, Joshua Horton and David Keith argue on distributive and consequentialist grounds that research into solar radiation management (SRM) geoengineering is justified because the resulting knowledge has the potential to benefit everyone, particularly the 'global poor.' I argue that this view overlooks procedural and recognitional justice, and thus relegates to the background questions of how SRM research should be governed. In response to Horton and Keith, I argue for a multidimensional approach to geoengineering justice, which entails that questions of how to govern SRM research should be addressed from the very outset – that is, now.

'It's an engineering problem, and it has engineering solutions.'

-Rex Tillerson, speaking on climate change at the Council on Foreign Relations, CEO speakers series, 27 June 2012[1]

'Does geoengineering raise any ethical issues not already considered by historical figures such as Aristotle, Hume, Kant, and so on? Isn't the ethics of making decisions that affect others not involved in making the decisions a problem as old as humanity? I just don't understand how there is anything new here for philosophy...'

-Stanford scientist Ken Caldeira, Geoengineering Google group, April 2012[2]

'Our government's first duty is to its people, to our citizens – to serve their needs, to ensure their safety, to preserve their rights and to defend their values. As President of the United States, I will always put America first, just like you, as the leaders of your countries will always, and should always, put your countries first.'

-Donald Trump, Speech to the United Nations General Assembly, September 19, 2017[3]

Introduction

A number of prominent proponents of research into solar radiation management (SRM) geoengineering argue on consequentialist grounds that such research is justified because the resulting knowledge has the potential to benefit everyone, particularly the least advantaged. Josh Horton and David Keith, for example, argue that 'a *prima facie* moral obligation exists to research SRM in the interest of developing countries, because SRM appears to be the most effective and practicable option available to alleviate a range of near-term climate damages that are certain to hurt the global South most of all' (Horton & Keith, 2016, p. 89). Horton and Keith provide a consequentialist, distributive justice argument for SRM research: research is justified – or even morally required – in order to save 'the global poor.'

Horton and Keith acknowledge that the obligation is defeasible. However, they offer little reason to think the *prima facie* moral obligation they assert might be overridden by countervailing considerations. In this paper, instead of accepting the *prima facie* obligation and considering concerns that might override it,[4] I want to challenge the very starting point of Horton and Keith's argument and its implicit suggestion that the decision to pursue research (or not) can be grounded primarily or exclusively in considerations of distributive justice. In the course of the paper, I will 'trouble' their position (in the sense of the Latin, *turbidus*, to make turbid – or to complicate)[5] and in so doing, show why I find the position troubling (in the worrisome sense) and offer an alternative approach.

My core concern is this: arguments such as Horton and Keith's often relegate to the background questions of how SRM research should be governed, implicitly suggesting that the decision to *pursue* research should happen first, *followed by* decisions about how to govern it. In my view, however, governance needs to begin at the beginning. Arguments such as Horton and Keith's, which focus on distributive and consequentialist considerations, should be considered not prior to, but in tandem with, concerns for participatory, recognitional, and multigenerational justice. This, in turn, entails that questions about how to govern SRM research – including questions of who decides whether such research should be pursued and how it should be pursued – should be addressed from the very outset – that is, now.[6]

The Limits of the Distributive Frame: Multidimensional Climate Justice and SRM

Before turning to Horton and Keith's argument specifically, I want to situate distributive justice within a broader multi-dimensional frame. I will argue for a multi-dimensional conception of justice – which incorporates procedural, participatory, and recognition justice – and show how the multi-dimensional view can broaden discussions of distributive justice in relation to SRM, moving beyond a narrow, consequentialist frame.

Although both academic discussions of climate justice and international climate negotiations through the UNFCCC process have emphasized distributive justice, the discourse around justice has long encompassed more than distribution alone. Some of the earliest accounts of justice in the Western philosophical tradition, for example, extend beyond distribution: Plato describes justice as a form of harmony within and among individuals (see, e.g., LeBar & Slote, 2016), and Aristotle discusses both distributive and rectifactory dimensions of justice. More recent accounts of environmental justice include both distributive and participatory dimensions (Figueroa & Mills, 2001; Shrader-Frechette, 2002), as well as additional dimensions such as recognition, intergenerational justice, ecological justice and capabilities (e.g., Figueroa, 2011; Nussbaum, 2009; Schlosberg, 2007; Whyte, 2011). Nevertheless, despite promising recent work arguing for a multidimensional approach to climate justice (see Kortetmäki, 2016), the ethics of climate mitigation, adaptation, and geoengineering remains centered on distribution. Who wins? Who loses? Who pays?

I will argue that this approach is inadequate, for multiple reasons. Some of these reasons reflect general concerns with conceptions of justice that focus only on distribution; others are more particular to the climate case, and to SRM specifically. Beginning with the general concerns, let's take a very simple example commonly used to conceptualize distributive justice, *the pie case*, where one is asked to consider how to fairly divide up a metaphorical pie among various parties. The pie case tends to provoke the intuition that when all else is equal, the pie should be divided equally. If you and I are sharing a pie, for example, it seems fair that we each get half. Of course, all else is rarely equal: for example, where did the pie come from, and who baked it? If I grew pumpkins for the pie and made it, then you stole it from me, it seems doubtful that justice will be achieved if you simply give half of it back. Context matters – and it often matters *a lot* – in assessing what sort of distribution would be fair. So one problem with the pie case, described in very abstract terms, is that it omits context, and what might at first seem like obvious distributive principles can turn out to be not so obvious after all, once the context is filled in. However, there is another problem with the pie metaphor: by focusing attention on the size of the slices in a fixed pie, the metaphor can distract from further questions about *who decides* and *through what process* distributive decisions are made.

One might reply that as long as the *principles of distribution* are fair, then the process doesn't matter. As long as the pie is fairly divided and everyone gets their share, it doesn't really matter who cuts and doles out the slices. Perhaps there are some cases in which it doesn't matter who divides the pie. For instance, if it were always fair to divide the pie equally among the relevant parties, and everyone agreed on this, then perhaps concerns about procedural and participatory justice would be less salient. The trouble is, in many complex situations of justice, *there is no ex ante general consensus on which distributive principle(s) to use*. Even in very simple, stylized cases, distributive principles can be controversial. And in most real-world cases, complex contextual considerations bear on the choice of relevant distributive principles. In these cases, *ex ante* agreement on the appropriate distributive principle(s) is unlikely, and certainly can't be assumed. But without consensus on what would be distributively fair, who has the authority to assert the applicability of a particular principle to a particular case? I want to suggest that in cases where the relevant distributive considerations are complex and controversial, it is morally problematic for an individual or a small group to select a distributive principle or determine a distributive outcome without taking seriously the need for participatory engagement. At minimum, then, it seems that where *ex ante* agreement is lacking, distributive justice cannot be achieved without a participatory component.

However, participatory justice also matters in its own right.[7] One way to ground the idea of participatory justice is in the ideal of *participatory parity*, which requires that people have the opportunity to engage as 'full partners in social interaction,' equal in status and standing to other participants (Fraser & Honneth, 2003, p. 36). There are multiple ways to operationalize this ideal from a participatory perspective, but one of the most basic approaches is through the *all-affected principle*: people should have opportunities to participate in decisions that significantly affect them, or put another way, 'all those affected by a political decision ought, directly or indirectly, to have a say in its making' (Näsström, 2011, p. 115).[8] Although some have objected to this principle on grounds that it is potentially *too* inclusive, Goodin (2007) argues that it is nevertheless 'the standard by which the adequacy of [other variations of the principle] is invariably assessed.' The all-affected principle thus represents an important starting point for participatory justice, and has significant implications for global scale issues like climate change. Adopting the all-affected principle in the case of SRM research and development would extend the community to which obligations of participatory justice apply to *all people*, given the technology's intended global reach.[9]

This line of argument shows that distributive considerations alone cannot be determinative in relation to SRM. Instead, arguments from distributive justice should be offered as fodder within a broader, more inclusive conversation about whether and how SRM research and development should take place, a conversation that gives voice to the perspectives of all those who would be affected by SRM, that is, to the perspectives of everyone.

The ideal of participatory parity, along with the associated all-affected principle, not only provides grounding for obligations of participatory justice; it also entails the need for justice as *recognition*. To enable people to engage 'full partners in social interaction' requires not only opportunities to participate in a formal sense in decisions that affect them; it also requires that people have similar status and standing *as* participants. Recognition justice focuses on these issues of status and standing, rejecting social arrangements and governance structures that silence or denigrate the perspectives of those in particular cultural groups, or with particular gender, racial, or class identities – or in the case of SRM, those who currently lack dominant voices in geoengineering debates. This requires respect – including respect for difference – and mutual acknowledgement. The importance of recognition appears clearly in the space of overlap with participatory justice: meaningful participation requires not just formal opportunities to offer one's views, but interlocutors, institutions, and processes that take seriously those perspectives.[10]

As Kyle Whyte (2011) explains, 'Recognition justice requires that policies and programs must meet the standard of fairly considering and representing the cultures, values, and situations of all affected parties.' This entails that the structure of participatory processes as well as the terms of the debate over SRM must take seriously the cultures, values, and situations of all involved. Recognition requires that processes, deliberations, and decisions about whether and how to proceed with SRM research and development engage all of the relevant perspectives without privileging a single normative perspective or worldview. Thus (to provide just one example), if non-consequentialist considerations matter to people in relation to SRM, any decision-making process that presupposes consequentialism will fail to provide adequate recognition by marginalizing perspectives that fall outside the consequentialist frame.

The view of justice just described – which incorporates distributive, participatory, and recognitive dimensions – has been called the *trivalent model of justice* (see, e.g., Kortetmäki, 2017; Schlosberg, 2007). The trivalent model not only incorporates three elements of justice that have value in their own right; it is a model in which each element supports the others. For example, distributive justice supports participatory justice and recognition by providing adequate resources to enable full participation and engagement of participants on an equal footing: significant disparities in knowledge, money, and power limit access to participatory processes (cf. Fraser & Honneth, 2003, p. 36). Conversely, participatory justice and recognition can play an important role in distributive justice by challenging certain assumptions about distribution and introducing a wider range of considerations and possibilities (Hourdequin, 2018). For example, robust, inclusive participatory processes may provoke important questions such as the following:

- What is being distributed? (Which pie is being divided?) Have all the relevant distributive dimensions been taken into account?
- Is a particular approach to remedying distributive injustice the only or best one? What alternatives exist?
- What principles of distribution are being presupposed?
- How are historical injustices taken into account? For example, do proposed distributive schemes or policies take history into account, and if so, how?
- How are future generations being considered?
- How will distributive decisions and frameworks be assessed and revisited over time?
- Are the values at stake fungible, as distributive approaches often presuppose?
- At what scale(s) are distributive frameworks being developed? What scale(s) are being overlooked?

Many of these questions are relevant to the case of SRM, and to the position articulated by Horton and Keith. For example, it *might* be the case that SRM could alleviate some of the anticipated harms of climate change for some of the world's most vulnerable people in the near term. However, participatory processes that take recognition seriously might raise the question of whether these are the only relevant harms. Perhaps some of the vulnerable people in question are also worried about the uneven distribution of power and control over the earth's climate, and see tradeoffs between alleviating physical/climatic harms through SRM and potentially exacerbating maldistributions of power through the development of this technology. Attention to participatory justice and recognition might also provoke a robust discussion about how research and development of SRM could affect future generations, and how best to address intergenerational justice. Thus, the trivalent model of justice can lead to fuller consideration of distributive questions in all their richness and complexity, and it can also make space for ethical concerns related to SRM that distributive arguments alone might overlook.

I have argued elsewhere for the importance of a trivalent model of justice, citing a number of reasons to pay particular attention to recognition:

> "[T]aking recognition seriously would encourage deeper dialogue about the principles of distributive justice and about what climate policy aims to distribute. It might prompt more careful thought about how climate change interacts with diverse cultures and ways of life, such that climate losses and vulnerabilities would not be reduced to economic losses and vulnerabilities [see McShane (2017), Preston (2017), Hourdequin (2015) (unpublished ms.)]. Additionally, it would require greater engagement with the perspectives and needs of those who lack significant political, economic, and cultural power." (Hourdequin, 2016, p. 37)

Participatory justice and recognition are important to climate mitigation and adaptation, but they play a *particularly* critical role in relation to SRM, for at least four reasons (Hourdequin, 2016): 1) SRM is global in scope and scale (in contrast to adaptation, which can occur on various scales); 2) SRM poses significant risks in comparison to mitigation (these risks include regional climate effects, but also geopolitical risks such as rogue or uncoordinated geoengineering); 3) SRM would represent a significant shift from past practices by exerting *intentional*, ongoing control over the climate system at the global scale; and 4) there is no consensus on the acceptability or desirability of SRM research, development, and deployment, and no settled mechanism for achieving an agreement on how to proceed. No international policy framework governs SRM research and planning, and both research on and discussions of SRM thus far have been relatively *ad hoc* and highly concentrated within a very small group of actors from wealthy countries such as the United States, England, and Germany.

For all of these reasons, the trivalent model of justice is critical to ethical discussions of SRM research and development, and to the establishment of governance frameworks for research. Distributive justice arguments for geoengineering research need to be embedded in a broader conversation that takes seriously issues of representation and recognition.[11]

Some Troubles with SRM to Save the Global Poor

> '[O]pposition to research on SRM threatens to violate principles of justice by effectively condemning developing countries to suffer the consequences of activities of which they have not been the primary beneficiaries.' – (Horton & Keith, 2016, p. 80)

The considerations offered above focus on the *general* insufficiency of arguments based on distributive justice alone to settle ethical questions regarding SRM research, development, and deployment. In this section, I want to return to the specific argument given by Joshua Horton and David Keith. After briefly summarizing their argument, I describe two key concerns – *paternalism* and *parochialism* – and describe potential links to the broader problem of *expertise imperialism*. I then suggest that a multidimensional approach to justice offers resources that can mitigate these concerns.

Before describing my concerns in more detail, it will be helpful to lay out Horton and Keith's argument, as I understand it. Here is a basic outline:

(1) **Global climate change will disproportionately burden the global poor**. This is because some of the most severe impacts of climate change are falling and will continue to fall on poorer regions and countries, and because the people in those areas have the fewest resources to adapt to climate change.

(2) **Global climate change is primarily the result of fossil fuel emissions that have primarily benefited the rich**.

(3) Thus, the **benefits associated with the use of fossil fuels have accrued primarily to the wealthy while the costs are being borne primarily by the poor**.

(4) **This is unfair**, because 'intuition tells us that the requirements of justice are violated when an activity benefits wealthy countries at the expense of poorer ones' (Horton & Keith, 2016, p. 80).

(5) **The unfairness of global climate change generates an obligation to 'take steps to reduce harms falling on the most vulnerable nations'** (80). There are three potential paths for reducing climate harms: mitigation (emissions reductions and/or carbon dioxide removal), adaptation, and SRM.

(6) **Mitigation won't effectively reduce climate harms in the near term** (81). Additionally, the near-term costs of mitigating will be borne more heavily by the

poor 'for whom energy is a larger fraction of expenses' (81). So mitigation may actually further burden those who are already least advantaged.

(7) **Adaptation can reduce climate harms, but it works locally (not globally) and is more expensive than SRM**. Because the benefits of adaptation are local, it is local actors who have the 'strongest incentives for action' but 'the poor...will have fewer resources' for adaptation than the rich. Adaptation's 'theoretical potential to help the poor most of all is undermined by the practical consequences of its costly, local character' (82).

(8) **SRM is preferable to adaptation as a way of addressing near-term harms to the least advantaged**, because its benefits are global in scale, and it is cheaper than adaptation. As Horton and Keith explain, 'Since a given unit of climate protection would benefit the poor disproportionately, the cost differential between adaptation and SRM imparts a comparative greater redistributive potential to the latter response option' (83).

(9) **SRM is the 'most effective and practicable option** to alleviate a range of near-term climate damages that are certain to hurt the global South most of all' (89).

(10) Therefore, **there is an obligation 'to research SRM in the interest of developing countries'** (89).

With this outline of the argument in view, I now turn to the concerns, the first of which is paternalism.

Paternalism

> 'Now, with these new technologies that evolve always come a lot of questions. Ours is an industry that is built on technology, it's built on science, it's built on engineering, and because we have a society that by and large is illiterate in these areas, science, math and engineering, what we do is a mystery to them and they find it scary. And because of that, it creates easy opportunities for opponents of development, activist organizations, to manufacture fear.'
>
> –Rex Tillerson (speaking on concerns regarding hydraulic fracturing), Council on Foreign Relations, CEO speakers series, 27 June 2012, https://www.cfr.org/event/ceo-speaker-series-conversation-rex-w-tillerson

Paternalism is a concept that has been central to discussions in bioethics, where questions arise about when and whether it is appropriate for physicians to make decisions on a patient's behalf, and relatedly, when and whether it is appropriate to withhold information from patients, on grounds that certain things are better for them not to know. Sjöstrand, Eriksson, Juth, and Helgesson (2013) characterize paternalism like this: 'Paternalism refers to courses of action (including decisions) that are done in the assumed interest of a person, but without or against that person's informed consent.' Gerald Dworkin offers a similar characterization, arguing that a person or institution acts paternalistically towards another when their act interferes with the other's autonomy, is done without consent, and is done with the intent of improving the welfare or advancing the interests of the other[12] (Dworkin 2017, section 2). By adding the 'interference with autonomy' condition, Dworkin's analysis has the virtue of circumscribing the scope of paternalism such that buying a gift for a friend or picking up a piece of trash from a neighbor's front lawn would not fall under its scope.

Horton and Keith's argument for SRM research seems to fit quite neatly within Sjöstrand et al.'s model of paternalism. The intent to advance others' interests is clear: Horton and Keith argue that SRM research ought to proceed in order to benefit 'the global poor,' and further, that there exists an *obligation* to conduct SRM research on their behalf (implicitly, it seems that this obligation is one they expect to be borne by researchers in wealthy countries[13]). However, Horton and Keith don't describe this obligation as contingent on the consent of those they seek to benefit: as noted above, they nowhere mention any data or consultation process that support the idea that the 'the global poor' share the view that SRM research is the best way to advance their interests in relation to climate change. By assuming it is possible to *know* that SRM research is in fact the best way to advance the interests of those in poorer parts of the world with respect to climate change, Horton and Keith make a paternalistic argument, insofar as paternalism involves acting to benefit another without their consent.

Horton and Keith's position also seems to fit Dworkin's more restrictive characterization of paternalism, with its 'interference with autonomy' condition. First, unless inclusively designed and governed, SRM research centered in wealthy countries (as it is now) will exacerbate power imbalances between powerful, wealthy countries and less powerful, less wealthy ones. The power to intentionally manipulate the global climate through the deployment of SRM is not inconsiderable, and SRM research thus far has amassed that power primarily within North America and Western Europe. While Horton and Keith might argue that they can sidestep concerns about interference with autonomy by advocating only for *research*, this is far from clear. In the case of nuclear weapons, for example, it is not only the *actual deployment* of these weapons that shapes the behavior of other nations, but the possession of the *power* to deploy them. The credible capacity of some countries to use nuclear weapons restricts the behavior of other countries; similarly, the developed capacity for wealthy nations to engage in global-scale climate engineering would constrain the actions and positions of less wealthy and less powerful countries, introducing another complex dimension into climate negotiations and another set of SRM-related justice concerns. Additionally, although Horton and Keith ostensibly argue only for an obligation to *research* SRM, this obligation gets its force directly from the assumed benefits of actually *using* SRM to address near-term climate harms. In other words, Horton and Keith seem to be arguing that we (in wealthy countries) should conduct SRM research so that we (in wealthy countries) can deploy SRM to help the global poor. They offer no reasons to think that SRM research *alone* will benefit the poor; without the link to deployment, it is not clear how SRM research discharges the obligation to 'take steps to reduce harms falling on the most vulnerable nations' (p. 80).[14] And even if research on SRM does not interfere with the autonomy of the global poor (which I have offered reasons to doubt), it is hard to see how even benevolent deployment of SRM by wealthy nations would avoid such interference.

Taking paternalism seriously highlights that even if SRM research *were* the best way to advance the interests of 'the global poor,' this doesn't entail that proceeding with such research without their consent or participation would be just (Jamieson, 1996). On the trivalent model of justice described above, arguments from distributive justice don't override the need for participatory justice and recognition.[15] Thus, from the perspective of multidimensional justice, paternalistic arguments for SRM are ethically problematic insofar as they attempt to circumvent the need for participatory processes and assume that the development of SRM can be justified based on assumed benefits to those most heavily burdened by the physical effects of climate change.

Although I want to acknowledge the limits my capacity to understand the diverse and context-dependent perspectives of the many people that Horton and Keith categorize as the 'global poor,' it seems that the following simple allegory might capture some aspects of the current situation in relation to SRM, seen in the context of the history of climate change more broadly[16]:

- I take actions that benefit me and (incidentally) harm you. I benefit a lot from these actions, which vastly increase my power and wealth.
- Later, I argue that you should not take similar actions to benefit yourself, because they might harm me, or others.
- At the very least, I say, if I am going to curtail my actions you should curtail yours, even though your actions are modest in comparison to mine and you've only recently started acting in the way that I've been acting for a long time.
- I continue to act in the ways that benefit me in the short term while harming you and others, and I determine that it is too expensive and inconvenient for me to stop doing so.
- Harms to you continue to accumulate, and your situation is becoming increasingly dire.
- I could provide resources to you that would enable you to better cope with your situation, but doing so would be expensive (though not so expensive as to significantly compromise way of life, which is luxurious compared to yours, or my power, which is vast compared to yours).
- So I argue that I should investigate a technology that I believe would benefit us both, alleviating some of the harm to you in the near term – though it might leave our grandchildren worse off.
- The technology is risky, but I am confident that my research will reveal the relevant risks, that I'll be able to control them, and that my research will be used to advance your well being most of all.
- Without consulting you, I argue that I have a *moral obligation* to investigate this technology, and that *not* investigating it would be *wrong*.
- I acknowledge the need to think about whether and how to use the technology I'm developing, but insist that we don't really need to work out those details now, because I'm just doing research, not actually advocating *use* of the technology.
- I embark on research (again, without consulting you, or at most, consulting you in a perfunctory way) and argue that it's ok to start experimenting with the technology in the real world, because the experiments I'm proposing don't really count as *testing or using* the technology (though the distinctions between research, testing, and deploying the technology are controversial).[17]

Recent empirical, social scientific research in climate-vulnerable regions supports the plausibility of this allegorical description in relation to climate engineering. In a study exploring the perspectives of people from the South Pacific, North American Arctic, and Sub-Saharan Africa, Wylie Carr and Christopher Preston found that although many of the interviewees did not categorically oppose geoengineering research or its consideration as one possible response to climate change, their '*overarching concern*...[was]...that climate engineering could further erode the already weakened self-determination of vulnerable populations due to a long history of oppression' (2017, p. 764, emphasis added). Interviewees not only expressed concern that they would be further disempowered by climate engineering, they worried that geoengineering research would not account for the climate variables most important to their lives and livelihoods (p. 767); that unintended and uncompensated impacts would fall on those already most vulnerable to climate change (p. 767); and that the costs of geoengineering research and deployment would be used by 'affluent technologically advanced societies [to] justify their control over the technologies' (p. 768). As one interviewee put it, 'It's scary as hell to be dependent on some other person to dictate the weather or climate change' (Carr and Preston, 2017, 770).

In the absence of a major reorientation that takes seriously these concerns, I submit that conducting SRM research in the current context is paternalistic, and not benignly so.

Cultural Parochialism

Because paternalism makes assumptions on others' behalf, it risks parochialism. Of course, all of us are positioned in ways that limit our view, and any argument is bound to reflect particular conceptions of the world. The trouble lies in a lack of reflexivity, and in assuming that one's own views are widely or universally shared (or ought to be so). It is in dialogue with others that we can come to recognize alternative perspectives, and this dialogue is *particularly* important for positions that prescribe courses of action – such as research on SRM – that are likely to have significant consequences for the world as a whole: for present and future generations, for people with diverse values and ways of life, for animals, plants, and ecosystems.

There are two particular aspects of Horton and Keith's argument that I believe reflect parochial assumptions that deserve further scrutiny. First, Horton and Keith assume a consequentialist framework in which risks and benefits can be traded off against one another.[18] For example, they assess mitigation, adaptation, and SRM in terms of their effects on climate risk and climate protection, focusing on the physical effects of climate change in various regions. They favor SRM over the other options because it is (they argue) the most cost effective way of reducing climate risk in the near future.[19] The language, throughout, reflects a consequentialist, cost-benefit frame. Although this frame is widely deployed throughout the world, it is arguably a culturally parochial one that encodes certain dominant Western presuppositions about the nature and fungibility of value.[20]

A second aspect of Horton and Keith's argument that embeds parochial assumptions has to do with what they take as 'fixed' and what they see as changeable. For example, Horton and Keith argue that in the near-term, the costs of both mitigation and adaptation will fall disproportionately on the poor, and therefore SRM – which is cheaper and whose (modest) costs are more likely to be borne by those in wealthier parts of the world – is a more just response to the near-term impacts of climate change. They further note that 'local actors pursuing local interests through the use of SRM might, if the intervention was properly designed, benefit the rest of the world (especially the global poor) as a virtual by-product of their otherwise self-interested use of solar geoengineering' (Horton & Keith, 2016, p. 83). There's a fair bit to unpack here. Clearly, the costs of adaptation and mitigation don't *necessarily* have to fall on the world's poorest people: what Horton and Keith take as fixed need not be so. If nations such as the United States truly cared about the well being of 'the global poor,' they could transfer resources to poorer countries burdened with the costs of adaptation or could shoulder a larger proportion of the global burdens of mitigation. Horton and Keith seem to assume that neither is likely to happen, thus SRM is a better option. But under a scenario in which the wealthy countries care too little about the poor to do their fair share with respect to mitigation and adaptation, how likely is it that research and development of SRM will prioritize the interests of 'the global poor'? And how confident do 'the global poor'[21] feel about the optimistic prospect that they will be the primary beneficiaries of an SRM program centered in and controlled by wealthy nations 'pursuing [their own] local interests'?

A broad and inclusive discussion of SRM research could illuminate the degree to which the assumptions and framings offered by Horton and Keith are shared by others throughout the world, and particularly by those their position purports to benefit.

Expertise Imperialism

The worries described above would be less troubling if Horton and Keith located their argument as one perspective among many on the complex and contested question of whether and how to pursue research on SRM. However, Horton and Keith instead present their argument as determinative, insisting that those who disagree and oppose SRM research are 'effectively condemning developing countries to suffer the consequences of activities of which they have not been the primary beneficiaries' (Horton & Keith, 2016, p. 80). They not only criticize the narrowness of perspective of their opponents – which seems to me a legitimate line of critique – they also question their opponents' 'deeper motives' (p. 90) and suggest that these motives eclipse concerns for those who are least well off. They claim that many of the opponents of SRM research are 'rich-country commentators criticizing solar geoengineering in an effort to shore up mitigation...while ignoring the potentially huge distributional advantages SRM might confer on the world's poorest in the global South,' and this, they hold, is 'ethically disturbing' (Horton & Keith, 2016, p. 90).

Yet it's not clear what gives Horton and Keith the epistemic position from which to understand their opponents' motives, and it seems a bit odd to hang both their arguments in favor SRM research and against those who oppose it on the importance of prioritizing the interests and needs of the least advantaged without consulting those same people about what their interests and needs really are. Despite arguing that the primary reason to engage in SRM research is to help the 'global poor,' Horton and Keith nowhere explicitly discuss the possibility of consulting or involving the 'global poor' or their representatives.[22] If, indeed, wealthy and privileged opponents of SRM research are resting their arguments on the suffering of those from other, less wealthy parts of the world, a robust conversation that includes those others may be the best way to bring this out. It might also expose limitations or lacunae in the arguments of Horton and Keith.

Paternalism and parochialism in discourse around SRM may reflect and be bound up with what Allen Buchanan calls *expertise imperialism*. Buchanan (2002) describes expertise imperialism as 'the tendency of experts to appeal to their genuine expertise in one area to justify their exercise of control in areas to which their expertise is in fact irrelevant.' Buchanan (2002) offers as an example the phenomenon of medical paternalism, which he suggests is enabled by various institutional factors, including the power and privilege associated with physicians' 'elite status,' which tend to 'insulate [them] from criticism' and undermine their own capacities for self-criticism (p. 133). However, Buchanan also notes that the phenomenon extends beyond the medical realm:

> "...not just medical professionals, but socially recognized experts generally, exhibit a combination of characteristics pregnant with the possibility of self-serving bias. On the one hand, the members of such groups have a common interest in sustaining and, if possible, expanding the privileges that they enjoy; this corporate interest can be in conflict with the interests of other individuals and of the community at large. On the other hand, the privileges that experts enjoy tend to insulate them from external criticisms that might serve as checks on the tendency to indulge in self-serving rationalizations." (Buchanan, 2002, p. 134)

It is difficult to judge whether and when 'self-serving rationalizations' are at work in any particular case, and I don't think it would be reasonable or fair to accuse Horton and Keith of expertise imperialism based on their article, 'Solar Geoengineering and Obligations to the Global Poor.' However, the structure of their argument is more confrontational than invitational, and seems to focus more on boxing out objectors to SRM research than on engaging seriously with their concerns.[23] Other scientists working on SRM have displayed a similar tendency. Take, for example, one prominent scientist's remark on the Google Geoengineering Groups listserv (also quoted above):

> "Does geoengineering raise any ethical issues not already considered by historical figures such as Aristotle, Hume, Kant, and so on? Isn't the ethics of making decisions that affect others not involved in making the decisions a problem as old as humanity?...In my conception, philosophers... develop theory and we all, when faced with moral problems, attempt to apply this theory in our moral reasoning. I drew the analogy with mathematics, where mathematicians develop theory and we all apply this theory when buying groceries....To me, a moral philosopher of geoengineering is like a mathematician of the grocery checkout line."[24]

The suggestion here is that the necessary ethical tools and considerations are already widely available and agreed upon, thus engaging philosophers in these debates is like having a theoretical mathematician supervising the clerks in the grocery checkout line: clearly unnecessary. Notice that not only does this scientist assume that it is straightforward to apply philosophical theories to the climate engineering context, he also seems to assume that canonical Western theories are sufficient to cover the ethical concerns, and that 'the ethics of making decisions that affect others not involved in making the decisions' is something that's been fully worked out. Moreover, the entire framework here is a technocratic one: philosophical experts (though perhaps only long-dead ones) develop ethical theories that they pass on to scientific experts who develop technologies to which these theories can then be straightforwardly applied.

So one strategy for limiting ethical discussion and calls for broader engagement in SRM governance is to argue that the relevant ethical considerations are available and at hand, and scientists can easily apply them to the climate engineering case. A second strategy suggests that there's nothing special or distinctive about SRM research at its current stage that warrants special attention or a special governance regime. The need for governance at a later stage is often acknowledged, though the criteria for determining when we will have arrived at the relevant stage are often unclear, and may shift. Note David Keith's comments at the 2017 international Climate Engineering Conference in Berlin on Harvard University's planned 2018 SRM field experiments:

> 'There are things that it's not. *It's not the first solar geoengineering experiment*, there have been at least two...*Second, it's not a field test*. To me, a test is something you do well down the engineering system, where there is a serious decision to deploy that system and you want to test that in a binary way. That's not what we're doing. What we're doing is something much earlier, *it's really applied atmospheric science*.' (Quoted in Dunne, 2017, emphases added)[25]

Is this expertise imperialism? Again, it's not easy to make that judgment in any particular case, nor is it necessarily helpful. But at the broader level, there are patterns worthy of attention. Much of the ethical discourse surrounding SRM – *particularly among those in the scientific community engaged in SRM-related research* – seems to focus on enumerating various ethical concerns primarily in order to develop replies to them that clear the way for further research and/or to provide reassurance that the ethics issues are covered, and there's no need for outsiders to worry, or intervene. Insofar as expertise imperialism has the potential to arise in any community of experts who are insulated from criticism, it is perhaps more important to take steps to guard against it than to diagnose it in any particular case. Such steps might address institutional structures, individual virtues, and their intersection, and could reveal ways in which existing institutional structures facilitate or hinder criticism and reflexivity within discussions of SRM.[26]

Conclusion: Multidimensional Justice and Governance – Developing Institutions and Cultures of Recognition

Paternalism, parochialism, and expertise imperialism are salient not only in relation to one particular position or argument, but in relation to discussions of SRM research and development more broadly. A multidimensional approach to justice, grounded in recognition and participatory parity, can mitigate against these troubles. Institutions that aim to instantiate recognition and participatory parity can directly combat insular or self-serving tendencies by self-consciously aiming to counteract the inequities that lead a few 'experts' to have outsized roles in guiding decisions involving normative concerns that deserve broader public deliberation.[27]

Taking seriously concerns about paternalism, parochialism, and expertise imperialism requires taking seriously their roles in knowledge production. Like all research, SRM research is not 'neutral,' and decisions about whether and how to pursue SRM research will affect not only the likelihood that SRM is deployed, but how it is developed, what concerns are taken into account, and what concerns are overlooked. SRM research raises questions of epistemic power and how asymmetries in knowledge and expertise can generate and exacerbate status inequality that in turn undermines recognition and participatory parity. Concerns of distributive justice in relation to SRM thus extend beyond the distribution of harms and benefits in relation to climate effects; they also encompass questions about the distribution of epistemic power, and relatedly, the power to direct and control responses to climate change (for further discussion, see Carr and Preston, 2017, Whyte, 2012a, 2012b). As Kyle Whyte (2012b, p. 175) explains in an article discussing indigenous sovereignty and research on SRM, the push for 'even early research represents an emerging crystallization of a commitment that will give some people greater capacity to impact the climate system...[T]his is an equity issue; it is also an issue of whose capabilities to dominate the environment and other people will be strengthened.'

A multidimensional perspective on climate justice suggests that the current concentration of power in relation to SRM research and development is untenable, and arguments from distributive justice do not preclude the need to address participatory justice and recognition, which can generate forms of accountability that the current system lacks. In imagining what multidimensional justice might mean for SRM research, it is critical that both the *terms of participation* and the *structure of research* be considered from the perspective of participatory parity. From this point of view, it seems unlikely that forms of participation in which researchers or governments merely 'consult' or survey members of the public about their views will be sufficient, especially if the terms of the information gathering are already set by the 'experts.' With respect to research, participatory parity will require addressing disparities in knowledge and power, and decentering research processes and institutions: if the justification for pursuing research on SRM is to benefit those who are most unfairly burdened by climate change, then why should SRM research be conducted primarily by those whose lived experience intersects relatively little with these burdens?

Multidimensional justice for SRM requires not only specific institutions and governance structures, but also certain practices and habits of mind among those engaged in geoengineering research, development, and governance. In this regard, I want to highlight the importance of cultivating *cultures of recognition*. As I have argued previously, cultures of recognition encompass 'particular habits of interaction that allow for mutual respect, which in turn rest on the development of individual and institutional virtues' (Hourdequin, 2018). Virtues that support recognition include patience, open-mindedness, care and concern, epistemic humility, and a commitment to sustaining relationships through disagreement. Recognition also requires self-awareness and receptivity, particularly on the part of the wealthy and powerful, who – as noted above – may be particular vulnerable to certain forms of *in*sensitivity, though they often control the development of critical discourses, bodies of scientific knowledge, and governance systems. Multidimensional justice requires a multifaceted approach: the development of virtues, practices, and cultures of recognition can work synergistically with efforts to re-envision both research and governance institutions in ways that constructively respond not only to the direct harms of climate change, but to the marginalization of those harmed in critical decisions that shape their and others' futures.

Notes

1. Available at: https://www.cfr.org/event/ceo-speaker-series-conversation-rex-w-tillerson [accessed 5 March 2018].
2. The quotes above are drawn from a thread with multiple posts and replies; they can be found at https://groups.google.com/d/topic/geoengineering/hECCEn5AG4E/discussion and https://groups.google.com/d/topic/geoengineering/TnO6Cce66fE/discussion [accessed 5 March 2018].
3. Available at: https://www.whitehouse.gov/the-press-office/2017/09/19/remarks-president-trump-72nd-session-united-nations-general-assembly [accessed 5 March 2018].
4. Though some of the concerns adduced here could also be relevant under this frame. It is worth noting also that the assumption that SRM is 'the most effective and practicable option available to alleviate a range of near-term climate damages' that would fall most heavily on the poor (Horton & Keith 2016, p. 89), on which the *prima facie* obligation is based, is itself subject to challenge on distributive and consequentialist grounds.
5. For further discussion, see *Staying with the Trouble* (Haraway, 2016, 1).

6. I am far from the first to advocate for wide-ranging, participatory engagement in geoengineering governance, including research governance. For other examples, see: Carr et al. (2013) (a paper on which David Keith is a co-author), Corner, Pidgeon, and Parkhill (2012), Carr, Yung, and Preston (2014). What this paper seeks to add is a critique of pro-research arguments that focus solely on consequentialist and/or distributive concerns, and an argument for a broader, multi-dimensional framework for geoengineering justice.
7. I recognize that I am making a normative argument for participatory justice here, the basis of which is contestable on the same grounds as I contested the legitimacy of an individual's or small group's choice of distributive principles without broader participatory engagement. My argument can thus be seen as one among many normative arguments that others might bring to the table. The same is true of the normative arguments made in this paper more generally. As Sophia Näsström (2011) explains with respect to the all-affected principle, discussed below, 'When we appeal to the all-affected principle we are therefore not only doing political philosophy. We are also engaged in a conflict with other citizens about the future boundaries of democratic life.'
8. The all-affected principle, as I understand it, does not entail that every person be directly consulted on every decision that affects them: it permits delegation of authority to others to make decisions through various systems of representation.
9. The all-affected principle, as described here, is not the same as a requirement of *consent* – it is, instead, a principle requiring opportunities for those affected to substantively engage in decisions that affect them. Nevertheless, concerns that Stephen Gardiner (2013a, pp. 28–29) has raised regarding consent-based justifications for geoengineering might be raised in relation to the all-affected principle: if the principle is construed as requiring engagement or representation of present people only, then the principle excludes future generations and non-human beings, who also matter morally. To this concern, I offer two replies: 1) in my view, honoring the all-affected principle, if interpreted to require engagement or representation of present people, is an important starting point for participatory justice, but is not necessarily sufficient to justify SRM. As I argue throughout, *multiple* dimensions of justice should be considered. What's more, there are normative considerations other than those of justice that are relevant to SRM; 2) it is worth considering how the all-affected principle might be elaborated so as to take account of future generations and non-human beings. If the all-affected principle were taken seriously for present persons, it seems likely that concerns for future generations and non-humans would enter into the dialogue more fully than they have to date. However, this is clearly contingent on present persons' concerns for future generations and diverse forms of life, and a more robust instantiation of the principle would seek to represent directly these constituencies.
10. Although I focus here on the way in which recognition and participatory justice interact, recognition extends beyond the participatory realm. Recognition, as I understand it, is fundamentally relational, with respect as its normative core. Thus, recognition requires relations of respect and non-domination, and misrecognition occurs when these relations fail at the interpersonal level, as well as where institutions and structures exclude, marginalize, or ignore those they purportedly serve. Participatory justice plays an important role in correcting failures of recognition, but misrecognition occurs outside of contexts involving participatory justice. For instance, misrecognition manifests frequently in everyday interactions, particularly where certain stereotypes or unwarranted assumptions block respectful engagement with others. Claudia Rankine's recent book, *Citizen: An American Lyric*, provides numerous examples.
11. I want to acknowledge other dimensions of justice not foregrounded by this paper. For example, a fourth dimension of the multidimensional justice framework might focus on the temporal axis. We might call this multigenerational justice, and consider under this dimension justice to future generations as well as ways of considering and responding to historical injustices in the present. This fourth dimension deserves treatment in a separate paper, which might also address the question of whether a multigenerational justice lens can sufficiently account for structural injustices in the present that grow out of historic patterns

of marginalization and oppression. Questions of multigenerational justice might also be considered as involving the *scope* of justice, rather than a distinct *dimension* (I thank Augustin Fragnière for this point), but for the purposes of this paper, I leave this issue aside.

12. According to Dworkin's analysis, a paternalistic act is one done '*only* because [the actor] believes [the act] will improve the welfare of [the other]...or in some way promote the interests, values, or good of [the other' (Dworkin 2017, emphasis added). But the stipulation that the other's interests or welfare be the *sole* motive for a paternalistic act may be too strong, as Dworkin himself acknowledges, so I have slightly loosened the third condition here.
13. See p. 88, where Horton and Keith argue that 'to close off research into SRM is to shirk the Northern responsibility to address the full range of climate risks destined to affect the global South most of all.'
14. One could imagine an argument for SRM research to benefit the 'global poor' that was decoupled from deployment and preserved autonomy. Such an argument might emphasize that SRM research has the potential to provide a greater array of *options* for responding to climate change, but it seems that this argument would work only if SRM research expanded the range of options *available to the most vulnerable.*
15. On a related note, Stephen Gardiner (2013b) argues that acting beneficently (to benefit another) can be morally problematic when such actions impinge on that other's rights. Breaking into your neighbor's house in order to clean it is problematic, even if your neighbor would like a cleaner house and you leave the house in significantly better condition than that in which you found it.
16. Stephen Gardiner (2013c) also offers an allegory for thinking through the decision context for climate engineering. This is the case of 'Wayne's Folly,' where a chronic philanderer takes steps to address the risk of spreading sexually transmitted disease (STD) to his wife and others by investing in a drug company working to improve treatments for STDs. Wayne's response to his promiscuity, argues Gardiner, bears important resemblance to certain contemporary arguments for climate geoengineering: both involve moral schizophrenia, a divergence between reasons and motives for action, and more specifically, 'creative myopia,' where 'an agent invokes a set of strong moral reasons to justify a given course of action, but this course of action is supported by these reasons *only because* the agent has ruled out a number of alternative courses of action strongly supported by the same reasons, and where this is due to motives she has that are less important, and are condemned by those reasons' (Gardiner, 2013b, 19).
17. Note David Keith's comments at the Berlin Conference (fall 2017) on Harvard's planned SRM field experiments, (quoted below, p. 14), which he insists should be categorized not as 'field testing' but rather as 'applied atmospheric science.' See: https://www.carbonbrief.org/geoengineering-scientists-berlin-debate-radicaly-ways-reverse-global-warming [accessed 5 March 2018].
18. For example: 'The ethical implications of these responses [to climate change] turn on the particular distribution of benefits and harms associated with each' (Horton & Keith 2016, p. 81).
19. With respect to mitigation, they argue that 'in the near term, mitigation has significant costs compared to only modest benefits, and a disproportionate share of the costs may fall on the poor' (p. 81), and in relation to adaptation, they explain, 'Since a given unit of climate protection would benefit the poor disproportionately, the cost differential between adaptation and SRM imparts a comparatively greater redistributive potential to the latter response option' (Horton & Keith 2016, p. 83).
20. What's more, even if one accepts the cost-benefit frame, Horton and Keith's conclusions about the cost effectiveness of SRM in comparison to other climate responses is controversial. For a succinct and helpful discussion of the costs of SRM, see (Lenferna, 2017). Lenferna points out, for example, that cost estimates that focus only on the costs of delivering aerosols to the stratosphere fail to consider the full panoply of costs associated with SRM.
21. Throughout the paper, I typically refer to 'the global poor' with scare quotes. Although it sometimes makes sense to speak in terms of broad categories ('the global North,' 'the global South,' or 'the global poor'), it is important to recognize that these sweeping terms obscure vast diversity within these categories. So the question, as phrased above, 'how confident do "the global poor" feel...?', is a rhetorical one that mirrors the categories

through which Horton and Keith frame their argument, though truly addressing the concerns of those who Horton and Keith see as falling into this broad category warrants a much more nuanced and fine-grained discussion.

22. Horton and Keith mention procedural justice very briefly at one point in the paper: this is in discussing the possibility of a global cost-benefit analysis (CBA) for SRM. After pointing out that some of the negative impacts of SRM on vulnerable regions might be offset by 'gains attributable to other avoided damages from climate change,' Horton and Keith acknowledge the difficulties in conducting a CBA for SRM, noting that 'any utilitarian assessment would need to ensure robust procedural and substantive protections so as not to violate fundamental principles of justice.' (Horton & Keith 2016, p. 86). However, they offer no further specification of what such 'procedural protections' would entail, and these protections are not highlighted in the context of their argument as a whole, but only in association with this brief discussion of the challenge of quantifying SRM's costs and benefits.
23. For example, Horton and Keith repeatedly impugn the motives of their opponents, insisting that those who advocate against SRM research care more about their 'rich-world political agendas' than about the suffering of the poor (2016, p. 91).
24. Available at: https://groups.google.com/d/topic/geoengineering/hECCEn5AG4E/discussion [accessed 5 March 2018].
25. https://www.carbonbrief.org/geoengineering-scientists-berlin-debate-radicaly-ways-reverse-global-warming, emphases added.
26. In this regard, see Allen Buchanan's (2002) work on social moral epistemology and José Medina's writing on the ways in which the privileged are susceptible to the vice of meta-blindness, an 'insensitivity to insensitivity' which makes the privileged not only inattentive to the limits of their own perspectives, but also unaware of their own inattentiveness. Medina further describes meta-blindness as 'a special difficulty in realizing and appreciating the limitations of [one's] horizon of understanding' (2013, p. 75).
27. I should note that moral philosophers are not exempted here – though in my view the greatest current risk is not that SRM policy is set by philosophers, but by a small group of technical experts working at the interface of geoengineering research and policy.

Acknowledgments

I am grateful to Stephen Gardiner, Augustin Fragnière, Alex Lenferna, and Thomas Ackerman, as well as participants at the fall 2017 University of Washington conference on Geoengineering, Political Legitimacy, and Justice for very helpful questions and comments on an earlier version of this paper. I would also like to thank students and faculty at University of Colorado-Boulder and University of Washington for discussion and feedback on additional presentations of this work.

Disclosure statement

No potential conflict of interest was reported by the author.

References

Buchanan, A. (2002). Social moral epistemology. *Social Philosophy & Policy, 19*, 126–152.

Carr, W., & Preston, C. J. (2017). Skewed vulnerabilities and moral corruption in global perspectives on climate engineering. *Environmental Values, 26*(6), 757–777.

Carr, W., Yung, L., & Preston, C. (2014). Swimming upstream: Engaging the american public early on climate engineering. *Bulletin of the Atomic Scientists, 70*(3), 38–48.

Carr, W. A., Preston, C. J., Yung, L., Szerszynski, B., Keith, D. W., & Mercer, A. M. (2013). Public engagement on solar radiation management and why it needs to happen now. *Climatic Change, 121*(3), 567–577.

Corner, A., Pidgeon, N., & Parkhill, K. (2012). Perceptions of geoengineering: Public attitudes, stakeholder perspectives, and the challenge of 'upstream' engagement. *Wiley Interdisciplinary Reviews: Climate Change*, *3*(5), 451–466.

Dunne, D. (2017, October 11). Geoengineering: scientists in Berlin debate radical ways to reverse global warming. *CarbonBrief*, Retrieved March 5, 2018, from https://www.carbonbrief.org/geoengineering-scientists-berlin-debate-radicaly-ways-reverse-global-warming

Dworkin, G. (2017). Paternalism. In: E. N. Zalta (ed.), *The Stanford encyclopedia of philosophy* (Winter 2017 Edition), https://plato.stanford.edu/archives/win2017/entries/paternalism/. Accessed December 26, 2018.

Figueroa, R., & Mills, C. (2001). Environmental justice. In: D. Jamieson (Ed.), *A companion to environmental philosophy* (pp. 426–438). Malden, MA: Blackwell.

Figueroa, R. M. (2011). Indigenous peoples and cultural losses. In: J.S. Dryzek, R.B. Norgaard, and D. Schlosberg (Eds.), *The Oxford handbook of climate change and society* (pp. 232–250). New York, NY. Oxford University Press

Fraser, N., & Honneth, A. (2003). *Redistribution or recognition: A political-philosophical exchange*. London: Verso.

Gardiner, S. M. (2013a). The desperation argument for geoengineering. *PS: Political Science & Politics*, *46*(1), 28–33.

Gardiner, S. M. (2013b). Why geoengineering is not a 'Global Public Good', and why it is ethically misleading to frame it as one. *Climatic Change*, *121*(3), 513–525.

Gardiner, S. M. (2013c). Geoengineering and moral schizophrenia. In: W. Burns & A. Strauss (Eds.), *Climate change geoengineering: Philosophical perspectives, legal issues, and governance frameworks* (pp. 11–38). Cambridge: Cambridge University Press.

Goodin, R. E. (2007). Enfranchising all affected interests, and its alternatives. *Philosophy & Public Affairs*, *35*(1), 40–68.

Haraway, D. J. (2016). *Staying with the trouble: Making kin in the chthulucene*. Durham, NC: Duke University Press.

Horton, J, & Keith, D. (2016). Solar geoengineering and obligations to the global poor. In: C. J. Preston (Ed.), Climate justice and geoengineering: Ethics and policy in the atmospheric Anthropocene (pp. 79-92). Lanham, MD: Rowman & Littlefield International.

Hourdequin, M. (2015). Recognizing climate losses. *Conference paper (unpublished)*, Presented at the *Second Buffalo Workshop on Ethics and Adaptation: Loss, Damage, and Harm*. University at Buffalo, Buffalo, NY.

Hourdequin, M. (2016). Justice, recognition and climate change. In: C. Preston (Ed.), *Climate justice and geoengineering: Ethics and policy in the atmospheric anthropocene* (pp. 33–48). Lanham, MD: Rowman & Littlefield International.

Hourdequin, M. (2018). Geoengineering justice: the role of recognition. *Science, Technology, And Human Values*. 10.1177/0162243918802893

Jamieson, D. (1996). Ethics and intentional climate change. *Climatic Change*, *33*(3), 323–336.

Kortetmäki, T. (2016). Reframing climate justice: A three-dimensional view on just climate negotiations. *Ethics, Policy & Environment*, *19*(3), 320–334.

Kortetmäki, T. (2017). Justice in and to nature: An application of the broad framework of environmental and ecological justice. *Jyväskylä Studies in Education, Psychology and Social Research*. Vol. 587. Retrieved from http://urn.fi/URN:ISBN:978-951-39-7127-4.

LeBar, M., & Slote, M. (2016). Justice as a virtue. In: E. N. Zalta (Ed.). *The Stanford encyclopedia of philosophy* (Spring 2016 Edition), https://plato.stanford.edu/archives/spr2016/entries/justice-virtue/. Accessed March 5, 2018.

Lenferna, A. 2017. The costs of solar geoengineering. *Ethics & International Affairs* online blog https://www.ethicsandinternationalaffairs.org/2017/costs-of-geoengineering/. Accessed March 6, 2018.

McShane, K. (2017). Values and harms in loss and damage. *Ethics, Policy & Environment*, *20*(2), 129–142.

Medina, J. (2013). *The epistemology of resistance: Gender and racial oppression, epistemic injustice, and the social imagination*. New York: Oxford University Press.

Näsström, S. (2011). The challenge of the all-affected principle. *Political Studies*, *59*(1), 116–134.

Nussbaum, M. C. (2009). *Frontiers of justice: Disability, nationality, species membership*. Cambridge, MA: Harvard University Press.

Preston, C. J. (2017). Challenges and opportunities for understanding non-economic loss and damage. *Ethics, Policy & Environment, 20*(2), 143–155.

Schlosberg, D. (2007). *Defining environmental justice: Theories, movements, and nature*. New York: Oxford University Press.

Shrader-Frechette, K. (2002). *Environmental justice: Creating equality, reclaiming democracy*. New York: Oxford University Press.

Sjöstrand, M., Eriksson, S., Juth, N., & Helgesson, G. (2013). Paternalism in the name of autonomy. *The Journal of Medicine and Philosophy: A Forum for Bioethics and Philosophy of Medicine, 38*(6), 710–724.

Whyte, K. P. (2011). The recognition dimensions of environmental justice in Indian country. *Environmental Justice, 4*(4), 199–205.

Whyte, K. P. (2012a). Indigenous peoples, solar radiation management, and consent. In: C. Preston (Ed.), *Engineering the climate: The ethics of solar radiation management* (pp. 65–76). Lanham, MD: Lexington Books.

Whyte, K. P. (2012b). Now this! indigenous sovereignty, political obliviousness and governance models for SRM research. *Ethics, Policy & Environment, 15*(2), 172–187.

Indigeneity in Geoengineering Discourses: Some Considerations

Kyle Powys Whyte

Introduction

Indigenous peoples are referenced at various times in communication, debates, and academic and policy discussions on geoengineering (i.e. geoengineering discourses). The discourses I have in mind focus on ethical and justice issues pertaining to some geoengineering research and (potential) implementation. The issues include concerns about potential inequalities in the distribution of environmental risks, research ethics, and abuses of social power. These issues are critical for many reasons, including the fact that some people advocate that certain kinds of geoengineering projects should figure as part of the portfolio of solutions pursued by people who stand to be most harmed by climate change impacts. I have been following some of the references to Indigenous peoples in geoengineering discourse in light of the work that I do. I have been invested for many years in the goal of advancing Indigenous peoples' self-determination in climate justice, both in mitigating and adapting to climate change.

Working primarily in North America, I have worked with dozens of Tribes, scientific organizations, and policy groups on climate change adaptation and resettlement planning. They include the Sustainable Development Institute at the College of Menominee Nation, the U.S. Global Change Research Program, and the U.S. Department of Interior's Advisory Committee on Climate Change and Natural Resources Science. In all of this work, I have rarely, if ever, heard Indigenous persons seek to organize any efforts to discuss geoengineering on their own terms. Yet Indigenous peoples have involved themselves in limited ways. Several Indigenous activists and organizations have denounced the very idea of geoengineering, including the Indigenous Environmental Network. The Haida Salmon Restoration Corporation recently participated in an ocean fertilization experiment at Haida Gwaii, which is considered to be a case of or analogous to some of the types of geoengineering projects being discussed today. There are also Indigenous peoples globally who have participated in or criticized the United Nation's (UN) Reducing Emissions through Deforestation and Forest Degradation (REDD) program. While I have not heard afforestation programs like REDD referred to as geoengineering, they are certainly similar or analogous. Indigenous peoples participated in a UN report, through the Convention on Biodiversity (CBD), that features Indigenous perspectives on geoengineering.

Based on my experience, I want to offer some considerations on the nature of geoengineering discourse as it pertains to Indigenous peoples. Indeed, Indigenous peoples and Indigenous issues are certainly considered as relevant topics in geoengineering discourse. Indigenous peoples have a long track record of taking leadership in addressing environmental issues, including climate change (Whyte, 2017, 2018). Indigenous peoples, now for many decades, have been leading advocates for the goal that they develop their own research, planning processes, public engagement and deliberative proceedings on issues that matter to them (Smith, 1999; Story & Lickers, 1997). However, I seek to raise some considerations in this essay that suggest that geoengineering discourse, as it stands now, does not articulate ethics and justice issues in ways that are salient to Indigenous peoples as they grapple with today's climate change ordeal. While some Indigenous persons may disagree with me, my tentative conclusion here is that most geoengineering discourses are not even setup in the first place to make it possible for Indigenous peoples to express and have an audience for many of their concerns about risk, research, and power.

Given how little Indigenous peoples come up, I hope in this essay to mainly contribute some basic considerations for why it is conceivable that geoengineering discourse is not entirely salient for at least some Indigenous peoples. I will cover these considerations in three different sections. While the sections overlap, they are also somewhat separate from one another. I will tie them together in the conclusion section. In the first section, 'Referencing Indigeneity in Geoengineering Discourse,' I discuss some examples from the discourse that resonate with me because they have to do with where Indigenous peoples (Indigeneity) have been explicitly or implicitly connected to topics of geoengineering research or (potential) implementation. I find that Indigeneity tends to be brought up when scholars and scientists are trying to learn about whether Indigenous perspectives offer insights or reasons that can weigh in on what is known or debated about the ethical acceptability and justice of geoengineering – especially the governance of geoengineering (research or implementation).

In the second section, 'Geoengineering and Colonialism,' I discuss the ways in which Indigenous voices frame climate change and how that relates to how some Indigenous peoples might relate to geoengineering if they created their own discourse on it. Indigenous histories and perspectives on climate change suggest the possibility that at least some Indigenous peoples would not see geoengineering as a discrete topic or solution to consider in relation to climate justice. Rather, geoengineering can only be understood through the idea that colonialism – its history and linear continuance today directly from that history – is a major factor of Indigenous vulnerability to climate change. In the third section, 'Indigenous Consent,' I focus on how it is hard to discuss Indigenous consent in relation to geoengineering precisely because Indigenous peoples never consented to any of global or local structures of colonial power that have generated the topic of geoengineering in the first place. I will conclude by making the point that for geoengineering discourse to be more salient to some Indigenous peoples, scholars and scientists will have to take up what Indigenous peoples have already conveyed about climate change and colonialism and the solutions for climate justice.

Referencing Indigeneity in Geoengineering Discourse

Discourse on ethics and justice issues is critical for geoengineering for key reasons. Scholars including Christopher Preston, Holly Jean Buck, Wiley Carr, Andrea Gammon, Martin Bunzl, Toby Svoboda, Nancy Tuana, Klaus Keller, and Marlos Goes, among others, demonstrate some of these issues on environmental risk. They show that the potential harmful environmental impacts of geoengineering projects will likely, across a number of cases, be suffered by people who already live under conditions of economic exploitation, racial and gender discrimination, and political powerlessness (Buck, 2012, 2018; Buck, Gammon, & Preston, 2014; Bunzl, 2009; Carr & Preston, 2017; Preston, 2013; Svoboda, Keller, Goes, & Tuana, 2011; Tuana, 2013). Nancy Tuana et al. raise a number of ethical concerns pertaining to research practices themselves (Tuana et al., 2012). Scholars such as Holly Jean Buck, Petra Tschakert, Stephen Gardiner, Ben Hale and Lisa Dilling, among others just referenced, have discussed moral challenges of social power. They have raised worries about careless argumentation about disaster avoidance, proposals for democratic engagement, and the possibility of genuine global consent (Buck, 2012, 2018; Carr et al., 2013; Gardiner, 2010; Hale & Dilling, 2011; Tschakert, 2010).

I first became aware of how Indigeneity was being discussed in geoengineering discourse when I went to a workshop at the University of Montana on the ethics of solar radiation management in Fall 2010. The workshop, sponsored by the National Science Foundation and producing two collections of publications, convened ethicists, social and physical scientists, and engineers (Preston, 2012; Scott, 2012). A main point for discussion at the workshop was whether it was ethically acceptable and just for governments and other parties to support early research on one specific type of approach to geoengineering: solar radiation management (SRM). If ethical and just in conception, many participants were also interested in what approaches to research ethics would be best given empirical testing could actually or perceivably alter local environments. While focused on SRM, the implications of the workshop's discourse on geoengineering relate to other approaches, such as ocean fertilization and carbon capture.

I remember, without recalling any specifics, several conversations about how research in SRM could pose disproportionate risks on Indigenous peoples living in the global south. There were also conversations, including after my own presentation, about whether Indigenous peoples would endorse SRM research given its connection to the possibility of the implementation of SRM in the future. In 2010, my main concern at the workshop was whether Indigenous leaders in North America were being adequately informed about the developments in different geoengineering technologies and the emerging research projects. I heard similar concerns from other parts of the world. Petra Tschakert claimed how 'So far, geo-engineering has been an almost exclusive debate in rich countries of the North' (Tschakert, 2010). In light of my main concern at the time, I learned a lot at the workshop about what geoengineering plans were unfolding, what is known about their range of potential impacts on the earth system, and what some of the benefits and burdens would be for many different populations. But what I also gathered was that there was, at least among some of the participants, something further at stake regarding Indigenous peoples beyond their vulnerability to either climate change or the rollout of SRM research. Of course, this is just my opinion having interpreted my interactions at the workshop. It seemed to me that references to Indigeneity seemed largely to turn on whether Indigenous peoples would find geoengineering ethically acceptable and just. There was interest in determining and debating whether Indigenous peoples, taken as diverse groups or as a global movement, would seem to favor, critique, or engage further on the importance of early research on SRM.

For me, it seemed like the underlying stake had to do with what implication Indigenous acceptance would have for a variety of pro- and anti-geoengineering positions (whether these positions are about research or potential implementation). In cases where Indigenous peoples, say, go along with a geoengineering-related project or vision, that would provide support to extrapolate more broadly that Indigenous peoples have reasons for favoring it. In cases where acceptance is lacking, that would point out key issues pertaining to values, power, and other matters that are important to take into consideration for governance. Or, in cases where Indigenous peoples might express an interest in engaging further, that interest would suggest that ethical and justice issues about geoengineering itself are not pertinent to Indigenous peoples, but are pertinent to research and implementation processes affecting Indigenous peoples.

Since the workshop, I continually hear of people referring to the Haida Salmon Restoration Corporation's experiment in ocean fertilization at Haida Gwaii. Holly Jean Buck was among the first people I knew who was discussing this issue in some of the geoengineering scientific and scholarly communities, posing key questions about how it came to be framed as geoengineering in the first place. Due to media attention in 2012, including a piece in the Guardian, non-Indigenous persons are discussing this case more widely (Buck, 2014). In one study, Kate Elizabeth Gannon and Mike Hulme interviewed local residents of the area about their views on ocean fertilization. Their purpose for doing so was that 'the case of ocean fertilization off the islands of Haida Gwaii may, therefore, provide a useful benchmark for reflexivity in geoengineering governance. Our case study shows that engaging with the situated beliefs and values that underpin human attitudes and responses towards novel geoengineering technologies is a *sine qua non* for good governance' (Gannon & Hulme, 2018, p. 1). The purpose of the research then is to gain greater clarity on Indigenous 'beliefs and values' on geoengineering governance, which includes the ethics and justice issues, as they discuss inclusivity, equity, and social power (16-17).

Interestingly, Gannon and Hulme cite an argument by Howitt et al (2012, p. 48) that states 'global environmental challenges like climate change "should be addressed as opportunities for decolonization"'. Gannon and Hulme claim that: 'In the case of geoengineering, this [decolonization] can only be realized through a clearer focus on the beliefs and values that underpin different attitudes and responses towards different technologies. Such a focus would provide the opportunity for geoengineering interventions to be governed in a more creative, inclusive and equitable manner. Yet the case of the Haida Salmon Restoration Corporation shows how difficult this will be and how easy it will be for geoengineering technology deployments to perpetuate or reinforce existing asymmetrical power relations. This is a salutary lesson given the rapidly growing attention now being given to new carbon dioxide removal technologies in light of the hugely ambitions goals of the Paris Agreement' (Gannon & Hulme, 2018, pp. 16–17). In this work, Indigenous voices and perspectives are valuable for weighing in on a range of ethics and justice issues associated with possible geoengineering governance, serving as 'benchmarks,' 'lessons,' and 'opportunities.'

In a recent study by Wiley Carr and Christopher Preston, they sought to see whether populations vulnerable to climate change expressed differing or similar ethical perspectives on geoengineering to those of ethicists. The perspectives of the ethicists occur largely in what I am referring to as geoengineering discourse. They interviewed people in three different regions, the Solomon Islands, Arctic (Alaska) and Kenya, some of whom are Indigenous. They write that 'One goal of this project was to bring the ethics literature into dialogue with empirical data documenting the hopes and fears of members of vulnerable populations, it is notable that the perspectives expressed by the interviewees corresponded to a surprising degree with many of the concerns articulated in the ethical literature examined above.' Carr and Preston see the results as '[indicating that] members of vulnerable populations shared concerns about their own particular vulnerabilities and about potential moral corruption in developed nations. However, interviewee perspectives also extended the arguments found in the ethics literature by revealing an overarching concern, namely that climate engineering could further erode the already weakened self-determination of vulnerable populations due to a long history of oppression. Indigenous peoples expressed diverse answers about whether they, in fact, endorsed different types of geoengineering or not' (Carr & Preston, 2017, p. 764). For Carr and Preston, there are a diversity of Indigenous perspectives, among the other persons they interviewed. What comes through in their essay is that collective self-determination in the face of climate change, an issue I will discuss further later in this essay, is at the forefront of their concerns.

The work of Gannon and Hulme and Carr and Preston fits in with the general guidance offered by the United Nations on emerging technologies and Indigenous peoples. A 2012 report from the Subsidiary Body on Scientific, Technical and Technological Advice, pertaining to the Convention on Biodiversity, claims that 'It is necessary for decision-makers and scientists to understand the wider multidisciplinary concerns expressed by indigenous peoples, to root their geoengineering proposals within this broader framework and to set aside part of their investigation to understanding how to incorporate a holistic approach into their work' (Subsidiary Body on Scientific, 2012). The idea in this report is that now that geoengineering is inevitably on the table as a potential solution, there needs to be more 'understanding' among those who can advance research and implementation about Indigenous perspectives, which are portrayed as adding a 'broader framework' and '[holism]' to geoengineering plans.

Another place in which Indigenous peoples are brought up is in discussions of the history of the very idea of geoengineering. Advocates of early research on some types of geoengineering often seek to move beyond hasty negative reactions to the idea of humans exercising landscape and earth systems scale environmental interventions. David Keith, for example, has appealed to Indigenous histories of fire management and hunting to show that human-alterations of the environment to suit pressing needs are not unusual for all humanity. In an online published version of a manuscript that was revised later and published elsewhere, he writes that 'Humans transform their environment. While global-scale transformations are a recent consequence of industrial civilization, human transformation of nature is ancient. Some transformations are deliberate, such as the use of fire by aboriginal peoples who altered landscapes to suit their needs or even the modern use of dams to create new lakes. Other transformations occur as an unintended side effect of resource use, such as the mass extinctions of indigenous fauna by early hunters in Australia and the Americas or the more recent threat of climate change caused by our use of fossil energy' (Keith, 2010).

Keith's position is used to suggest that it is precisely Indigenous peoples' histories (among others) that show us how normal it is to deliberately change the environment at large scales. Apart from geoengineering discourses, Indigenous peoples have long advocated for the idea that their societies cultivated landscapes in ways that promoted their sustainability and resilience in the face of environmental change (Whyte, 2017, 2018). Indigenous leaders and scholars point out that it is not always correct to say that there is such a thing as 'wilderness' since Indigenous economies and cultures had regional impacts on ecosystems and landscapes (Trosper, 2002). Myself, among others, certainly reject Keith's tone if it is taken to level off different types of human-induced environmental change, which may or may not be Keith's intention (Harkin & Lewis, 2007). Yet, nonetheless, the point being made here, by Keith, is that geoengineering should not seem like such a stretch given humans, including Indigenous peoples, alter the environment as business as usual and to respond to environmental crises.

Expressing different views, some Indigenous leaders and organizations have engaged in geoengineering discourse and denounced all forms of geoengineering. These denunciations arise in a context where Indigenous peoples globally, from activists to (Indigenous) government leaders, have criticized the failure of the international community to adequately respond to the need to lower carbon footprints for the sake of future generations (Tauli-Corpuz, 2017). The Indigenous peoples' 'Anchorage Declaration' calls geoengineering a 'false solution' (The Anchorage Declaration, 2009). The declaration states 'We challenge States to abandon false solutions to climate change that negatively impact Indigenous Peoples' rights, lands, air, oceans, forests, territories and waters. These include nuclear energy, large-scale dams, geo-engineering techniques, "clean coal", agro-fuels, plantations, and market-based mechanisms such as carbon trading, the Clean Development Mechanism, and forest offsets. The human rights of Indigenous Peoples to protect our forests and forest livelihoods must be recognized, respected and ensured.' Ben Powless, Mohawk from Six Nations, representing the Indigenous Environmental Network, explains that 'For too long our peoples' bodies and lands have been used to test new technologies. Now, in response to climate change, these same people want to put Mother Earth at risk with geoengineering technologies. We cannot afford to threaten our planet in this way, especially when simple, just and proven solutions are at hand' (ETC Group [ETC], 2010).

At the same time, and this is conveyed in Carr and Preston, but also in Gannon and Hulme, there are diverse perspectives on geoengineering among Indigenous peoples. The United Nations report on geoengineering and Indigenous peoples for the Convention on Biodiversity claims that 'Geoengineering has received little support from indigenous and local communities who are acknowledged as being among the world's most vulnerable populations to climate change. Indigenous participants have called for greater involvement of indigenous and local communities in the development of proposals for geoengineering. Not all indigenous and local communities have called for a total ban or for modeling work or controlled in-laboratory experimentation to cease. In fact, some see it as useful in further understanding the complexities of the Earth's ecosystems and in better understanding the potential benefits and harms of geoengineering proposals. On the other hand, there is certainly a strong reluctance to see geoengineering experiments being carried out on a significant scale in the natural world.' (Subsidiary Body on Scientific, 2012)

Scientific reports, including the Arctic Climate Assessment (in 2004), Intergovernmental Panel on Climate Change (AR4 and AR5), and multiple U.S. National Climate Assessments (including 2001, 2014, and 2018), state that Indigenous peoples are among populations most vulnerable to climate change. This work is important given that advocates of geoengineering research often cite, as a strong reason in favor of doing so, the benefits to vulnerable populations. In this way, references to the vulnerability in geoengineering discourse can be taken as referencing some Indigenous peoples by implication. Though often not named, Indigenous peoples include, as evidenced in the above reports, many millions of people living in the global south. Advocates of early geoengineering research, such as Joshua Horton and David Keith, '... contend that a prima facie moral obligation exists to research SRM in the interest of developing countries, because SRM appears to be the most effective and practicable option available to alleviate a range of near-term climate damages that are certain to hurt the global South most of all' (Horton & Keith, 2016, p. 89).

Horton and Keith claim that "Fundamental principles of justice require that, all things being equal, the disadvantaged should not suffer from the results of actions benefiting the better off. Opponents of research into the possible benefits (*and harms*) of solar geoengineering threaten to violate this requirement in at least two ways. First, failing to conduct research puts the global South at risk of paying the highest near-term price for rich-world industrialization and the historical emissions associated with it. And second, stopping research may advance some rich-world political agendas in which geoengineering is at most a tangential issue, but it would come at the cost of assured suffering for poor countries confronting immediate threats that are largely absent from such agendas. Supporting research on solar geoengineering offers the best way to avoid these unjust outcomes (Horton & Keith, 2016, pp. 90–91). For these authors, this is important because 'One of the very few things that nearly all participants in the climate change debate agree on is that the effects of climate change will disproportionately affect the poor, for the simple reason that poorer people will have fewer resources available to them to manage climate risks and adapt to unavoidable changes compared to their wealthier neighbors. This simple fact applies both to disadvantaged people in every country and more broadly to the developing world in relation to rich, industrialized nations' (Horton & Keith, 2016, p. 79). They go on to say that 'The rich have got richer doing things that hurt the poor most of all' (80). In this way, the situations of Indigenous peoples and other groups – taken as urgent and pressing – factor into geoengineering discourse as reasons in favor of doing research on different types of geoengineering.

The references to Indigeneity in geoengineering discourse that I have just discussed, whether explicit or implicit, show how Indigenous perspectives and histories figure into discussions about whether to invest in early research and potential implementation. Gannon and Hulme suggest that understanding Indigenous perspectives in the case of

Haida Gwaii support decolonization if we take it to mean getting a 'clearer focus on the beliefs and values that underpin different attitudes and responses towards different technologies.' Carr and Preston see it as important to compare the views of ethicists with Indigenous persons and persons of other groups who are vulnerable to climate change, and show how Indigenous and other persons add substantial points to geoengineering discourse, such as the importance of self-determination. Keith seeks to contextualize concerns that geoengineering is novel (and hence not controllable), by showing how humans, including many Indigenous peoples, have always modified the environment, for better or worse. Well-meaning advocates of geoengineering could be seen as referencing Indigenous peoples indirectly when they claim that perhaps the primary goal of geoengineering is to support people who are vulnerable to some types of climate change impacts, and where vulnerability is primarily looked at as rooted in economic inequality.

Geoengineering and Colonialism

I cannot think of any Indigenous peoples who have ever had a workshop, event, or major meeting on how to address the prospect of geoengineering research or implementation. That Indigenous peoples do not have their own geoengineering discourse is not because Indigenous peoples never grasp or embrace nonIndigenous concepts and projects. Indeed, Indigenous peoples have not had a problem gravitating around 'climate change' and 'climate justice.' In North America, and beyond too, Indigenous peoples fund their own scientific staff and collaborate with nonIndigenous scientific institutions on climate change planning. They lead and actively participate in local and global advocacy on climate justice. The U.S. 4th National Climate Assessment in chapter 15 documents over 800 Indigenous-led climate actions just in the U.S. sphere. Indigenous peoples often want to highlight their heritages and histories of cultivating landscapes and shaping ecosystems. So it is not as if Indigenous people have some kind of intolerance to science, engineering or the idea anthropogenic environmental change (and taking responsibility for such change). I have my suspicions regarding the broad radio silence, which has to do largely with how geoengineering discourse is not yet open to Indigenous engagement. The key point I want to highlight in this section is that, I would argue, Indigenous peoples do not approach the topic of geoengineering through precisely the same narrative as I have seen in geoengineering discourse.

It is important to begin with where something like geoengineering shows up in certain Indigenous accounts of climate change. While Indigenous peoples certainly appeal to their own heritage, such as in the case of fire management, what often goes missing is an account of the recent wave of European and settler colonialism (e.g. U.S. and Canada) that has produced, in a short period of time, a dramatic terraformation and hydrological engineering of Indigenous territories. Heather Davis and Zoe Todd call such terraforming and engineering 'seismic' in its impact, to highlight both its rapidity and physical effects (Davis & Todd, 2017). Candis Callison, relating to Arctic Indigenous peoples, writes that we need to recognize what 'climate change portends for those who have endured a century of immense cultural, political and environmental changes' (Callison, 2014, p. 42). Callison's work emphasizes that the harms many nonIndigenous persons dread most of the climate crisis are ones that Indigenous peoples have endured already due to different forms of colonialism: ecosystem collapse, species loss, economic crash, drastic relocation and cultural disintegration. Dan Wildcat claims that Indigenous vulnerability to climate change today is part of previous removals occurring through U.S. colonial expansion: 'geographic' (displacement, e.g. Trail of Tears and the forced occupation of reservations); 'social' and 'psycho-cultural' (such as through removal of children to boarding schools) (Wildcat, 2009, p. 4). Leanne Simpson discusses how 'Indigenous peoples have always been able to adapt, and we've had a resilience. But the speed of this – our stories and our culture and our oral tradition doesn't keep up, can't keep up... Colonial thought brought us climate change' (Klein, 2013). Sheila Watt-Cloutier claims 'Climate change is yet another rapid assault on our way of life. It cannot be separated from the first waves of changes and assaults at the very core of the human spirit that has come our way' (Robb, 2015; see also Watt-Cloutier, 2015)

As these voices express, Indigenous peoples do not always see climate change as a future of potential environmental impacts that will threaten their current ways of life such that those impacts must be curtailed as much as possible. Rather, climate change impacts are an intensification of entangled processes of colonialism, capitalism and industrialization that continue to inflict violence and harm on Indigenous peoples. Consider, in more detail, what I mean by this. Different forms of colonialism, from forced relocation, to the creation of reservations, to property dispossession and land grabs, can be looked at as human-induced (anthropogenic) environmental or even climate change (in some cases) (Whyte, 2016). For these changes either forced Indigenous peoples to adapt rapidly to new climate regions (in the case of relocation) or they shrunk or fractionated Indigenous land tenure to situations where it is impossible to plan effectively how to anticipate environmental change. In these examples, there is really no such thing as an isolated climate change impact (Haalboom & Natcher, 2012; Whyte, 2016). Colonial domination, for example, continues to be the problem that generates – to a large but not exclusive extent – the risk. Moreover, solutions to climate change, when they do not deal with colonial domination, also inflict harm. Current hydropower or forest conservation solutions to climate change still displace Indigenous peoples, for example (Beymer-Farris and Bassett, 2012; Campbell, 2015; Cooke, Nordensvard, Saat, Urban, & Siciliano, 2017).

In line with what I am saying, in Carr and Preston, an anonymous Alaska Native interviewee discussed how: 'Eighty-five percent of our communities here are coastal communities, and... people really are having to consider moving their communities. A lot of these villages and their location where they're at now – they're in those particular areas not of their choosing. The Bureau of Indian Affairs located them, because they're on navigable waters. It was easy to barge freight and other stuff into those communities. Our communities were either nomadic or semi-nomadic, moved depending on seasons, on whatever food sources were available. Living by the water is not one of the things that we would have probably chosen for a lifetime commitment, because we recognized that there are issues that are bound to happen... the problem is that no one accepts the responsibility for having located these communities in the areas that they now find themselves' (Carr & Preston, 2017, pp. 765–766).

For this interviewee, colonialism and other forms of oppression are at the heart of the problem. Yet often geoengineering discourses isolate geoengineering as a topic and only add in colonialism, capitalist exploitation, imperialism and other forms of domination later as governance challenges or stakeholders' values or views that must be understood and weighed. Yet for many Indigenous peoples, similar to what the person cited by Carr and Preston shows, colonialism is among the central topics. Colonial domination is not a frame for geoengineering. Geoengineering is itself an ethical or justice issue that arises in situations where Indigenous self-determination is disrespected via ecologically disruptive colonialism, among other forms of domination. So, for each Indigenous people, there is usually a rather clear path forward that they have been calling for, and for years, to address colonialism. For example, the Treaty Tribes of Western Washington associate many different environmental threats to salmon populations, including climate change, as stemming from decades of failures of the U.S. and state of Washington to honor their treaty rights (Treaty Indian Tribes in Western Washington, 2011). Indigenous peoples of the Isle de Jean Charles in Louisiana have long worked to secure recognition as sovereigns in the eyes of the state of Louisiana and the United States. Some people from the island cite both its vulnerability to sea level rise, but also injustices in the resettlement process, as due to continued lack of respect for Indigenous self-determination (Maldonado, Shearer, Bronen, Peterson, & Lazrus, 2013). Many other environmental injustices are traceable precisely to ongoing laws, policies and practices of colonialism (Grijalva, 2008; Whyte, 2011).

In this way, the focus on geoengineering itself as an isolated topic must be contextualized. Some Indigenous peoples would not, I would argue, isolate geoengineering so discretely. They have not put on the table geoengineering, no matter what form, as a solution that must be deliberated on in terms of its pros and cons and in terms of how it relates to localized interests, values or beliefs. Consider what this means. Regarding the ocean fertilization project at Haida Gwaii, among those Native people who have heard of it, they would prima facie associate it more with the decades of environmental injustices at the hands of the Canadian settler state that, among other factors, have affected salmon and other habitats. Gannon and Hulme document some of the histories of colonialism in the region and the current challenges. Buck brings up the issue that the Nation's rights have not been extinguished in the area, the relationship between the two councils (Old Massett and Skidegate), and that the Nationhas explored numerous scientific and political solutions, including a multi-use marine planning process lead by the Council of the Haida Nation (Buck 2014). The Old Massett Village Council's engagement in the ocean fertilization project occurs, for an important part, in the context of colonially induced declining salmon runs. The declining salmon runs do not arise only or primarily from the looped back effects of recent anthropogenic climate change. They are due to factors including land dispossession, disrespect of rights and ecological degradation. Many of these factors can still be addressed by settler Canadians today in ways beyond ocean fertilization, and the Haida are known for some of their recent efforts in protecting their lands, such as forest protection. There is still much political reconciliation work that needs to be done between Indigenous peoples and Canada. So to say that those Haida persons who advocate for ocean fertilization see it as a potential solution to domination is a strange proposition since there are so many other solutions they have been advocating for across generations that historically would have curtailed habitat degradation and are still relevant solutions today.

Indigenous peoples, then, often share perspectives on climate change that emphasize issues of systematic injustice that are local and global. Buck sheds light on this in her work on people in Finland's perception of geoengineering. Though this work does not pertain to Indigenous peoples, it is nonetheless illustrative. She writes that the 'assumption is that people will look at geoengineering as a local concern, or through an individual utilitarian lens. However, these respondents – in part because of a systems view, but also due to empathy – understand the interconnectedness of the world's economies and peoples. (This might seem counterintuitive in a time that seems to be marked by rising nationalism, but these could be simply two sides of a coin.). These respondents did not read the issue through the prism of their local interests' (Buck, 2018, p. 85). Buck's work opens up that how geoengineering is constructed as a discreet topic needs further interrogation.

In my analysis so far, it is important to consider some of the differences in the narrative that Indigenous peoples might have. It is not a given that today's social-ecological systems are ones that are important to conserve. For the state of these systems today is already, for some, an Indigenous dystopia. So, what are Indigenous peoples being asked, then, when some people try to persuade them to adopt geoengineering as Indigenous people's best available solution to climate change impacts (that Indigenous peoples did not cause)? It is also the case that Indigenous peoples have been arguing for decades and even centuries for certain reforms from colonial nations, such as the U.S. or New Zealand, that would improve Indigenous peoples' capacities to adapt to climate change and would hasten mitigation efforts. These solutions should be on the table in discussions about geoengineering, even if many who are involved in geoengineering discourse do not initially see these solutions as on topic. For Indigenous peoples, I would argue, they are exactly the topic.

If I extrapolate more speculatively, I think the concern is that the construction of geoengineering as an issue to be debated regarding its governance and acceptability can create a powerful form of obfuscation of ethics and justice. For decades, Indigenous and allied scholars in areas like Indigenous studies have argued for 'Indigenous erasure' as one of the strategies of colonial domination. Erasure includes the erasure of colonialism in discourse. Scholars and writers such as (among many others) Lee Maracle, Glen Coulthard, Audra Simpson, Megan Bang, Ananda Marin, Tsianina Lomawaima, Theresa McCarthy, Eve Tuck and K. Wayne Yang have shown how conversations about Indigenous participation may occur in the name of ethics and justice, but actually serve to obscure (erase) the full implications of what colonial nations' responsibilities actually are to Indigenous peoples (Bang et al., 2014; Coulthard 2014; Lomawaima & McCarty, 2006; Maracle, 2015; Simpson, 2014; Tuck & Yang, 2012). In geoengineering discourse more recently, Stephen Gardiner refers to an analogous moral problem. The problem occurs when people 'emphasize and endorse strong ethical concerns that we are otherwise unwilling to act on, and which would, if earnestly and coherently embraced, lead us to approach both climate policy in general and geoengineering in particular in very different ways. In short, the worry is that even if ethically serious people have reason to support (some forms of) geoengineering research and perhaps even deployment in the abstract, their approach would look very different from anything currently under consideration, let alone actually likely to transpire' (Gardiner, 2013a, p. 12).

When confronted with the issue of geoengineering, it should not be surprising that Indigenous peoples have different views on its acceptability regarding ethics and justice. Calling for greater attention to Indigenous issues on geoengineering involves centering and acting to address relevant forms of domination, instead of obscuring them. And, if it is true that Indigenous peoples have long legacies of explicitly seeing their histories and cultures as modifying the environment, then they would have a lot to say about climate change as a form of environmental modification that has rendered us into the situation we are in today. If we look at colonialism, we can see that it is not accurate or factual to see Indigenous peoples, and perhaps other populations too, as primarily burdened by lacking financial resources for adapting to climate change. It is the legacy and ongoing practices of colonialism, and other forms of domination, that hamper mitigation of and adaptation to climate change.

Indigenous Consent

The topic of consent has been raised as a major issue in geoengineering. Consent is considered to be an important topic for at least two reasons, one general for any actions, and one more specific to geoengineering. First, consent invokes the idea that those parties, human or nonhuman, affected by the actions of others should have opportunities to self-determine their acceptance of or opposition to those the actions. In cases where affected parties oppose harmful effects on them, their wishes should be honored with the cessation of the harmful or risky actions or the modification of the actions to end harm and reduce risk. Affected parties' opposition or provisional acceptance can often be followed by their engagement in processes that seek to end harms and reduce risks. Second, the particular context of geoengineering is rife with scenarios where more powerful or sovereign parties, from scientific organizations to nations, could – in morally problematic ways – engage in research or implementation without securing the consent of the affected parties. Moreover, geoengineering is a global scale issue that cannot possibly be directly consented to by everyone who will be affected. Indeed, some ways of thinking about potential geoengineering implementation affect everyone on earth. For this reason, there are approaches to consent that do not require the securing of direct acceptance, such as tacit or implicit consent. Scholars such as Benjamin Hale, Lisa Dilling, and Pak-Hang Wong document a range of work and issues in the literature on consent in relation to geoengineering that explores some of these options for consent, such as implicit consent in cases where direct consent is impossible (Hale & Dilling, 2011; Wong, 2016).

In the literature cited previously in the last two sections, one of the issues that keeps arising is that some Indigenous peoples endorse geoengineering research and implementation. But I am not convinced that there is much of an issue to discuss regarding *whether* Indigenous peoples, some or all, consent or dissent to geoengineering. Rather, we need to reflect on what it means to even narrow in on the idea of Indigenous consent to geoengineering, to begin with. I have two points to make here. First, when Indigenous consent or acceptance is discussed, I often do not see very much clarity in what it means in the first place to say that Indigenous peoples can even participate in the politics of science and international relations meaningfully in relation to geoengineering, and many other topics too. Second, Indigenous peoples never consented to the various global and local orders in which geoengineering emerges as an issue anyways. This raises the question of what it means for groups of people to have an opportunity to consent when the very opportunity itself arises from conditions they did not consent to. While it is perhaps the case that all consent situations are ones in which it is imaginable that the consenter did not consent to the surrounding conditions, there is still a difference here. For the conditions, I am discussing are conditions of domination, as described, in part, in the previous section. These conditions were not setup or intended to support Indigenous well-being or self-determination. Unfortunately, in this essay, I cannot focus more space on discussing when the status of consent to any conditions is morally relevant and when it is not.

Starting with the first point, the problem concerns the fact that Indigenous peoples globally are not in the position relative to any potential geoengineering research or future implementation to give meaningful consent or dissent. Much of why this is the case has to do with the impacts of colonialism on Indigenous diplomacy both locally and globally. Indigenous worlds are diverse. While all Indigenous peoples trace their sovereignty and collective self-determination to origins prior to or separate from the formation of nation-states, they have been denied ethical and just capacities to represent themselves in international, regional, national and local fora. For example, the United Nations does not include Indigenous peoples as members, though, based on years of activism, Indigenous peoples have become a major 'non-state' force in the U.N. In terms of political representation, it is hard to convey just how different Indigenous peoples can be from one another, which I will discuss briefly in what follows.

In many regions, neighboring nation states do not recognize Indigenous peoples as sovereigns or semi-sovereigns. Indigenous peoples in this situation often form or work with non-governmental organizations and send representatives to human rights bodies at the United Nations. Often the U.N. or the International Labour Organization are their main avenues for gaining relief from human rights abuses given they are not recognized as sovereigns locally. In other regions, settler states like the U.S. recognize *some* Indigenous peoples as having political sovereignty, for example, the 573 U.S. federally recognized Tribes. But there are many Indigenous peoples the U.S. does not recognize. Moreover, many Indigenous persons do not feel that Tribal governments represent them because many federally recognized Tribes are forms of government the U.S. forced Tribes to adopt to promote extractive industries. Hence, they are often times not governments that some people feel represent some of their constituents, interests in morally acceptable ways. There are many Indigenous NGOs that represent interests politically that are often ignored by 'recognized' Tribal governments. In the case of the U.S., colonialism has also diminished Indigenous capacities to participate on equal terms in consultative (e.g. consent) processes. Often U.S. agencies and corporations have a larger staff and greater financial resources to participate in consultative processes, whereas the Tribes who may be affected by the U.S. and corporate actions have far fewer staff and financial resources. Historically, we know that North American Indigenous peoples, among others, had tremendous diplomatic capacities, as evidenced in eras such as the fur trade (White, 1991; Witgen, 2011). But over time colonialism works to strip Indigenous peoples of these capacities, which then opens them up to criticism when they cannot follow through in participatory processes at the same level as a nation or corporation.

In terms of geoengineering research or future implementation, then, there is no way to suggest something like Indigenous consent or dissent. Imperialism and settler colonialism have rendered a situation in which many Indigenous peoples do not have local leverage to consent or dissent to any projects that might affect them. Indigenous leadership at the U.N. cannot possibly cover the different forms of political representation that Indigenous peoples have to use to put their issues on the table. Moreover, while the U.N. Declaration on the Rights of Indigenous peoples enshrines *free, prior and informed consent* (FPIC), it is unclear whether most nation-states are institutionally setup to make that possible. For example, with U.N. REDD policies, a forest conservation program, there may be many of Indigenous peoples living near a nation-state that have not enjoyed good diplomatic relationships with the nation-state and local communities, making jurisdiction unclear. So to suggest that it is even doable to implement policies of FPIC is deeply problematic. For the needed relationships are not present to facilitate FPIC. Or in the U.S. and Canada, while both nations have consultation policies with Indigenous peoples, they do not actually ensure Indigenous peoples can dissent in terms of a veto power. Consultation policies that would provide the possibility of Indigenous dissent have been dismissed as not being politically feasible given Indigenous peoples rarely have population numbers to sway national voting trends.

So, in this context, it is interesting to hear discussions about Indigenous consent or dissent to geoengineering and attempts to persuade vulnerable populations like Indigenous peoples for several reasons. At one level, the ways in which the world is stacked up politically against Indigenous representation and diplomacy makes it inevitable that whoever has the capacity to propose geoengineering research or implementation will not require Indigenous support. While there are certainly groups of people out there who could oppose geoengineering with political force, Indigenous peoples are unlikely to be one of those groups. At another level, I have yet to see advocates of geoengineering learn about the different aspects and problems of Indigenous political representation, and attempt to offer a solution to securing consent that at least tries to grapple with the very real challenges Indigenous peoples have to endure every day regarding their consent.

Second, Indigenous peoples did not consent anyways to the global and local orders for which geoengineering becomes an issue in the first place. Some Indigenous peoples do not invest in geoengineering qua geoengineering. That is, if an Indigenous people goes along with a geoengineering research project or potential implementation, that is not the same as saying that group or many other Indigenous peoples consent or endorse geoengineering more broadly. For Indigenous peoples have not consented to the very conditions that give rise to the opportunity to consent to something like geoengineering. Let me explain, in more detail, what I mean by this. In the climate justice literature, it is often stated that groups like Indigenous peoples have done little to cause anthropogenic climate change, yet they will bear many burdens from climate change impacts. Moreover, scholars, such as Chris Cuomo, note that anthropogenic climate change emerges from the crucible of inequality. As Cuomo would add, this is just the very beginning of understanding the relationships between power and climate change relating to Indigenous peoples (Cuomo, 2011).

For Indigenous peoples, there is a continuing and very direct connection between anthropogenic climate change and their struggles with colonialism. The fossil fuel industry and climate-altering land use change are largely possible because of Indigenous dispossession from the lands where those activities first took place. Those activities were made possible through laws, policies, and practices that opened up Indigenous territories; they were also made possible through colonial cultures that made colonialists feel it was morally acceptable (or morally inevitable) to do so. Critically, current laws, policies, practices and cultures that descend from the original ones (or are literally still the same ones today) are largely responsible for what makes Indigenous peoples vulnerable to certain climate change impacts today (Whyte, 2016). Over the years, Indigenous peoples have been offered a range of solutions to climate change and environmental problems, including the privatization of their lands, the installation of large-scale dams, among others. Each of these solutions rarely engaged with the ultimate change of the laws, policies, practices and cultures that are responsible for the bad aspects of Indigenous peoples' situations today. Indigenous peoples did not consent to these conditions, whether historically or today, where conditions refer to the laws, policies, practices and cultures I have just described – the local and global orders based on colonial domination.

Gardiner has also covered the argument about desperation. 'In conclusion, the desperation argument misses much of what is at stake, ethically speaking, in geoengineering policy. As far as justification is concerned, neither the consent nor the self-defense interpretations clearly license geoengineering, and they may even count against it. Moreover, the contextual question reveals more about the live threats and ethical import of geoengineering...' (Gardiner, 2013b, p. 31). Gardiner's views echo Marilyn Frye's original contributions to defining oppression, which is an ethical problem of people being put in a situation where the optional actions each have consequences (Frye, 1983). Here, context refers to the conditions that put somebody in a dilemma. Gardiner claims that "The argument suggests that we can approach the plight of the desperate from some distance, thinking only 'who can blame them?', 'wouldn't we do the same?', and 'shouldn't we help out?' However, if desperation is over invoked, the powerful countries are unlikely to be in this situation. Instead, they are likely to have played a substantial role in reducing the desperate to the point of begging. This is a horrifying moral territory. In my view to bring others to the point of desperation, and especially to put them in a situation here they are forced to make a tragic choice, constitutes a special kind of moral wrong (32)."

For Frye and Gardiner, what matters are the conditions under which a situation arises in the first place in which somebody has the option of consenting or dissenting. Some Indigenous peoples, such as those cited in this essay, claim it is the laws, policies, practices and cultures of colonialism, among other forms of domination, that play a crucial role in exacerbating climatic vulnerability. The literature on climate change resettlement demonstrates that it is not so much about having the financial resources to relocate that supports adaptation; rather it is territorial mobility and collective self-determination. Territorial mobility and collective self-determination, which for many Indigenous peoples have been curtailed exponentially, are issues of colonialism. Hence, to suggest that some Indigenous peoples should consent to geoengineering because it is their best option to adapt to climate change is to, again, take off the table the anti-colonial and decolonial reforms that Indigenous peoples have been calling on nations and corporations to do for years. Indigenous peoples did not consent to the conditions of colonial domination having such a sway over their lives.

Conclusion

Holly Jean Buck asks whether it is 'possible to imagine geoengineering being driven by a desire for social change, as a means to transform society? This does not place geoengineering in the box of "environmental issues", but acknowledges that geoengineering is coupled to complex social, cultural and economic systems' (Allenby, 2010; Buck, 2012, p. 266). Indigenous voices *should* be involved in scientific and policy discussions of different types of geoengineering. But, context matters. Geoengineering discourses cannot just be associated with geoengineering to the exclusion of topics and solutions that Indigenous peoples value. A conversation about geoengineering that, say, disallows or is silent on, treaty rights or colonialism, is not a space for Indigenous voices to matter, in my opinion. Or a discussion where Indigenous peoples are asked to trust non-Native people again, this time, is problematic if there are not direct reasons given for why trust is an appropriate attitude. For the conversation must address *why* distrust occurred in the first place, which has to do with legal and policy frameworks, social and culture norms and economic systems that are anti-Indigenous.

Forces of domination render even the most well-intentioned solutions ineffective. Indeed, Indigenous peoples face environmental risks from all sides, from encroachment on their territories, pollution and pollution drift, land-grabs and global climate change, among many others. Whether Indigenous peoples accept any geoengineering research or implementation will mean very little if Indigenous territories are not rematriated (Maracle, 1996) and political reconciliation with colonial states remains lacking. In some ways, the future of geoengineering discourse should be to bring back into focus why it may be problematic in the first place to isolate geoengineering as a discrete topic. Ethics and justice issues pertaining to geoengineering perhaps emerge most clearly when we question how it came to be that some people see the best path forward as involving the weighing of different perspectives, beliefs, values, and interests.

Disclosure statement

No potential conflict of interest was reported by the author.

References

Allenby, B. (2010). Climate change negotiations and geoengineering: Is this really the best we can do? *Environmental Quality Management*, *20*(2), 1–16.

Bang, M., Curley, L., Kessel, A., Marin, A., Suzukovich, E. S., III, & Strack, G. (2014). Muskrat theories, tobacco in the streets, and living Chicago as Indigenous land. *Environmental Education Research*, *20*(1), 37–55.

Beymer-Farris, B. A., & Bassett, T. J. (2012). The REDD menace: Resurgent protectionism in Tanzania's mangrove forests. *Global Environmental Change*, *22*(2), 332–341.

Buck, H. (2014). Village science meets global discourse: The Haida Salmon restoration corporation's ocean iron fertilization experiment. *Case Study, Engineering Our Climate Working Paper and Opinion Article Series*. Retrieved from http://wp. me/p2zsRk-9M

Buck, H. J. (2012). Geoengineering: Re-making climate for profit or humanitarian intervention? *Development and Change*, *43*(1), 253–270.

Buck, H. J. (2018). Perspectives on solar geoengineering from Finnish Lapland: Local insights on the global imaginary of arctic geoengineering. *Geoforum, 91*, 78–86.

Buck, H. J., Gammon, A. R., & Preston, C. J. (2014). Gender and geoengineering. *Hypatia, 29*(3), 651–669.

Bunzl, M. (2009). Researching geoengineering: Should not or could not? *Environmental Research Letters, 4*(4), 045104. Retrieved from http://stacks.iop.org/1748-9326/4/i=4/a=045104

Callison, C. (2014). *How climate change comes to matter: The communal life of facts*. Raleigh-Durham, NC: Duke University Press.

Campbell, C. (2015). Implementing a greener redd+ in black and white: Preserving wounaan lands and culture in panama with indigenous-sensitive modifications to redd+. *American Indian Law Review, 40*(p), 193.

Carr, W., & Preston, C. J. (2017). Skewed Vulnerabilities and moral corruption in global perspectives on climate engineering. *Environmental Values, 26*, 757–777.

Carr, W. A., Preston, C. J., Yung, L., Szerszynski, B., Keith, D. W., & Mercer, A. M. (2013). Public engagement on solar radiation management and why it needs to happen now. *Climatic Change, 121*(3), 567–577.

Cooke, F. M., Nordensvard, J., Saat, G. B., Urban, F., & Siciliano, G. (2017). The limits of social protection: The case of hydropower dams and indigenous peoples' land. *Asia & the Pacific Policy Studies, 4*, 437–450.

Coulthard, G. S. (2014). Red Skin, White Masks: Rejecting the Colonial Politics of Recognition. Minneapolis, MN: University of Minnesota Press.

Cuomo, C. J. (2011). Climate change, vulnerability, and responsibility. *Hypatia, 26*(4), 690–714.

Davis, H., & Todd, Z. (2017). On the importance of a date, or, decolonizing the anthropocene. *ACME: an International Journal for Critical Geographies, 16*(4), 761–780.

Declaration, T. A.The Anchorage Declaration. (2009, April 24).*Indigenous peoples' global summit on climate change*. Anchorage, Alaska.

ETC Group. (2010). Hands off mother earth! Retrieved 2018, from http://www.etcgroup.org/content/hands-mother-earth.

Frye, M. (1983). *The politics of reality: Essays in feminist theory*. Trumansburg, NY: Crossing Press.

Gannon, K. E., & Hulme, M. (2018). Geoengineering at the "edge of the world": Exploring perceptions of ocean fertilisation through the Haida Salmon restoration corporation. *Geo: Geography and Environment, 5*(1), 1–21.

Gardiner, S. (2013a). Geoengineering and moral schizophrenia: What is the question? In: W. C. G. Burns & A. L. Strauss (Eds.), *Climate change geoengineering: philosophical perspectives, legal issues, and governance frameworks* (pp. 11–38). Cambridge, UK: Cambridge Univesity Press.

Gardiner, S. M. (2010). Is "arming the future" with geoengineering really the lesser evil? Some doubts about the ethics of intentionally manipulating the climate system. In: S. M. Gardiner, S. Caney, D. Jamieson, & H. Shue (Eds.), *Climate ethics: Essential readings* (pp. 284–314). Oxford, UK: Oxford University Press.

Gardiner, S. M. (2013b). The desperation argument for geoengineering. *PS: Political Science & Politics, 46*(1), 28–33.

Grijalva, J. M. (2008). *Closing the circle: Environmental justice in Indian country*. Durham, NC: Carolina Academic Press.

Haalboom, B., & Natcher, D. C. (2012). The power and peril of "vulnerability": Approaching community labels with caution in climate change research. *Arctic, 65*(3), 245–366. Retrieved from http://arctic.synergiesprairies.ca/arctic/index.php/arctic/article/view/4219

Hale, B., & Dilling, L. (2011). Geoengineering, ocean fertilization, and the problem of permissible pollution. *Science, Technology & Human Values, 36*(2), 190–212.

Harkin, M. E., & Lewis, D. R. (Eds.). (2007). *Native Americans and the environment: Perspectives on the ecological Indian*. Lincoln, NE: University of Nebraska Press.

Horton, J., & Keith, D. (2016). Solar geoengineering and obligations to the global poor. In: C. Preston (ed.), *Climate justice and geoengineering: Ethics and policy in the atmospheric anthropocene* (pp. 79–92). Lanham, MD: Rowman and Littlefield.

Howitt, R., Havnen, O., & Veland, S. (2012). Natural and unnatural disasters: Responding with respect for indigenous rights and knowledges. *Geographical Research*, *50*(1), 47–59.

Keith, D. (2010). Engineering the planet. In: S. Schneider & M. Mastrandrea, Eds., *Climate change science and policy* (pp. 1–11). Washington DC: Island Press. Retrieved from https://www.researchgate.net/profile/David_Keith/publication/228552593_Engineering_the_planet/links/549233b40cf2991ff5560b2c/Engineering-the-planet.pdf. Online, Draft Not Reflective of Final Published Version.

Klein, N. (2013, March 5). Dancing the world into being: A conversation with Idle No More's Leanne Simpson. *Yes! Magazine*.

Lomawaima, K. T., & McCarty, T. L. (2006). *" To Remain an Indian": Lessons in democracy from a century of Native American Education*. New York, NY: Teachers College Pr.

Maldonado, J. K., Shearer, C., Bronen, R., Peterson, K., & Lazrus, H. (2013). The impact of climate change on tribal communities in the US: Displacement, relocation, and human rights. *Climatic Change*, *120*(3), 601–614.

Maracle, L. (1996). *I am woman: a native perspective on sociology and feminism*. Toronto, Ontario: Press Gang.

Maracle, L. (2015). *Memory Serves*. Edmonton, AB: NeWest Press.

Preston, C. J. (2012). *Engineering the climate: The ethics of solar radiation management*. Lanham, MD: Lexington Books.

Preston, C. J. (2013). Ethics and geoengineering: Reviewing the moral issues raised by solar radiation management and carbon dioxide removal. *Wiley Interdisciplinary Reviews: Climate Change*, *4*(1), 23–37.

Robb, P. (2015). Q and A: Sheila Watt-Cloutier seeks some cold comfort. *Ottawa Citizen*.

Scott, D. (2012). Introduction to the special section, 'the ethics of geoengineering: Investigating the moral challenges of solar radiation management'. *Ethics, Policy & Environment*, *15*(2), 133–135.

Simpson, A. (2014). *Mohawk interruptus: Political life across the borders of settler states*. Durham, NC: Duke University Press.

Smith, L. T. (1999). *Decolonizing methodologies: Research and indigenous peoples*. London, UK: Zed books.

Story, P., & Lickers, F. (1997). Partnership building for sustainable development: A first nations perspective from Ontario. *Journal of Sustainable Forestry*, *4*(3–4), 149–162.

Subsidiary Body on Scientific,Technical and Technological Advice (2012). Impacts of climate-related geoengineering on biodiversity: Views and experiences of indigenous and local communities and stakeholders. Montreal, PQ: Canada.

Svoboda, T., Keller, K., Goes, M., & Tuana, N. (2011). Sulfate aerosol geoengineering: The question of justice. *Public Affairs Quarterly*, *25*(3), 157–179.

Tauli-Corpuz, V. (2017). U.N. Special Rapporteur: Indigenous Peoples' rights must be respected in global climate change agreement. Retrieved, 2017 from http://unsr.vtaulicorpuz.org/site/index.php/press-releases/61-clima-change-hrc.

Treaty Indian Tribes in Western Washington. (2011). *Treaty rights at risk: Ongoing habitat loss, the decline of the Salmon resource, and recommendations for change*. Retrieved from http://nwifc.org/w/wp content/uploads/downloads/2011/08/whitepaper628finalpdf.pdf

Trosper, R. L. (2002). Northwest coast indigenous institutions that supported resilience and sustainability. *Ecological Economics*, *41*, 329–344.

Tschakert, P. (2010). "Whose Hands are Allowed at the Thermostat?" Voices from Africa. Retrieved from http://www.umt.edu/ethics/ethicsgeoengineering/Workshop/articles1/tschakert.pdf

Tuana, N. (2013). The Ethical Dimensions of Geoengineering: Solar Radiation Management through Sulphate Particle Injection." Working Paper, Geoengineering Our Climate Working Paper and Opinion Article Series (pp. 1–20). Available at: http://wp.me/p2zsRk-7B. Accessed January 6, 2019.

Tuana, N., Sriver, R. L., Svoboda, T., Olson, R., Irvine, P. J., Haqq-Misra, J., & Keller, K. (2012). Towards integrated ethical and scientific analysis of geoengineering: A research agenda. *Ethics, Policy & Environment*, *15*(2), 136–157.

Tuck, E., & Yang, K. W. (2012). Decolonization is not a metaphor. *Decolonization: Indigeneity, Education & Society*, *1*(1), 1–40.

Watt-Cloutier, S. (2015). *The right to be cold: One woman's story of protecting her culture, the Arctic and the whole planet*. Toronto, ON Penguin.

White, R. (1991). *The middle ground: Indians, empires, and republics in the Great Lakes region, 1650-1815*. Cambridge, U.K.: Cambridge University Press.

Whyte, K. P. (2011). Environmental justice in Native America. *Environmental Justice*, *4*(4), 185–186.

Whyte, K. P. (2016). Is it Colonial Déjà Vu? Indigenous peoples and climate injustice. In: J. Adamson, M. Davis, & H. Huang (Eds.), *Humanities for the environment: Integrating knowledges, forging new constellations of practice*, 88–104. New York, NY: Earthscan.

Whyte, K. P. (2017). Indigenous climate change studies: Indigenizing futures, decolonizing the anthropocene. *English Language Notes*, *55*(1–2), 153–162.

Whyte, K. P. (2018). Critical investigations of resilience: A brief introduction to indigenous environmental studies & sciences. *Daedalus*, *147*(2), 136–147.

Wildcat, D. R. (2009). *Red alert! saving the planet with indigenous knowledge*. Golden, CO: Fulcrum.

Witgen, M. (2011). *An infinity of nations: How the native new world shaped early North America*. Philadelphia, PA: University of Pennsylvania Press.

Wong, P.-H. (2016). Consenting to geoengineering. *Philosophy & Technology*, *29*(2), 173–188.

Recognitional Justice, Climate Engineering, and the Care Approach

Christopher Preston and Wylie Carr

ABSTRACT
Given the existing inequities in climate change, any proposed climate engineering strategy to solve the climate problem must meet a high threshold for justice. In contrast to an overly thin paradigm for justice that demands only a science-based assessment of potential temperature-related benefits and harms, we argue for the importance of attention to *recognitional justice*. Recognitional justice, we go on to claim, calls for a different type of assessment tool. Such an assessment would pay attention to neglected considerations such as relationships, context, power, vulnerability, narrative, and affect (amongst others). Here we develop a care-ethics related tool for assessing the justice (or injustice) of climate engineering with stratospheric aerosols, and suggest that qualitative social science methods may be required for its effective application. We illustrate the use of this tool with a case study involving interviews about stratospheric aerosol injection conducted in Kenya, the Solomon Islands, and the North American Arctic. Having shown through this case study the efficacy of the care approach for spotting recognitional injustice, we suggest that a care approach is not only sensitive to the considerations that count, it can also be powerfully normative.

Disruption to established weather patterns brought on by climate change has the potential to create enormous economic – as well as non-economic – burdens (IPCC, 2014a). These burdens fall disproportionately on individuals and populations least responsible for causing them and raise serious concerns about justice (IPCC, 2014b; Shue, 2014). With climate harms increasingly evident and highly likely to escalate, preventing as many of these harms from occurring as possible is a clear moral priority for people of good conscience.

The hope of avoiding some of these injustices through intentionally engineering the climate is one of the strongest reasons in favor of pursuing what would otherwise be an outrageous attempt at a technological fix (Horton & Keith, 2016). Although not a preferred option for dealing with the harms of climate change, a number of commentators have become convinced that hoping for a better option to materialize is nothing more than a 'pious wish' (Crutzen, 2006). Stacked up against the extensive array of concerns levelled against climate engineering (Burns & Strauss, 2013; Gardiner, 2013, 2010; Preston, 2012), the potential to alleviate significant injustices is one of the most compelling moral arguments to take climate engineering seriously.

For many forms of climate engineering, however, there remains a heated disagreement about whether the 'cure' might not be worse than the 'disease.' A considerable amount of uncertainty remains over the immediate and the long-range impacts of such dramatic interventions into the climate system. Concerns about disruption to precipitation patterns, impacts on vegetative productivity, harmful changes in land use, and the increased likelihood of extreme weather events all haunt the discussion (Barrett et al., 2014; Ferraro, Highwood, Charlton-Perez, & Meehl et al., 2014; Ito, 2017; MacMartin, Ben Kravitz, Long, & Rasch, 2016).

In addition to these biophysical worries, concerns about the nature of the politics that might be used to legitimize these schemes compound the overall sense of uncertainty (Cairns and Stirling 2014; Hulme, 2014; Macnaghten & Szerszynski, 2013). Other commentators have expressed apprehension about the ethical visions embedded in climate engineering, arguing that the technology seems destined to profoundly impact social and political relationships as well as changing our fundamental understanding of the relationship between humans and the surrounding world (Gardiner, 2010; Hamilton 2013; Jamieson, 1996).

The international dimensions of climate engineering justice are particularly fraught. Jane Flegal and Aarti Gupta are worried that as currently framed, justifications of climate engineering as a way for wealthy nations to help the vulnerable may turn out to disempower the very groups advocates claim to be concerned about (Flegal & Gupta, 2018). With these concerns in mind, it is clear that for any type of climate engineering deployment to be morally acceptable, not only must it be just, it must be just in the relevant ways.

Taking Flegal and Gupta's concerns about disempowerment seriously, we argue below that an important type of justice is *recognitional justice* and that recognitional justice requires a different set of tools for its assessment. We argue that a 'care approach' can provide a useful corrective to the often incomplete approaches to justice. A care approach and the qualitative methods it demands can fill out important elements of justice revolving around recognition, thereby providing a fuller moral assessment of climate engineering than some of the current discussion displays.

To make this case, we begin by highlighting how recognitional justice is often glossed over by researchers advocating for climate engineering. We go on explain why the care approach is the right tool for identifying this type of justice (and injustice). We then illustrate how qualitative social science is the appropriate method for operationalizing the care tool, drawing on a case study to illustrate our claims. The conclusion we come to is that justice in climate engineering cannot be achieved unless it is first sanctioned by the care approach.

Recognitional Justice

To identify the type of injustice that has routinely been missed by climate engineering commentators, it helps to identify the approach to climate justice which tends to dominate. Flegal and Gupta find that the ethics of climate engineering (in its research stage) is typically assessed too narrowly around 'distributional outcomes' and 'comparative risk assessment' (Flegal & Gupta, 2018, p. 49–52).

One problem with this, as advocates of climate engineering themselves acknowledge, is that predicting risks and outcomes remains scientifically problematic due to the inherent uncertainty about climate outcomes at regional levels. A bigger problem is that, even if this uncertainty was not present, the focus on distributional outcomes has arguably bent the ethics discussion too far towards a narrow quantitative cost-benefit analyses of potential physical harms that might be forestalled by the use of stratospheric aerosols. In bending the ethics this way, considerable sources of injustice can be missed.

This narrowing of the ethics is particularly evident when scientific modelling is allowed to set the direction for the ethical conversation. Horton and Keith, for example, adopt the position that the ethical implications of different responses to climate harms 'turn on the particular distribution of benefits and harms associated with each' (Horton & Keith, 2016, p. 81). The emphasis on the distribution of benefits and harms leads to the consideration of justice being reduced to the question of 'are temperatures and precipitation likely to be normalized by the climate engineering strategy?'

The focus on distributional outcomes is arguably characteristic of a larger trend within political justice in recent decades. According to some critics, the distributional paradigm has taken over the justice discussion (Fraser, 2007; Honneth, 2001; Schlosberg, 2007; Young, 2006). Thinking of justice only in terms of the distribution of benefits and harms has a number of shortcomings. One of them is that this approach makes justice appear to be too much about what benefits and burdens must to be shared and not enough about how societies and cultures are structured. What David Schlosberg identifies as problematic is, '...the sole emphasis on distribution without an examination of the *underlying* causes of maldistribution' (Schlosberg, 2013, p. 14). Looking more carefully at that those structures opens up questions of power and vulnerability that are integral to a complete picture of justice.

Critics such as Schlossberg, Frazer, and Young insist that those who seek a just society must be concerned about much more than simply the distribution of physical harms. The narrow paradigm of thinking about justice in terms of physical (or economic) harms and benefits simply misses too much. By focusing only on how physical goods and harms stack up, it fails to see both *how* and *why* the interests of certain groups have been systematically neglected throughout history. It fails to identify the underlying causes of the disproportionate burden of climate harms borne by some populations. And it fails to spot a number of less tangible, but highly significant, forms of injustice.

Taking these other considerations seriously has led to a new focus on recognition as a key accompaniment, or even a precursor, to questions of just distribution. Recognition – and its absence, misrecognition – are not only requirements for distributional justice, but also key components of justice in themselves (Schlosberg, 2007). At the core of recognition is the straightforward idea of an individual or a group being adequately acknowledged. Adequate acknowledgment means respecting and noticing people for *who* they are and *where* they are. Kyle Powys Whyte says recognition means 'fairly representing and considering the cultures, values, and situations of all affected parties' (Whyte, 2011, p. 200).

Although recognition and misrecognition clearly have personal and psychological components to them, Fraser thinks it important to emphasize their social and political dimensions. She places a large part of the blame for failures of recognition on the institutionalization of subordination exhibited in 'patterns of representation, interpretation, and communication' displayed across society (Fraser, 1999, p. 7). Groups, in other words, can be victims of recognitional injustice. The social status of particular groups is routinely diminished not just by individuals but by civic arrangements and relations. Particular kinds of oppressive social arrangements, or 'structural injustices' as Young has called them (Young, 2006), repeatedly create misrecognition for disadvantaged groups and the individuals within them.

Highly pertinent for this social, structural, and historical perspective, is the notion of an 'environmental heritage' developed by Robert Figueroa. An environmental heritage of a group is a product of a distinct and particular past and it creates 'an environmental identity in relation to the community viewed over time' (Figueroa, 2006, p. 371–372). This heritage frames the decisions, values, and practices of a particular group. One of the conditions for adequate recognition is whether this environmental heritage is fully taken into account. David Schlosberg affirms this by calling for 'recognition and preservation of diverse cultures, identities, economies, and ways of knowing.' One of the biggest threats to recognition, according to Schlosberg, is a 'growing global monoculture' that attempts to erase the significance of different lifeways, practices, and knowledge systems (Schlosberg, 2007, p. 86).

In sum, recognition is a key component of political justice that a narrow framing of justice in terms of physical harms and benefits neglects. On the one hand, recognition is necessary for diverse groups to participate equally in political processes and to be recipients of a fair distribution of benefits and harms. On the other hand, recognition is simply a component of justice in its own right. Ensuring recognition means working to understand the different starting assumptions, values, and knowledge practices brought to a particular situation. Gaining such an understanding requires careful attention to historical, cultural, and institutional factors. Put simply, it demands the ability to appreciate who people are, where they are coming from, and what they are saying about their situation.

A Care Based Assessment

As a relatively new component of political justice, recognition demands new frameworks for assessment. The previous emphasis on distributive elements of justice meant that policies were likely to be assessed primarily through a consequentialist lens. If a particular policy is in place, how do the resulting benefits and burdens stack up? Thinking about benefits and burdens in a certain way invariably points towards particular disciplines and approaches as being most relevant to their appraisal. Flegal and Gupta suggested that the cost-benefit frame ensures that '...concerns about equity are treated as empirical matters, requiring scientific assessment of feasibility, risks ... distributive outcomes and optimizations'(Flegal & Gupta, 2018, p. 56). In the case of climate engineering, this means considering questions such as: What would be the temperature change? How would this change influence precipitation? What positive and negative impacts will it have on the economy? What additional risks of harm does it generate? The methods through which such questions are answered are scientific, not political or cultural. Flegal and Gupta point out that, as far as methods for assessing the justice of prospective climate engineering, 'science is [considered] the institution most capable of steering [its] technological emergence' (Flegal & Gupta, 2018, p. 54).

The scientific answers to questions about the distribution of the benefits and burdens are clearly important for considering the justice of any climate engineering strategy. Temperature and precipitation are important. These don't, however, give much of a sense of whether a strategy will provide adequate recognition to a particular people. Recognitional justice requires a different type of lens through which to assess justice. Rather than relying on framings focusing on welfare-related benefits and harms, it requires a framing more sensitive to the kinds of things that recognition is all about.

It is our contention that a 'care approach' proves itself to be particularly appropriate for identifying recognition or misrecognition. This approach is not primarily about developing a caring attitude – though developing this attitude might help – it is about using particular lenses deemed important by care ethics theorists. For example, an interest in considering the historical circumstances that have disadvantaged certain groups, the requirement to carefully listen to marginalized voices, and the necessity of identifying the structural injustices that sustain subordination have all been hallmarks of the care approach (Held, 2006; Lindemann, 2006). Digging in a bit deeper, it is possible to identify several distinctive features of a care approach that make it a suitable lens for identifying crucial aspects of justice that have often been missed.[1]

A care approach starts with a *relational worldview*, one that prioritizes relationships among people – and relationships between people and institutions – as much as it prioritizes people themselves. In matters of governance, a care approach will pay attention to shifts or ruptures in relationships brought about by a particular policy. It stresses the importance of sustaining and maintaining healthy relationships within social and ecological communities and is on the lookout for the creation of new relationships that might be harmful.

A care approach also recognizes that *context* matters. Different individuals, communities, settings, and situations have their own unique characteristics and challenges. Understanding the specifics of a situation is a shift away from generalized forms of assessment that assume uniformity or commonality across contexts. Failures of recognition will occur if it is assumed that 'one size fits all.' As Whyte observed, the cultures, values, and situations of people vary considerably. Taking these differences seriously is a feature of care.

Care-based approaches to ethics and politics are also sensitive to the presence of *dependence* in society, acknowledging that there are often asymmetries in capabilities between individuals and groups who interact with each other. The existence of dependence in a particular setting need not always be negative. Children, for example, can depend upon their parents in appropriate ways. Many of the personal and social relationships in which we participate do not start from positions of absolute equality.

A care approach will be on the lookout for how relationships of dependence might emerge from a particular policy decision. It will encourage questions about whether, for example, a particular proposal to address the climate problem will enable relationships to become more nurturing and enabling or whether it will lead them to become more dependent and limiting.

With its awareness of dependence, a care-based approach pays heed to the distribution of *power* and *vulnerability* within any network of relationships. Relationships can be empowering or disempowering. They can ameliorate or accentuate vulnerability. A care approach will scrutinize whether a particular policy proposal will improve or worsen the situation of vulnerable actors. It will investigate whether the new power dynamics created by the technology might lead to the degradation of autonomy. It will ask whether a disempowered culture will be unwillingly forced to reject their own ways of knowing and knowledge practices, something that will result in a loss of 'espistemic self-determination' (Werkheiser, 2017).

Unlike some ethical frameworks, a care approach recognizes that *emotional reactions* are a legitimate component of moral assessment. Responses to injustices can be appropriately informed (and motivated) by affective considerations. The affective component of a response to a policy proposal can sometimes say a lot about its fairness. Structural injustices wear people down. Affective responses can provide a clue about whether a policy meets the conditions of recognitional justice. A care approach recognizes emotional as well as physical harms.

Finally, a care approach embraces the useful role that *narrative* can play in the assessment of justice. Narratives are valuable tools for highlighting different understandings of the world and different experiences of stakeholders. Narrative is an important way for people to define and communicate their identity and to make sense of the challenges they face. Margaret Urban Walker claims that narrative illuminates 'the location of human beings' feelings, psychological states, needs, and understandings as nodes of a story (or of the intersection of stories) that has already begun, and will continue beyond a given juncture...' (Walker, 2007, p. 18). Narratives are a suitable vehicle for revealing an environmental heritage at risk. Taking narratives seriously is an acknowledgment of the importance of people being able to tell (and to adequately control) their own story.

With its emphasis on *relationships, context, dependence, power/vulnerability, affect*, and *narrative*, a care approach draws attention to areas that traditional assessments of political justice have tended to bypass. By asking about more than the distribution of physical goods and harms, a care approach is highly alert to instances of misrecognition. Through its attention to *narrative*, it recognizes the importance of environmental heritage and how identity is created through communal practices over time. By being sensitive to *power, dependency*, and *vulnerability* it is responsive to structural injustices and the ways that subordination can be institutionalized. By focusing on *relationships*, a care approach is attentive to how injustices affect not just individuals but also groups and communities. By seeking out *context*, it is grounded in real world situations rather than abstracted ideals. A care lens pays close attention to the concrete details of how people actually live their lives. It respects what the sciences can tell us about benefits gained and physical harms averted. But it insists that these narrowly quantitative matters are not the only justice issues in play.

With a new framework for assessment in hand, it will help to consider next the methods through which this assessment lens can be operationalized. Whyte points out that recognitional justice by its nature demands different methods. These methods, he suggests, must include 'the development of creative participatory processes...that aim as much as possible for the inclusion of ...values and...particular situations into policy' (Whyte, 2011, p. 204–205). To detect when the specter of misrecognition is approaching, qualitative research that probes the views, opinions, and concerns of those who are most vulnerable to harm is required. Marion Hourdequin expresses this need when thinking about how best to judge the losses associated with climate change.

> If we are to truly recognize diverse forms of loss, it seems to me that there is no way around the time intensive work of actually listening to and trying to understand the perspectives of those who will be affected. This, inevitably, requires not just quantification but qualitative work – understanding the narratives, the world views, the conceptual frameworks of those living in particular places.[2]

Recognitional justice insists that quantification of biophysical and economic impacts is not enough. It demands sitting down with those who are, or may be, affected by different policies to appreciate potential impacts from their perspective. These impacts include social, political, and cultural impacts. Only when these impacts are adequately understood do they stand any chance of being included meaningfully in the policy process.

Recognition and Climate Engineering: A Case Study

In the light of the potential for uneven impacts from climate engineering, social scientists, ethicists, humanitarian and environmental organizations, and members of the public have all advocated for the inclusion of more geographically and culturally diverse perspectives in future research and decision-making (Preston, 2012; Suarez and van Aalst, 2017; Whyte, 2012). Suarez et al., for example, have argued, 'There is a moral imperative to facilitate involving the most vulnerable in decision-making about [climate engineering] ... to help inform a more inclusive and nuanced conversation about what can go wrong – and what must go right' (Suarez, Banerjee, & de Suarez, 2018). The following section draws on qualitative research designed to listen attentively to the concerns that vulnerable populations might have about climate engineering. The results provide a startling illustration of how appropriate the care lens can be for revealing the presence (or absence) of recognitional justice.

The research findings presented here are based on in-depth interviews conducted in the Solomon Islands, Kenya, and Alaska (with Alaska Natives) about the prospect of stratospheric aerosol injection. The interviews were designed to better understand the hopes and fears of those who have been cast as the potential beneficiaries of climate engineering.[3] Applying a care lens to the data reveals the critical importance of recognition for vulnerable populations and the effectiveness of the care approach at sniffing out misrecogntion. The method employed a listening posture in conducting and analyzing the interviews. Here we do not attempt to speak for interviewees but rather present their concerns in their own words.

All of the interviewees grounded their perspectives on climate engineering within their experience of climate change. In many cases, significant climate harms were already being felt. These harms included disruptions to agriculture, damage to traditional cultural practices, and threats to public health. Nearly all interviewees indicated some desperation for solutions to the tangible harms of climate warming. They expressed what might be characterized as a deeply reluctant willingness to consider climate engineering (Carr & Preston, 2017). A large majority of interviewees indicated that, all other things being equal, climate engineering would not be their preferred response. Nevertheless, given the lack of global political will for effective mitigation and the severity of the impacts already being felt, many of the interviewees were at least willing to consider some of these controversial techniques (Carr & Yung, 2018).

This reluctant willingness to consider climate engineering, however, came with a number of caveats about the possible injustices that might accompany it. It is here that the care approach starts to shine. For instance, interviewees expressed deep concerns about how climate engineering might create imbalances of *power*. One interviewee said, 'I'm just afraid that something this powerful could be used as the way to affect different parts of the world without their approval.' Another respondent had similar worries: 'And where would the power be in terms of who decides what to do? In the past, countries with not as much wealth and the indigenous populations always get put on the back burner and don't get to decide these things. Would that be the same case?' Interviewees across all three research sites came to climate engineering sensitized by centuries of exploitation.

Concerns about power were often accompanied by comments about *vulnerability*. Many interviewees worried that climate engineering would exacerbate rather than ease their vulnerability.

> Who has the resources to invest in doing it and then as a result, who gets to make the decision about who's going to feel the worst impacts? Generally it's the poor people, the brown people and black people on the planet who will not have the resources to pony up for this kind of work.

The sense of disempowerment expressed here linked quickly to concerns about the creation of new relationships of *dependency*. In the quote above, worries about dependency were expressed in terms of race and ethnicity. A different interviewee framed similar concerns in the context of political and economic relationships between countries:

> I think my question is, for climate engineering, countries that are for it might be going to do it, but within developing countries like Solomon Islands, there's no money to do the different things they're talking about. If we have support from your country to help us, maybe it will work. We do not have money to do things like that.

Many interviewees were also acutely attuned to the historical relationships between the nations likely to be developing the technologies and those whom they were supposed to help. They expressed concerns about how climate engineering could follow a storyline already experienced too many times. In the words of one interviewee, 'As a Native person ... there has been too much horrible stuff done to us in the name of science to trust it. To us personally, to us as a culture, to us as peoples, to us as inhabitants of the environment.' Similar concerns about historical circumstances were expressed by others: 'I fundamentally struggle with the concept of humans messing around even more than they already have. Because we've borne the brunt of too many of these scientific experimentations – from physical and medicinal aspects, to landscape and natural parts of it.'

In the light of this past, considerable skepticism remained about who, in the end, would see the most benefit from climate engineering. Vulnerable populations were concerned that the technology would amount to another excuse for the rich nations take control of poorer countries' weather, echoing centuries of colonial exploitation.

Interviewees were cautious about ways in which well-meaning outsiders could end up doing them harm. 'Regardless of what happens with climate engineering, we'll be taking the brunt of the problem again.'

The types of self-determination under threat were not just perceived to be political and economic, but also epistemic. Interviewees zeroed in on how a technology like climate engineering threatened to ignore distinctive ways of knowing and the importance of *narrative* to traditional knowledge. One interviewee spoke at length about the importance of narrative to indigenous science and cosmologies, comparing that approach with typical western scientific approaches to technologies such as climate engineering.

> Western science looks at one thing very closely. And indigenous science often looks at thousands of things in a different way. It's not that deep quantitative longitudinal thing. It's personal experience, it's relationships, it's listening to stories, it's talking to people. But you are looking at a thousand different things. You're not looking at one thing very closely, you're taking a broad sweep of a thousand things.

This same interviewee went on to speak about how difficult it can be to integrate indigenous science, with its basis in experience, relationship, and storytelling, into Western science.

> Storytelling, and what does that mean exactly? How is that actually a transmission of knowledge and creation of knowledge? How do you incorporate that into Western science? Does Western science want that? Even if you've got a scientist on board with you, how do you communicate that to funding agencies? How do you tell the NSF, 'Wait, story is important!'

Clearly there was a significant worry that distinctive ways of knowing expressed in unique cultural narratives would be silenced by the epistemology embedded in climate engineering.

For example, one interviewee broadened the point about different ways of knowing by adding a discussion of different practices and traditions that highlight the importance of environmental heritage.

> Don't forget we are living on one planet and we have our own ways of life. When you come in from the university with the concept of climate engineering, you're coming in with knowledge that scientists have put in place. But don't forget our traditional life, our traditional skills, our traditional taboos, things that we value. They work for us in the Solomon Islands, because this is our local environment. This is where we have been living, our forefathers shared this information to I don't know how many generations. It's good to value these things as well.

Closely related to these comments about distinctive practices and ways of knowing were concerns about whether climate engineering could pay sufficient attention to *context*, given the varying needs of people living in radically different circumstances. For instance, one interviewee spoke at length about important differences between weather patterns in the temperate climatic zones (where the majority of climate engineering research is taking place) and the tropical and subtropical climatic zones (where most of the world's subsistence farming occurs). This respondent drew attention to the relevance of local micro-climates:

> The higher yield agricultural areas, especially the highlands both east and west of the Rift Valley, they have got a local system, and especially to the west of the Rift valley, they have one long growing season. So the variety of maize they grow, the variety of other crops they grow, is long-maturing, because they have got enough rainfall and the rainfall system is long. It means sometimes when I want the rains in the short rainfall season areas, it's actually when they don't want the rains in the long rainfall area because they want a dry season so that the crops can dry up for harvesting.

The need for self-determination sought by this individual was contingent on important contextual details about the types of technologies and tools that might be deployed and about the particular geographical and climatic situation in which he and others lived.

This sense of a loss of control of one's destiny coupled with the prospect of additional harms created the frequent expression of anxiety. In the words of one interviewee, 'It's scary as hell to be dependent on some other person to dictate the weather or climate change.' Fear, unease, and a sense of vulnerability to the whims of the richer nations were broadly expressed across the populations interviewed. One interviewee even expressed the need for members of vulnerable populations to make space for a 'primal scream' when contemplating the injustices of climate change and potential implications of climate engineering. The *emotional* response of pessimism and despair that this sense of vulnerability and dependency created weighed on several respondents.

Recognizing the emotional stakes, one respondent suggested that one of the ways to reduce this sense of injustice would be to ensure local self-determination within a climate engineering strategy:

> I want something that I can actually take home and actually get engaged in it and practice it. The only thing I can ask a climate engineer is, what are some of the tools, technologies, and ideas that you can give me that I can actually go and implement at very minimal costs, so the communities, the people I interact with on a daily basis can easily adopt it, and get the process going?

Even though they might have been reluctantly accepting that something had to be done, many of those interviewed struggled to get over the sense of emotional unease that the whole prospect created. 'My initial reaction is, I just don't have trust. I don't have trust for the process. It just makes me very leery.'

Taken together, what is striking about the responses is how pertinent are many of the themes prioritized by a care approach. Concerns about the expression of *power* were revealed by interviewees already sensitized to centuries of exploitation. Nation-to-nation and people-to-people *relationships* were a key orienting lens. The *emotional* responses of pessimism and despair that this sense of vulnerability and dependency created were clearly in evidence. The need for self-determination sought by the respondent in Kenya was contingent on important *contextual details* about the types of technologies and tools that might be deployed and about the particular geographical and climatic situation in which Kenyans lived. The importance of *narrative* to expressing distinctive ways of knowing that might conflict with the epistemology embedded in climate engineering was highlighted in the responses of others.

As the above quotes collectively demonstrate, interviewees were at pains to emphasize recognitional concerns in addition to concerns about the distribution of physical harms and benefits. They were certainly interested in whether climate engineering might help distribute physical climate harms and benefits more justly, but they were also concerned about social, political, and cultural harms getting downplayed by those who view climate engineering only through the lens of physical benefits and burdens. Injustices that respondents worried a great deal about included the ways in which traditions, practices, and ways of knowing might be ignored or made impossible. They included relationships of power, dependence, and vulnerability that might be established. They also involved the likelihood of not being equal participants in decisions that would impact them. These worries provide compelling evidence that adequate recognition is not only a necessary condition for just procedural and distributive conditions. It is also a key component of justice in itself. Using the lens provided by the care approach promises to key a careful listener into these critical prerequisites for justice.

Making Policy with the Care Approach

If the care approach is helpfully tuned in to considerations of recognitional justice, can this attentiveness contribute to good policy-making? One of the concerns often raised about the care approach in other areas of ethics is that it is insufficiently normative. The approach asks policy makers to be sensitive to power and dependence, for example, but it does little to determine how much power or how much dependence is too much. It says that emotions are important but it does not provide clear procedures for what to do when emotions conflict. It values relationships but does not give clear enough guidance on when relationships need to be curtailed. If these are accurate criticisms of care ethics, then one might reasonably ask where normativity is going to originate from if recognitional justice is to be served.

The care themes articulated above certainly focus more on bringing attention to particular areas of concern than they do on providing definitive answers about what is just or unjust. It should be clear, however, that when attention is focused on the right areas, injustices are more likely to be identified. A series of targeted questions about relationship, context, power, etc. can provide answers to help assess whether a particular climate policy is likely to be morally acceptable.[4]

On *relationship* it should be asked: How might social and ecological relationships shift if this climate engineering strategy is introduced? How have interconnections within and between communities been considered in the development of this strategy? Is the development or deployment of this climate engineering technology likely to create significant ruptures in relationships? Can it create new types of beneficial relationships? Is the developer of the technology oblivious to the relationships it will create and destroy?

On *context* one might inquire: What are the important particularities of this context? What is the unique history, ecology and culture of this place? What specific actors or groups will be most affected by the proposed climate engineering technology? Does the proposed technology, for example, give sufficient attention to the microclimates and agricultural practices of the affected regions? Does it pay enough attention to the historical details or does it risk re-inscribing elements of an undesirable colonial past?

On *dependence* one might wonder: Where are the current relations of dependence in this situation (e.g. people dependent on each other, on companies, on infrastructure, on ecological processes etc.), and how might these change due to the use of this particular climate engineering technology? What is the character of the relations of dependence in play (e.g. are they experienced as nurturing and empowering or extractive and destructive by those involved)? Does the development and use of this technology exacerbate dependencies or lessen them? Are the economic dependencies created desirable or not? Are they entered into willingly?

On *power and vulnerability* one might question: How does the development, deployment and use of this technology affect the distribution of power? Are any actors/groups favored or granted more power than others? How will the technology affect the level of control the impacted actors have over their own future? Would the interest of indigenous populations be put on the back-burner again? Who are the most vulnerable actors – both human and non-human – and what measures are in place to prevent their abuse? Will this technology lead to the concentration of power or its redistribution?

On *affect* one might investigate: Does the development, introduction, or use of this technology evoke strong emotions among those impacted? Are these emotional reactions positive or negative? Is it the case that those who will experience the technology find the prospect 'scary as hell'? Can the technology settle the mind of someone who initially states 'I just don't have trust.' Is affect being granted a legitimate role in the decision-making processes or are the affective dimensions of this technological change being excluded from consideration?

On *narrative* one might seek to find out: What are the narratives being told by those promoting and by those who will be subjected to this climate engineering technology? What worldviews, values, assumptions and beliefs are being expressed in these different narratives? What alternative visions, strategies, and technologies do the different stories reveal as available and important for the assessment process? Are the different ways of knowing displayed by a stakeholder who implores 'Wait, story is important!' being recognized? Are culturally significant narratives being suppressed, dismissed or excluded?

This sample list of questions is not definitive nor exclusive. Other questions may be more appropriate with of each of the various climate strategies. Nevertheless, answers to these sorts of questions have the potential to be highly instructive about potential injustices within a particular policy pathway. They might reveal, for example, whether a given climate strategy will shatter valuable relationships, concentrate power, cause negative affective reactions, impose new narratives from the outside, and increase dependence. On the other hand, they might indicate whether a strategy is sensitive to context, ensures a fairer distribution of power, decreases vulnerability, and allows a community to continue to determine its own story. The answers to these questions will reveal whether or not recognitional justice is likely to be served by a given deployment of climate engineering.

Progress will only be made on answering these questions by listening to local accounts, taking first person perspectives seriously, and by not imposing overly narrow understandings of justice from the outside. Getting satisfactory answers will require the funding of significant ethics and social science research developed in concert with the populations of concern. It will also require the governments funding climate engineering research to take ethics oversight seriously, as happened, for example in the case of the UK's Stratospheric Particle Injection for Climate Engineering (SPICE) project (Stilgoe, 2015).

There might still be reason to hesitate about the care approach. One of the appeals of the narrower, distributional paradigm is that physical and economic goods and bads are much more readily quantifiable. Quantification can mostly be completed by cost-benefit analyses employing methods developed by economists over more than half a century (Hicks, 1943).

One might suggest in response that, while quantifiability is appealing when considering the just distribution of economic and material goods, it can be very difficult to achieve in other equally significant domains of justice. Insisting on quantification when discussing justice marks an attempt to be falsely definitive. One might define too narrowly the relevant concerns in play and fail to allow the sphere of justice and injustice to be adequately opened up. If considering the just distribution of benefits and harms is straightforwardly consequentialist and quasi-economic, recognitional justice is more care-based and qualitative. Due to its qualitative nature, the care approach may provide the kind of 'plural and conditional' policy advice that is most appropriate for this domain (Stirling, 2010).

The Challenges of Recognition

What should be clear by now is that it is not enough to consider only the potential benefits and harms of climate engineering in terms of welfare (physical, economic or otherwise). What Flegal and Gupta call an 'expert driven, outcome-oriented, and risk-based understanding of equity' (Flegal & Gupta, 2018, 56) tells only part of the justice story. When Horton and Keith cite multi-model studies (Kravitz et al., 2014) that consider the impacts on temperature and precipitation of stratospheric aerosol deployment they provide important information but offer an incomplete picture of the moral terrain. Other dimensions of justice must also be included if climate engineering is to be politically legitimate. Climate change is a complex and multi-dimensional problem and recognitional justice is an important element of the ethical territory. Avoiding misrecognition is likely to require qualitative methods that acknowledge the complexities of justice across geographical, temporal, and cultural boundaries.

Our suspicion is that ethicists and policy-makers will require better tools to detect all of the subtle forms of injustice that attend climate change and the various attempts to fix it. Some of these tools are relatively new to the justice discussion and are likely to take concentrated and sustained work to become fully operational. Nevertheless, given the 'crucible of inequality' (Cuomo, 2011) in which climate change was created, decision-makers must make exceptional effort to avoid compounding past injustices as they seek desirable solutions to what is unprecedented crisis.

The care approach looks like an effective tool to have in the toolbox when setting out on such a path.

Notes

1. See Preston and Wickson (2016) for an account of the care approach applied to biotechnology assessment.
2. Marion Hourdequin, 'Recognizing Climate Losses' (draft). Presented at 2nd Buffalo Workshop on Adaptation: Loss, Damage, and Harm, May 8–9th, 2015.
3. The data collection was conducted by W. Carr while a Ph.D. candidate at the University of Montana. Between 2012 and 2014 he conducted interviews with 33 Solomon Islanders, 29

Alaska Natives, and 38 Kenyans. For a more detailed description of the research methods involved see (Carr & Yung, 2018). The fieldwork was funded by the National Science Foundation (SES-0958095) and an Environmental Protection Agency STAR Grant (Assistance Agreement No. FP917316).

4. These questions are adapted from (Wickson et al., 2017).

Acknowledegments

This publication has not been formally reviewed by either the US Fish and Wildlife Service (FWS) or the EPA. The views expressed in this document are solely those of the authors, and do not necessarily reflect those of the FWS or EPA. Neither FWS nor EPA endorse any products or commercial services mentioned in this publication. Thanks also to the participants in the workshop on Geoengineering, Political Legitimacy, and Justice at the University of Washington, Nov 2–3, 2017.

Disclosure statement

No potential conflict of interest was reported by the author.

Funding

Research for this paper was supported by the US National Science Foundation [Grant Number SES 0958095]; and the US Environmental Protection Agency (EPA)[Assistance Agreement No. FP 917316].

References

Barrett, S., Lenton, T. M., Millner, A., Tavoni, A., Stephen Carpenter, J. M., Stuart Chapin, A. F., et al. (2014). Climate engineering reconsidered. *Nature Climate Change*, *4*(7), 527–529. Nature Publishing Group.

Burns, W. C. G., & Strauss, A. (2013). *Climate change geoengineering: philosophical perspectives, legal issues, and governance frameworks*. Cambridge, UK: Cambridge University Press.

Cairns, R., & Stirling, A. (2014). 'Maintaining planetary systems' or 'concentrating global power?' High stakes in contending framings of climate geoengineering. *Global Environmental Change*, *28*, 25–38. Elsevier Ltd.

Carr, W., & Preston, C. J. (2017). Skewed vulnerabilities and moral corruption in global perspectives on climate engineering. *Environmental Values*, *26*(6). doi:10.3197/096327117X15046905490371

Carr, W. A., & Yung, L. (2018, January). Perceptions of climate engineering in the South Pacific, Sub-Saharan Africa, and North American arctic. *Climatic Change*, 1–14. Springer Netherlands. doi:10.1007/s10584-018-2138-x

Crutzen, P. J. (2006). Albedo enhancement by stratospheric sulfur injections: A contribution to resolve a policy dilemma? *Climatic Change*, *77*(3–4), 211–220.

Cuomo, C. J. (2011). Climate change, vulnerability, and responsibility. *Hypatia 26*(4), 690–714.

Ferraro, A. J., Highwood, E. J., Charlton-Perez, A. J., Meehl, G., et al. Crutzen P J, The Royal Society, Tilmes S, Turco R P, Robock A, Oman L, Chen C-C, Stenchikov G L and Garcia R L Rasch P J, et al. 2014. Weakened tropical circulation and reduced precipitation in response to geoengineering. *Environmental Research Letters 9* (1).IOP Publishing. 014001.

Figueroa, R. (2006). Evaluating environmental justice claims. In J. Bauer (Ed.), *Forging Environmentalism: Justice, Livelihood, and Contested Environments* (pp. 360–376). New York: M. E. Sharpe.

Flegal, J. A., & Gupta, A. (2018). Evoking equity as a rationale for solar geoengineering research? Scrutinizing emerging expert visions of equity. *International Environmental Agreements: Politics, Law and Economics, 18*(1), 45–61. Springer Netherlands.

Fraser, N. (1999). Social justice in the age of identity politics. In L. Ray & A. Sayer (Eds.), *Culture and economy after the cultural turn* (pp. 25–52). Thousand Oaks, CA: Sage. https://books.google.nl/books?hl=en&id=epacb_Iz6VAC&oi=fnd&pg=PA72&dq=social+justice+in+the+age+of+identity+politics&ots=4vq3EId3BO&sig=16b6lpoXbfN1hvzoN2X7mhO3QnA#v=onepage&q=social justice in the age of identity politics&f=false

Fraser, N. (2007). Feminist politics in the age of recognition: A two-dimensional approach to gender justice. *Studies in Social Justice, 1*(1). Retrieved from https://search.proquest.com/openview/0be2ebf16992eabd2e0f195d0fe5248d/1?pq-origsite=gscholar&cbl=1636338

Gardiner, S. M. (2010). Is 'arming the future' with geoengineering really the lesser evil? Some doubts about the ethics of intentionally manipulating the climate system. *Climate Ethics: Essential Readings*, 284–314. Retrieved from http://ssrn.com/abstract=1357162

Gardiner, S. M. (2013). The desperation argument for geoengineering. *PS: Political Science & Politics, 46*(01), 28–33.

Hamilton, C. (2013). *Earthmasters: The dawn of the age of climate engineering*. New Haven, CT: Yale University Press.

Held, V. (2006). The ethics of care: Personal, political, and global. New York, NY: Oxford University Press. doi:10.1093/0195180992.001.0001

Hicks, J. R. (1943). The four consumer's surpluses. *The Review of Economic Studies, 11*(1), 31.

Honneth, A. (2001). Recognition or redistribution? *Theory, Culture & Society, 18*(2–3), 43–55. SAGE PublicationsLondon, Thousand Oaks and New Delhi.

Horton, J., & Keith, D. (2016). Solar geoengineering and obligations to the global poor. In C. J. Preston (Ed.), *Climate justice and geoengineering: Ethics and policy in the atmospheric anthropocene* (pp. 79–92). London: Rowman & Littlefield International.

Hulme, M. (2014). *Can science fix climate change: A case against climate engineering*. New York, NY: Wiley & Sons.

IPCC. 2014a. Climate Change 2014: Impacts, Adaptation and Vulnerability. Part A: Global and Sectoral Aspects. Contribution of Working Group II to the Fifth Assessment Report of the Intergovernmental Panel on Climate Change. Cambridge, UK. http://www.ipcc.ch/report/ar5/wg2/

IPCC. 2014b. Climate change 2014 synthesis report summary chapter for policymakers. *Ipcc*, 31. doi:10.1017/CBO9781107415324.

Ito, A. (2017). Solar radiation management and ecosystem functional responses. *Climatic Change, 142*(1–2), 53-66.

Jamieson, D. (1996). Ethics and intentional climate change. *Climatic Change, 33*(3), 323–336.

Kravitz, B., MacMartin, D. G., Robock, A., Rasch, P. J., Ricke, K. L., Cole, J. N. S., Curry, C. L., et al. (2014). A multi-model assessment of regional climate disparities caused by solar geoengineering. *Environmental Research Letters, 9*(7), 074013. IOP Publishing.

Lindemann, H. (2006). *An invitation to feminist ethics*. New York, NY: McGraw-Hill.

MacMartin, D. G., Ben Kravitz, J. C., Long, S., & Rasch, P. J. (2016). Geoengineering with stratospheric aerosols: What do we not know after a decade of research? *Earth's Future, 4*(11), 543–548. Wiley Periodicals, Inc.

Macnaghten, P., & Szerszynski, B. (2013). Living the global social experiment: an analysis of public discourse on solar radiation management and its implications for governance. *Global Environmental Change, 23*(2), 465–474. Elsevier Ltd.

Preston, C. J. (2012). *Engineering the climate: The ethics of solar radiation management*. C. J. Preston, Ed.. Lanham, MD: Lexington Press.

Preston, C. J., & Wickson, F. (2016). Broadening the lens for the governance of emerging technologies: Care ethics and agricultural biotechnology. *Technology in Society, 45*, 48–57.

Schlosberg, D. (2007). Defining environmental justice: Theories, movements, and nature. *9780199286*. New York: Oxford University Press. doi:10.1093/acprof:oso/9780199286294.001.0001

Schlosberg, D. (2013). Theorising environmental justice: the expanding sphere of a discourse. *Environmental Politics*, *22*(1), 37–55.

Shue, H. (2014). *Climate justice : vulnerability and protection* (1st ed.). Oxford, UK: Oxford University Press.

Stilgoe, J. (2015). Experiment earth: Responsible innovation in geoengineering. New York, NY: Routledge. doi:10.4324/9781315849195

Stirling, A. (2010). Keep it complex. *Nature*, *468*(7327), 1029–1031. Nature Research.

Suarez, P., Banerjee, B., & de Suarez, J. M. (2018). Geoengineering and the humanitarian challenge: What role for the most vulnerable. In J.Blackstock & S. Low (Eds.), *Geoengineering our climate? Ethics, politics, and governance* (pp. 193-197). New York: Earthscan.

Suarez, P, & van Aalst, M.K. (2017). Geoengineering: A humanitarian concern. *Earth's Future*, *5*, 183–195.

Walker, M. U. (2007). *Moral understandings : A feminist study in ethics*. New York, NY: Oxford University Press.

Werkheiser, I. (2017). Loss of epistemic self-determination in the Anthropocene. *Ethics,Policy, and Environment*, *20*(2), 156–167.

Whyte, K. P. (2011). The recognition dimensions of environmental justice in indian country. *Environmental Justice*, *4*(4), 199–205.

Whyte, K. P. (2012). Indigenous peoples, solar radiation management, and consent. In C. J. Preston (Ed.), *Engineering the Climate: The Ethics of Solar Radiation Management* (pp. 65–76). London: Rowman & Littlefield International.

Wickson, F., Preston, C., Binimelis, R., Herrero, A., Hartley, S., Wynberg, R., & Wynne, B. (2017). Addressing socio-economic and ethical considerations in biotechnology governance: The potential of a new politics of care. *Food Ethics*, *1*(2), 193–199.

Young, I. M. (2006). Responsibility and global justice: A social connection model. *Social Philosophy and Policy*, *23*(01), 102.

Institutional Legitimacy and Geoengineering Governance

Daniel Edward Callies

ABSTRACT
There is general agreement amongst those involved in the normative discussion about geoengineering that *if* we are to move forward with significant research, development, and certainly any future deployment, legitimate governance is a must. However, while we agree that the abstract *concept* of legitimacy ought to guide geoengineering governance, agreement surrounding the appropriate *conception* of legitimacy has yet to emerge. Relying upon Allen Buchanan's metacoordination view of institutional legitimacy, this paper puts forward a conception of legitimacy appropriate for geoengineering governance, outlining five normative criteria an institution ought to fulfill if it is to justifiably coordinate our action around geoengineering.

1. Introduction

Geoengineering is a controversial topic.[1] On the one hand, many of the technologies under the geoengineering umbrella boast characteristics that make them appropriate for continued research and possible future deployment. On the other hand, many of these same technologies are dangerous. Take stratospheric aerosol injection, for example. It boasts relatively strong leverage on the climate system. The fact that with relatively little effort we could generate huge impacts on global climate gives us reason to look into the technology. However, that same strong leverage makes the technology troubling. In the hands of an irresponsible actor, such technology could have devastatingly negative effects across the world population and could irreparably damage natural ecosystems and the species that comprise them. In a similar vein, the relatively inexpensive price tag attached to the technology means that we could alleviate some future climatic harms without having to divert scarce resources away from mitigation, adaptation, and other valuable goals such as the eradication of global poverty. However, that same relatively inexpensive price tag makes the technology vulnerable to unjustified unilateral action since a multilateral cost-sharing arrangement is not strictly necessary.[2]

Given that geoengineering technologies have the potential to bring about both significant benefits and drastic burdens, regulation is important. Nearly everyone involved in the normative discussion about climate engineering agrees that *if* we are to move forward with significant research, development, and certainly any future implementation of geoengineering technologies, legitimate governance is a must.[3]

Despite this agreement that further research and certainly any future deployment should be accompanied by legitimate governance, there has been relatively little discussion to date among political philosophers about what would constitute legitimate governance of such a technology.[4] That is, while we all might recognize that the abstract *concept* of legitimacy ought to guide geoengineering governance, agreement surrounding the appropriate *conception* of legitimacy for geoengineering governance has yet to emerge.

The main point of this paper is to introduce a framework out of which an appropriate conception of legitimacy can spring. To do so, I'll begin in the next section by exploring specific conceptions of legitimacy that have been developed in the philosophical literature to date. I'll argue that what we need is a general concept of legitimacy that is appropriate for the diversity of institutions that occupy the international realm – the realm where geoengineering governance would take place. Drawing on the recent work of Allen Buchanan, Section 3 then puts forward a general concept of institutional legitimacy. This general concept will pave the way for a more specific normative conception of legitimacy (explored in Section 4) that will serve to coordinate judgements about the legitimacy of an institution set up to oversee climate engineering. Section 5 offers a quick recap and then concludes the paper.

2. Conceptions of (State) Legitimacy

Before looking at specific *conceptions* of legitimacy, we need to be clear about the general *concept* of legitimacy.[5] In the philosophical literature, the concept of legitimacy has generally denoted either the justification of political power – understood as coercive power backed by government sanctions (Rawls, 2005, p. 136) – or the justification of political authority – understood as a right to rule and a correlating obligation to obey (Buchanan, 2003, p. 147). Aiming at these understandings of the *concept* of legitimacy, various *conceptions* have been developed that spell out exactly when it is that political power or political authority is justified. For example, according to Rawls, '... political power is legitimate only when it is exercised in accordance with a constitution the essentials of which all citizens, as reasonable and rational, *can* endorse in the light of their common human reason' (2001, p. 41, as cited in Peter, 2014). Here we see Rawls grounds legitimacy (understood as the justification of coercive power) in the processes whereby political power is exercised.[6] As long as political power is exercised in accordance with a constitution that citizens could hypothetically endorse, then the exercise of such political power is justified. A. J. Simmons has a more demanding conception of legitimacy. For Simmons, a legitimate government has a right to rule and citizens of a legitimate government are under an obligation to obey. The only way, according to Simmons, for a government to derive such a right and for citizens to be under such an obligation to obey is if the citizens have actually expressed their consent (Simmons, 2001, p. 136). In the absence of such consent, the government has no right to rule and citizens are under no obligation to obey. Given that the legitimacy of the government is conditioned upon voluntary consent, consent theorists like Simmons are called 'voluntarists.'

Now, the Rawlsian or the voluntarist conception of legitimacy may be an appropriate conception for the legitimacy of state-like institutions.[7] But these conceptions do not apply to international institutions very well. The Rawlsian conception places a priority on the publicly recognized conception of justice embedded in the constitution. And the voluntarist conception places a priority on the express consent of those within the purview of the state. But international institutions exercise political power without any publicly recognized constitution and without the consent of all of those within their purview. Rather than reach the conclusion that international institutions are illegitimate because they fail to meet either the Rawlsian or the voluntarist standards, I contend that we should rely upon a different conception of legitimacy for two reasons.

First, international institutions are beneficial. For example, the world is a better place with the International Atomic Energy Association (IAEA) and the United Nations Framework Convention on Climate Change (UNFCCC) wielding the power they do, notwithstanding their imperfections. And this is the case despite the fact that they are wielding this power in the absence of any international or global constitution and despite the fact that they do not enjoy the consent of every individual over whom they wield such power. To conclude that these institutions are illegitimate because they fail to meet the standards of state legitimacy outlined by Rawls and Simmons would be counterproductive.[8] But, second, it makes sense to say that how high we set the bar for an institution to be considered legitimate should be sensitive to the characteristics and the function of the particular institution in question. If an international institution is wielding significant power like a state, then perhaps it should meet the demanding requirements we expect of states for them to be considered legitimate. On the other hand, if the institution is merely providing suggestive guidelines and has no ability to enforce any of its directives, then we may want to relax the criteria it needs to fulfil to be considered legitimate.

To be more explicit, what we need is a general concept of institutional legitimacy; a concept that is malleable enough to give rise to appropriate conceptions of legitimacy for the variety of institutions that occupy our world. Clearly explicating a general *concept* of institutional legitimacy will prove invaluable in developing a more concrete *conception* of legitimacy that can be applied to the kind of climate engineering institution that ought to oversee research and possible development.

3. Institutional Legitimacy

To begin the outline of a general concept of institutional legitimacy, we can ask the following question: Why do we have institutions and what is it for an institution, in general, to be legitimate? We have institutions, according to Buchanan, in order solve coordination problems. They supply us with the kind of coordination we need to achieve certain desirable outcomes and do so without the excessive costs or inefficiencies associated with non-institutional alternatives. In order for institutions to effectively coordinate our action and deliver the desirable outcomes, they require us to grant them a certain kind of standing – the kind of standing they need to perform their functions. And in order for us to justifiably[9] grant them the kind of standing they need in order to perform this function, we first must converge (or coordinate) on particular normative criteria that institutions ought to fulfil to be worthy of that standing. So, legitimacy assessments serve to solve a metacoordination problem: they allow us to justifiably coordinate around normative criteria that institutions must meet if we are to grant them the kind of standing needed to solve the further problem of coordinating our collective action towards certain desirable outcomes. Given that legitimacy assessments allow us to solve this higher order coordination problem, Buchanan aptly refers to this as the *Metacoordination View* of institutional legitimacy.

3.1. The Metacoordination View of Institutional Legitimacy

The Metacoordination View of institutional legitimacy thus says, abstractly, that an institution is legitimate when it is worthy of the certain kind of standing needed for it to perform its institutional goals. Thus, a geoengineering governance institution would be legitimate if it is worthy of the kind of standing it needs to coordinate our action around research and development (or abandonment!) of technologies capable of modifying the planetary climate. Before moving on to a more specific conception of legitimacy for geoengineering governance, I want to highlight six aspects of the Metacoordination View.

First, this is an elucidation of the *concept* of institutional legitimacy, not a *conception* of institutional legitimacy. This general concept of legitimacy can be applied to institutions that rule in the strongest sense – that is, institutions such as the state that claim a monopoly over the use of force for a given jurisdiction. But it can just as coherently be applied to institutions that do not rule even in a weak sense of the word, but instead merely promulgate principles that can be used for coordination across either small or large populations. It may be helpful to think of the concept of institutional legitimacy as a large circle encapsulating smaller circles that articulate conceptions of legitimacy for particular institutions. For example, the particular conceptions of state legitimacy previously surveyed put forth different normative criteria that would count in favour of a *state* being legitimate. A particular conception for the legitimacy of, say, a national sports league would identify different – presumably, less demanding – normative criteria that would count in favour of the *sports league* being legitimate. And, similarly, a particular conception for the legitimacy of a geoengineering governance institution would identify normative criteria appropriate for the kind of institution it is. But all of these conceptions of legitimacy – the conception of state legitimacy, the conception of national-sports-league legitimacy, and the conception of geoengineering governance legitimacy – would be encapsulated by the broader concept of institutional legitimacy. The important point to take away is that the general *concept* of institutional legitimacy says that an institution is legitimate when it merits the kind of standing it needs to function (and it merits that standing when it sufficiently fulfils context-specific normative criteria). A particular *conception* of legitimacy will then identify and outline the normative criteria appropriate for the kind of institution under consideration. So, while it doesn't outline specific normative criteria for all kinds of institutions, the Metacoordination View of institutional legitimacy is flexible enough to accommodate various specific conceptions of legitimacy.

Second, the special standing we grant to legitimate institutions can be considered a kind of respect. Thus, if a state is legitimate, it means that the state is worthy of our respect. Similarly, if a sports league is legitimate, it means it is worthy of our respect. And if a geoengineering governance institution is to be considered legitimate, it means it is an institution that is worthy of our respect. Admittedly, 'respect' is a complex term with many different variations. Showing respect for a person is certainly distinct from showing respect for, say, the law or another institution. Following Buchanan, we can say that the kind of respect one accords an institution will vary depending upon one's relationship to the institution. In the case of those who are addressees of the institution's directives, showing respect need not mean uniformly complying with those directives or regarding oneself as duty-bound to comply with them. Rather, for those who are addressees of the institution, showing respect implies that institutional directives should be taken seriously, independently of the particular content of the directive (Buchanan, 2013, p. 184). In more technical words, addressees of the institution, when showing it proper respect, would take one of its directives as a '(defeasible) content-independent, exclusionary reason to comply' (Buchanan, 2018, p. 74). However, for non-addressees, the appropriate form of respect for the legitimate institution will generally equate to at least some kind of presumption in favour of non-interference.[10]

Third, there is a difference between an institution being *worthy* of respect and an institution actually *gaining* respect. This is the distinction between normative and descriptive legitimacy. The descriptive or sociological sense of legitimacy refers to the actual attitudes or beliefs of agents (Weber, 1997). For instance, if people believe an institution to be legitimate, their general belief is what confers (descriptive) legitimacy onto the institution – regardless of whether or not it actually *ought* to be granted such standing. Take North Korea as an example. It could be the case that the North Korean government enjoys descriptive or sociological legitimacy from the standpoint of its people. It is possible that the people of North Korea believe that their government is legitimate. But even if we grant that the people of North Korea believe their government to be legitimate, we can still doubt its *normative* legitimacy. In other words, we can doubt whether the people actually *should* consider the government legitimate, that is, whether the government is actually deserving of such a predicate. To bring this back to the Metacoordination View, if an institution is worthy of respect, regardless of whether or not it actually gains the respect it deserves, it has normative legitimacy. Thus, we will be exploring when it is that we *should* consider a geoengineering governance institution worthy of our respect. However, the Metacoordination View also 'explains why achieving sociological legitimacy is important from a normative standpoint, namely, because it enables us to empower institutions so that they can provide their distinctive benefits' (Buchanan, 2018, p. 76). That is, we want an institution to not only be *worthy* of our respect; we also want it to actually *garner* our respect, since such actual respect is often necessary for it to deliver its beneficial coordinating functions without relying upon excessive coercion (coercion that could make the institution dangerous). So, the fact that a geoengineering governance institution is worthy of our respect is not enough; we'll want it to actually garner respect too.

Fourth, the proper standpoint from which to judge the legitimacy of institutions is social rather than individual. When declaring an institution worthy of respect, we are not declaring that it is worthy of *my* individual respect, in the sense of the first-person *singular* possessive pronoun. Rather, we are declaring that is worthy of *our* respect socially, worthy of our respect in the sense of the first-person *plural* possessive pronoun (Buchanan, 2013, p. 180). This is due to the fact that institutions are attempting to coordinate not my or your action alone, but our action as a group. Whether or not an institution is able to solve the coordination problem it is meant to address will depend upon its success in coordinating the action of the *group* to a sufficient degree. Therefore, to judge a geoengineering governance institution as legitimate means that it is worthy of the respect of the group, not any particular individual within the group.

Fifth, it is possible for an institution to be legitimate without every single one of its rules or directives being legitimate. That is, a geoengineering governance institution may, in general, be worthy of the respect of those within its purview (and those outside) notwithstanding the fact that certain of its directives are unworthy of respect. This has direct implications for how we ought to respond to a deficient institution. 'Generally speaking, the proper response to deficiencies in legitimate institutions is reform, not revolution, to attempt to modify it so as to remedy its flaws rather than to scrap it' (Buchanan, 2013, p. 185). To require that a geoengineering governance institution only be labelled legitimate if every single aspect of its existence is worthy of respect would be too demanding.[11] As is commonly said, this would make the Best the enemy of the Good – something I'll say more about in the penultimate section of the paper.

This leads directly to the final clarification of the concept. Institutional legitimacy is not a bivalent concept, with institutions either being absolutely legitimate or absolutely not. Rather, it is a concept of degree; an institution can be more, or less, legitimate. Thus, the more and/or the better a geoengineering governance institution fulfils the appropriate normative criteria, the stronger the reason we have to consider it worthy of respect – in other words, the more legitimate the institution is. The normative criteria identified as relevant for geoengineering governance will serve as what Rawls called 'counting principles' rather than necessary and sufficient conditions (Rawls, 1999, p. 364). '[T]he more of them an institution satisfies, and the higher degree to which it satisfies them, the stronger its claim to legitimacy' (Buchanan & Keohane, 2006, p. 424). There will be institutional arrangements that are closer to the end of the spectrum that characterizes clear legitimacy – we can call this *robust legitimacy*. And there will be institutional arrangements that are closer to the end of the spectrum that characterizes clear illegitimacy, while nonetheless sufficiently satisfying the relevant criteria to be worthy of respect – we can call this *weak legitimacy*. Where a geoengineering governance institution falls upon this spectrum will be determined by how well it satisfies the normative criteria that are salient for the kind of institution it is.

3.2. The Normative Criteria

Now, the general concept of institutional legitimacy as adumbrated by the Metacoordination View does not necessarily specify any particular normative criteria. However, there are some things we can say about the normative criteria simply from an analysis of the general concept. First, we know that the normative criteria used in legitimacy assessments will vary depending upon the specific form and function of the institution being assessed. Whereas democracy may be a salient normative criterion for state-like institutions, it may not be entirely relevant when making legitimacy assessments of, say, a national sports league.

Second, we know that the normative criteria will lie somewhere on a continuum between 'the excessively demanding requirements of full justice or optimal efficacy and the excessively forgiving requirement of bare advantage relative to the non-institutional alternative' (Buchanan, 2013, p. 193). Moreover, we know that where on this continuum the normative criteria of a given institution fall will depend upon the particulars of the institution and the environment in which the institution is situated. The more important the need for the coordination the institution is providing, the more we should relax our criteria. However, the greater the risk we run by empowering the institution, the more demanding we should make our criteria. There are at least two ways that we can be put at risk by empowering an institution. The first, and perhaps most obvious way, is that the institution could abuse its power. For instance, an institution endowed with significant coercive power could wield that power either through questionable means or to undesirable ends. But, second, we can even be put at risk by institutions that wield mere coordinative power. For instance, an institution setup to coordinate action around geoengineering could suggest that resources be directed towards a particular (perhaps comparatively less optimific) technology or could have a bias in favour of development. In such a situation, even if the institution didn't back its suggestions with coercive power, even its mere coordinative power could put us at risk of being locked into the development of a suboptimal technology or could even lead us towards objectionable deployment scenarios. The upshot here is that we'll want to make sure the normative criteria identified as appropriate for judging a given institution as legitimate are sensitive to both the particulars of the institution and the environment in which the institution is situated. This entails that the more we need the coordination the institution can provide and the lower the risk we run by empowering it, the more relaxed we should make the normative criteria; conversely, the weaker the need for coordination and the greater the risk the institution poses, the more demanding the normative criteria ought to be.

Third, we know that the criteria we use to make legitimacy assessments will have to be translated into 'epistemically accessible standards' that can serve as proxies for the normative criteria (Buchanan, 2013, p. 193). Imagine that we identify transparency as a normative criterion for geoengineering governance. It may be difficult to agree upon whether or not the abstract normative criterion of transparency is being met. So, in order to coordinate our legitimacy judgements, we could establish a substantive standard relating to the normative criterion of transparency which requires the institution to release its meeting minutes and perhaps its voting record. This would provide a clear point around which addressees of the institution could determine whether or not the criterion of transparency was being fulfilled. Transparency is thus a normative criterion that plausibly can be translated into specific standards for myriad institutions. The point being made here is that if the abstract normative criteria that ought to be used to coordinate legitimacy assessments are incapable of being translated into substantive standards that agents can use to actually inform their attitudes of respect for the institution, the criteria are inadequate (Buchanan, 2013).

One final point to mention again about the normative criteria is that, according to Buchanan, they are not necessary and/or sufficient conditions for an institution being worthy of our respect. Rather, the more of the criteria that are satisfied, and the greater the extent to which they are satisfied, the more legitimate the institution is. Given that institutional legitimacy, as stated previously, is a concept admitting of degree, this makes sense. But I would depart from Buchanan and maintain instead that at least one normative criterion is a necessary condition for any institution to be worthy of our respect – namely, that of comparative benefit.[12] The normative criterion of comparative benefit is, in fact, a necessary condition, and not merely a counting principle. Why is it the case that comparative benefit is a necessary condition and not merely a point in favour of an institution being considered legitimate? The answer is conceptual. Insofar as we have institutions in order to help solve coordination problems and to do so without the costs and inefficiencies associated with a non-institutional alternative, an institution that fails to deliver a comparative benefit is not helping to solve the original coordination problem. Now, it may be the case that institutions that go significantly beyond the minimal satisfaction of the criterion of comparative benefit ought to be considered more legitimate than institutions that barely satisfy such a condition. But when assessing an institution that fails to meet the criterion of mere comparative benefit, we need not even examine the other normative criteria. Any institution that fails the test of comparative benefit ought to be judged illegitimate.

But I should make clear that the comparative benefit criterion is one that applies to the institution, and not to any of its particular policies. Institutions that deliver a comparative benefit on the whole can sometimes issue policies that are comparatively detrimental – this is most clear with patently unjust policies. An institution setup to oversee geoengineering would be making decisions about, most likely, various technologies at various stages of research and development. For instance, a geoengineering governance institution could be tasked with approving or denying field trials across a number of different geoengineering proposals. A particular *policy* of that institution, on the other hand, could, for example, place a moratorium on field trials of a specific technology (i.e. ocean fertilization). What we want to know is whether the institution as a whole is legitimate – that is, we want know whether the institution as a whole is justified in attempting to coordinate our collective action. And this can be the case even if some of its particular policies are not comparatively beneficial. Again, we don't want to set our bar of legitimacy so high as to make the Best the enemy of the Good.

4. Normative Criteria for Geoengineering Governance

What, then, are the appropriate normative criteria for an institution overseeing geoengineering? There are two methodological obstacles to proceeding with an adumbration of normative criteria appropriate for a climate engineering regulatory institution. First, without knowing the precise purpose of the institution and the kind of power it has at its disposal to achieve this purpose, nailing down clear and uncontroversial normative criteria for legitimacy assessments is difficult. Not only that, remember that the Metacoordination View sees legitimacy judgements as a *social* practice. The object of the practice is to reach justified agreement on shared normative criteria and related substantive standards. We'll need to decide together what the main goal of this institution is and what kind of power it ought to have to achieve such a goal.

What I offer here, then, are not *the* necessary and sufficient criteria that ought to be used to judge the legitimacy of geoengineering regulation. But given the characteristics of geoengineering technologies, the following proposed criteria seem relevant for assessing whether or not a governance institution is worthy of our respect. Thus, the following criteria are a good place to begin a discourse that will need to gather input from participants from all over the globe. With these words of caution in mind, the following five subsections explore tentative normative criteria for an institution charged with overseeing climate engineering. An institution empowered to oversee climate engineering ought to sufficiently satisfy the normative criteria of: (1) comparative benefit; (2) accountability; (3) transparency; (4) substantive justice; and (5) procedural justice.

4.1. Comparative Benefit

The first normative criterion of comparative benefit is actually a necessary condition for any justified claim to legitimacy. As mentioned previously, a climate engineering regulatory institution that fails to meet the criterion of comparative benefit ought not to receive our respect. This should be fairly straightforward. Insofar as we are not better off with a climate engineering institution than we would be without it, we have no reason to accord it the respect it would require to function.

There are a variety of ways in which a geoengineering governance institution could make the world a better place than the non-institutional alternative. For instance, if the institution made the threat of harmful unilateral deployment less likely, this would count as a benefit, since surely a world in which the threat of harmful unilateral deployment is less likely is better than a world in which it is more likely. Similarly, if the institution aided coordinated research so that resources could be more efficiently used, this would count as a comparative benefit. Or if the institution furthered other things we value – such as substantive and procedural justice – this too would count as a benefit compared to the non-institutional alternative. Conversely, if an institution increased the likelihood that geoengineering would be deployed prematurely, or exacerbated global injustices (again compared to the non-institutional alternative), then this would count against its fulfilment of the criterion of comparative benefit.

Now, the idea of comparative benefit has two readings. The first is the thought that the institution in question should lead to greater social benefit than the non-institutional alternative; that is, the world in which we have an institution to oversee climate engineering must be comparatively better than a world in which no such institution exists, *ceteris paribus*. I call this the *non-institutional alternative* reading. The second reading of comparative benefit is also counterfactual, but is not limited to merely a non-institutional alternative. Under this reading of the comparative benefit criterion, the climate engineering regulatory institution must deliver a social benefit compared to possible alternative institutions as well. This second – more demanding – reading, which I call the *institutional alternative reading*, has four conditions. A climate engineering institution will fail the comparative benefit criterion if there is an institutional alternative that: (a) provides significantly greater benefits, (b) enjoys similar or even greater feasibility, (c) is accessible without unacceptable transition costs, and (d) sufficiently fulfils the other normative criteria (Buchanan & Keohane, 2006, p. 422). Thus, a climate engineering institution will be worthy of our respect if it fulfils the comparative benefit criterion under the non-institutional alternative reading. But insofar as there is an alternative institution that satisfies conditions (a) – (d) mentioned above, it is that institution that should be coordinating our action around geoengineering. That is, a climate engineering regulatory institution ought to fulfil the *institutional alternative reading* of the criterion of comparative benefit.

4.2. Accountability

A second normative criterion for the legitimacy of geoengineering governance is the idea of accountability. Remember that empowering an institution to help us solve our coordination problem carries with it risks associated with the power the institution wields. In order to make sure that power is being used in the way we intend, we will want institutional agents and the institution itself to be accountable. The norm of accountability has, according the Buchanan and Keohane, three elements:

> First, standards that those who are held accountable are expected to meet; second, information available to accountability holders, who can then apply the standards in question to the performance of those who are held to account; and third, the ability of these accountability holders to impose sanctions – to attach costs to the failure to meet the standards (Buchanan & Keohane, 2006, p. 426).

To these three I'd like to add a fourth element that would clarify the appropriate group of accountability holders. It wouldn't be hard to imagine a scenario in which the three elements above are properly accounted for, and yet the group of accountability holders is insufficiently restricted so as to render the fulfilment of the first three elements hollow.

However, while accountability is a normative criterion that a climate engineering regulatory institution ought to satisfy if it is to warrant our respect, it is also possible for such an institution to be overly accountable in at least two ways. First of all, while we want institutional agents to be accountable to the right group, we also want them to exercise their own judgement and expertise to a certain degree. Presumably, institutional agents are better informed than the general group of accountability holders (Maskin & Tirole, 2004). If institutional agents are to act as responsible representatives for the accountability holders and not merely pander to their wishes, they will require a certain (perhaps small, perhaps larger) degree of insulation.[13] This leads to the second point, which is that – while they certainly should be part of the group of accountability holders – it is impossible to allow future generations to hold current institutional agents accountable. If the current group of accountability holders has too much sway over institutional agents, they can tilt the institution's functioning in their favour, disregarding important duties the current generation may have to future generations. This is what Stephen Gardiner has termed 'the tyranny of the contemporary' (Gardiner, 2011, ch. 5). For these two reasons, we will want to design our accountability mechanisms with some caution.[14]

4.3. *Transparency*

A third normative criterion relevant to geoengineering governance is that of transparency. Transparency is commonly defined as 'the principle of enabling the public to gain information about the operations and structures of a given entity' (Etzioni, 2010). While some have highlighted the costs associated with significant transparency (Bannister & Connolly, 2011), the norm has at least three significant functions. The first significant function correlates to the second element of accountability highlighted above. Transparency aids accountability holders in accessing information useful for determining how they should exercise their power of holding institutional agents accountable. But, secondly, broad transparency gives non-addressees of an institution the ability to analyse the institution's functioning and enables them to 'contest the terms of accountability' (Buchanan & Keohane, 2006, p. 427). Finally, transparency can serve as a kind of public justification and create more trust for institutions and the agents that occupy senior positions (O'Neill, 2006). And it is worth mentioning, given climate engineering's controversial nature, that transparency will be of the utmost importance in garnering descriptive legitimacy, aiding people in reaching concurrent judgements as to whether or not the institution is worthy of our common respect. However, given the complex and highly technical nature of climate change and especially geoengineering, true transparency will require that the information provided be accessible to both addressees and non-addressees of the institution. This information ought to be accessible not merely in the sense that it is available, but accessible in the sense that it is intelligible to institutional outsiders as well.

4.4. Substantive Justice

A fourth normative criterion that a climate engineering regulatory institution ought to sufficiently satisfy in order to warrant our respect is captured by the idea of substantive justice. Institutions can deliver certain benefits and often carry with them certain burdens as well.[15] This is certainly true of an institution overseeing geoengineering. It is universally recognized that geoengineering has the potential to create novel distributions of climatic benefits and burdens.

In the abstract, the idea of substantive justice refers to just substantive outcomes, or just distributions of the benefits and burdens produced by a geoengineering governance institution. We generally assume that everyone would like to secure as many of the benefits and as few of the burdens associated with the institution's functioning. We can imagine people having certain claims to some of the benefits and certain claims against the burdens associated with geoengineering governance. Substantive (or distributive) justice obtains when there is a proper balance between the competing claims to those benefits and against the accompanying burdens. While a perfect balance between these competing claims would have to obtain for us to determine the geoengineering governance institution was perfectly just, remember that legitimacy assessments admit of degree. We don't want to require a climate engineering institution to be perfectly just in order for us to grant it the kind of standing it needs to function. This is for at least the following two reasons. First, we may need such an institution in order to make progress on justice. So, 'refusing to regard an institution as legitimate unless it is fully just would be self-defeating from the standpoint of justice' (Buchanan, 2018, p. 55). Second, requiring that the institution be fully just in order for it to be considered legitimate would conflate 'legitimacy' and 'justice' and impoverish our moral lexicon. Thus, when making legitimacy assessments, we'll have to determine whether or not a geoengineering governance institution is sufficiently satisfying the demands of substantive distributive justice to be worthy of our collective respect, while nonetheless recognizing that it need not go all the way in satisfying such demands.

Exactly what would count as a just substantive outcome with respect to geoengineering (or climate change in general) is subject to ongoing debate. But there is a modest claim about substantive justice and geoengineering that seems to be on solid ground.[16] The fact that the least well off members of the global community have (a) contributed the least to the genesis of climate change, (b) have benefited the least from previous actions that have brought about climate change, and (c) have the weakest ability to respond to the burdens of climate change all point to the same conclusion regarding the distribution of benefits and burdens related to geoengineering. The three facts listed above lend support to the following conclusion: if a geoengineering governance institution is to minimally meet the normative criterion of substantive justice, then the distribution of benefits and burdens engendered by such an institution should (probably heavily) favour the least well off members of the global community. Exactly how much that distribution should favour them is difficult to say.[17] But what seems clear is that if a geoengineering governance institution were to lead to a world in which the least well off members of the global community were forced to shoulder even greater burdens than they already are in the face of anthropogenic climate change, such an institution would fail to satisfy the normative criterion of substantive justice.

4.5. Procedural Justice

The final normative criterion I put forward as applicable to geoengineering governance is that of procedural justice. As was noted in the previous subsection, we will want to make sure that our governance institution is conforming to norms of substantive justice. But the substantive distributive outcome with respect to geoengineering is not all we care about.[18] We are also concerned with how it is that we go about making decisions regarding research, development, and deployment. That is, along with substantive justice, we also care about procedural justice. The difference between substantive and procedural justice can be understood as the difference between fairness in the result and fairness in the process, respectively (Rawls, 2005, p. 424; Hampshire, 1993). If we think of substantive justice as a proper balance between the competing claims to the benefits and burdens associated with the institution, we can think of procedural justice as a proper balance between the competing claims to participate in the decision-making process that will determine the outcome. Just as in the case of substantive justice, the normative criterion of procedural justice can be satisfied to a greater or lesser degree. The more procedurally just the geoengineering governance institution is, the more confident we can be that we are justified in granting it the kind of respect it needs to coordinate our action. The more procedural injustice that exists within the institution – that is, the more there are people with claims to participation that are being ignored – the more that will count against us collectively appraising the institution as legitimate.

As in the case of substantive justice, it is difficult to say exactly what would count as a just procedure when it comes to making decisions about geoengineering. But in order for such a procedure to minimally fulfil the normative criterion of procedural justice, it would need to go some way towards providing what I call *fair terms of inclusion* and *fair terms of participation*.[19] By fair terms of inclusion, I mean that all those with legitimate claims to participate in the decision-making process are included. By *fair terms of participation*, I mean that all those included in the decision-making process are included on justifiable terms. Again, this doesn't provide us with specific instructions for securing procedural justice in geoengineering governance. But it does help us pick out clear instances of procedural injustice. For example, given the potentially beneficial and potentially catastrophic effects geoengineering could have on, say, those residing within small island states, any process that failed to include these people and allow their voices to be heard on reasonable terms of participation would be procedurally unjust (regardless of the substantive outcome).

Finally, whatever formal procedure ends up guiding the decision-making process around geoengineering, there is a good reason to think that such a procedure should include public participation in some way, shape, or form.[20] From a purely theoretical standpoint, public participation may not be strictly necessary provided that good representatives of all those with legitimate claims are included in the decision-making process. But including public participation in the decision-making process with respect to geoengineering would be valuable on two fronts. First, we know that even the best of representatives do not always fully represent the interests of their constituents – a problem exacerbated at the international level where (a) representatives of democratic states are often too far removed from their constituents to adequately represent their interests, and (b) representatives of non-democratic states often do not represent the interests of their constituents at all. Including public participation could go some minor way towards addressing this issue. Second, embedding public participation of some kind into the decision-making process can help a geoengineering governance institution secure descriptive or sociological legitimacy more easily. That is, having members of the public included in the institution could help a geoengineering governance institution actually garner the respect it needs to perform its function. And given geoengineering's controversial nature, a governance institution will need all the help it can get in gaining the actual support of the public.

5. Conclusion

A quick recap is probably in order. This paper began by noting that, given geoengineering's potential to deliver both great benefits and great harms to various populations throughout the world, continued research, development, and any possible deployment ought to be overseen by a legitimate governance institution. And, in order for such an institution to justifiably regulate geoengineering, it will have to be a legitimate institution. While perhaps everyone agrees that geoengineering governance ought to be legitimate, an agreed upon *conception* of legitimacy has yet to emerge. I posited that standard conceptions of (state) legitimacy may be inadequate to determine the legitimacy of a geoengineering regulatory institution. This is because geoengineering governance is something that will take place on the international stage, and international institutions vary dramatically in their characteristics and functions. Given this diversity, we should adopt the broader concept of institutional legitimacy, and then develop specific normative criteria that each kind of institution ought to fulfil. That is the route this paper has taken, identifying five normative criteria that could be appropriate for making legitimacy assessments of an institution setup to oversee climate engineering.

Now, it's clear that these normative criteria are somewhat ambiguous. And we need to remember that there may be other normative criteria that are also salient to legitimacy assessments of a geoengineering governance institution.[21] Furthermore, even upon reaching a justified agreement regarding the appropriate normative criteria for legitimacy assessments, we still need to explore how these normative criteria can be translated into substantive standards. These standards are to serve as proxies for the normative criteria and, thus, they will need to fall between two points. On the one hand, they need to capture the normative criteria as closely as possible. On the other hand, they need to be accessible to both addressees and non-addressees of the institution, and they need to be standards that could be widely accepted. Exactly how these standards are instantiated and the judgement of whether or not they are met must be done socially.

Some might find this conclusion unsatisfying for perhaps two reasons. First, one might think that geoengineering governance is going to require significant coercive power in order to effectively coordinate our action (which may be the case), and so such a loose conception of legitimacy is insufficient. But remember what was said at the end of Section 2. If a geoengineering governance institution is wielding significant coercive power, then it, just like states, ought to meet significantly more demanding normative criteria to be considered legitimate. Second, what we want from a conception of legitimacy, one might argue, are explicit necessary and sufficient conditions that unmistakably spell out exactly when it is that an institution is legitimate and when it lacks legitimacy. The Metacoordination View, clearly, does not do that. But to the extent that we recognize legitimacy assessments as social practices, what the Metacoordination View *does* do is provide us with the right framework from which to engage in that practice. The hope is that this paper has clearly explicated that general framework, and done some modest work towards generating a conception of legitimacy that will allow us to justifiably coordinate our assessments regarding geoengineering governance.

Notes

1. Throughout this paper, I'll be loose with language and refer to 'geoengineering' or 'climate engineering' broadly. But it is important to note that there are significant and morally relevant differences between the different proposals (Heyward, 2013; Jamieson, 2014, ch. 7).
2. For an argument against the prospect of unilateral deployment, see (Horton, 2011).
3. Note that one need not think that we *should* move forward with research and development to endorse this claim. Even if one thinks research should be abandoned today, this conditional claim could still be accepted. And note that by highlighting the importance of legitimate governance for geoengineering, no claim is being made about its relative importance. There are many other emerging technologies (i.e. gene drives, artificial intelligence, etc.) that deserve attention from normative theorists in order to help guide the development of legitimate regulatory institutions. The focus of this paper, however, is on the cluster of emerging technologies known as geoengineering.
4. For notable exceptions, see (Wong, 2016), (Morrow, Kopp, & Oppenheimer, 2013), (Morrow, 2017b), and (Gardiner & Fragniere, 2018). The conception of legitimacy I outline later in the paper is relatively similar to the conception outlined by Morrow, Kopp, and Oppenheimer – which shouldn't be surprising given that we're all relying upon work done by Buchanan and Keohane (2006). That being said, there are also significant differences. For example, Morrow, Kopp, and Oppenheimer would claim that if a climate engineering regulatory institution is legitimate that it (a) has a right to rule and (b) that those within its purview are under a duty to obey. The Metacoordination View of Buchanan's that I am relying upon claims instead that a legitimate institution is (a) justified in coordinating our action and (b) that those within its purview have a defeasible content-independent reason to comply, but not a duty to obey. Morrow, Kopp, and Oppenheimer also claim that the consent of all democratic states would be necessary but not sufficient to confer legitimacy. But what if all democratic states, except France, consented? Contrary to the view put forward by Morrow, Kopp, and Oppenheimer, the Metacoordination View *could* still consider such an institution legitimate (though this would of course depend upon how well it is satisfying the relevant normative criteria).
5. For explanations of the distinction between concepts and conceptions, see (Dworkin, 2013, p. 167; Hart, 1998, p. 160; Korsgaard, 2003, pp. 115–16; Rawls, 1999, p. 5).
6. Though, I should point out that even procedural conceptions like Rawls' recognize substantive constraints. See (Rawls, 2005, p. 428). I thank Augustin Fragnière for pointing this out to me.

7. I have serious doubts about the voluntarist conception of legitimacy for even state-like institutions, but I set that concern aside here.
8. I don't mean to suggest that merely being beneficial confers legitimacy onto an institution. Rather, I mean to suggest that we should recognize what we would be giving up by labelling all international institutions as illegitimate. We need international institutions both to make progress on the legitimacy of domestic institutions and on concerns of justice more globally. Thus, to label international institutions as illegitimate would be counter-productive to the causes of making sure that political power is only exercised legitimately and that it serves the cause of justice. See my fifth point in Section 3.1.
9. It should be noted that we are not looking for arbitrary or unjustified granting of this standing. As Buchanan writes, 'The goal is not simply agreement that we should support the institution, but rather agreement that is morally appropriate' (Buchanan, 2013, p. 179).
10. For a slightly different understanding of what it means to show an institution the kind of respect it needs to perform its function, see (Adams, 2018).
11. It could also be noted that sometimes institutions are required in order to further the cause of justice (the natural duty of justice). It would, ironically, undermine the cause of justice to require that all institutions be perfectly just in order for them to be considered legitimate.
12. Buchanan has altered his view and now agrees that comparative benefit is a necessary criterion. See (Buchanan, 2018, p. 60). But the reason for this isn't fleshed out. As I explain below, I think construing the comparative benefit criterion as necessary is conceptually entailed by the Metacoordination View.
13. Take the case of genetically modified mosquitoes, for instance. Imagine that politicians, speaking with informed scientists, recognize that genetic modification is the all-things-considered best way to stop the spread of infectious disease. If there is too much accountability, they may be reluctant to pursue such a policy option because their constituency may be averse to genetic modification. Adrienne LaFrance writes, 'A poll conducted by the Annenberg Public Policy Center in February found more than one-third of Americans believed genetically modified mosquitoes were to blame for the spread of Zika. (They're not.)' (LaFrance, 2016).
14. We will want to avoid 'narrow accountability' in favor of a more *broad accountability*. Narrow accountability is 'accountability without provision for contestation of the terms of accountability'. Broad accountability, however, would allow for public discourse and require a fifth element for the criterion, namely, that the terms of accountability themselves be publicly justified and open to revision. See (Buchanan & Keohane, 2006, p. 427).
15. Not only can institutions actually deliver certain benefits and burdens, they can also affect the possibility of certain benefits and burdens coming to fruition. Whether an institution has actually produced particular benefits or burdens can often only be assessed ex post. For instance, if we assume that a geoengineering regulatory institution would deliver a comparative benefit, we should grant it standing ex ante. But if it fails to deliver such benefits and is actually putting us at risk of catastrophe, then we should collectively revoke its standing. We could also recognize that an institution could put us at risk ex ante, in which case we should also withhold a positive legitimacy assessment. See Section 4.1 for more on risk and comparative benefit.
16. While it is not completely free from controversy, the following claim enjoys rather broad support. See (Gardiner, 2004, 2016a; Horton & Keith, 2016; Morrow, 2017a).
17. There are two ways to understand the claim that the least well off should be favoured most heavily: in an absolute sense and in a relative sense. I think that substantive justice would require any use of climate engineering to favour the least well off in both a relative and an absolute sense. This entails that if the use of climate engineering were to make everyone worse off, but were to make the least well off less worse off than others, that it would *not* further the cause of justice. That is, I am not entertaining ideas of levelling down. I thank the editors for bringing this to my attention.

18. For more about why we should care about more than the mere distributive outcome with respect to climate engineering, see the contributions of Marion Hourdequin, Patrick Taylor Smith, and Christopher Preston to this special issue.
19. While he doesn't use these terms and focuses more on what I call fair terms of inclusion, these terms are influenced by Charles Beitz's discussion of political equality. See (Beitz, 1989).
20. For more on what form that public participation could take, see: Callies (2018) and Jinnah (2018).
21. For other papers that address some salient normative criteria for geoengineering governance, see (Jamieson, 1996; Morrow et al., 2013; Rayner et al., 2013) along with the contributions of Christopher Preston, Kyle Powys White, and Marion Hourdequin to this special issue.

Acknowledgments

For helpful comments on earlier drafts of this paper, I'd like to thank Darrel Moellendorf, Catriona McKinnon, Clare Heyward, and especially Steven Gardiner and Augustin Fragnière. I'd also like to thank the participants of the following conferences and workshops for fruitful discussions: 'Geoengineering and Legitimacy Conference', University of Washington, Seattle (November, 2017); 'Science, Technology, and Public Policy Workshop', Harvard University (November, 2016); 'Climate Scholars' Conference', University of Reading, England (November, 2016).

Research for this paper was supported by the DFG-funded Excellence Cluster, The Formation of Normative Orders (Goethe-Universität Frankfurt), and the Bernheim Foundation's Hoover Chair for Social and Economic Ethics (Université catholique de Louvain)

Disclosure statement

No potential conflict of interest was reported by the author.

References

Adams, N. P. (2018). Institutional legitimacy. *Journal of Political Philosophy*, *26*(1), 84–102.

Bannister, F., & Connolly, R. (2011). The trouble with transparency: A critical review of openness in e-government. *Policy & Internet*, *3*(1), 1–30.

Beitz, C. R. (1989). *Political equality: An essay in democratic theory*. Princeton, NJ: Princeton University Press.

Buchanan, A. (2003). *Justice, legitimacy, and self-determination*. Oxford: Oxford University Press.

Buchanan, A. (2013). *The heart of human rights*. Oxford University Press. doi:10.1093/acprof:oso/9780199325382.001.0001

Buchanan, A. (2018). Institutional legitimacy. In: D. Sobel, P. Vallentyne, & S. Wall (Eds.), *Oxford studies in political philosophy* (Vol. 4, pp. 53–78). Oxford: Oxford University Press.

Buchanan, A., & Keohane, R. O. (2006). The legitimacy of global governance institutions. *Ethics & International Affairs*, *20*(4), 405–437.

Callies, D.E. (2018). The slippery slope argument against geoengineering research. Journal of Applied Philosophy, doi:10.1111/japp.12345

Dworkin, R. (2013). *Taking rights seriously*. London: Bloomsbury.

Etzioni, A. (2010). Is transparency the best disinfectant? *Journal of Political Philosophy*, *18*(4), 389–404.

Gardiner, S. M. (2004). Ethics and global climate change. *Ethics*, *114*(3), 555–600.

Gardiner, S. M. (2011). *A perfect moral storm: The ethical tragedy of climate change*. New York, NY: Oxford University Press.

Gardiner, S. M. (2016a). *Debating climate ethics*. New York, NY: Oxford University Press.

Gardiner, S. M., & Fragniere, A. (2018). The tollgate principles for the governance of geoengineering: Moving beyond the oxford principles to an ethically more robust approach. *Ethics, Policy and Environment*. Forthcoming.

Hampshire, S. (1993). Liberalism: The new twist. *New York review of books* 40(14), 43–47.

Hart, H. L. A. (1998). *The concept of law* (2nd ed.). Oxford: Oxford University Press.

Heyward, C. (2013). Situating and abandoning geoengineering: A typology of five responses to dangerous climate change. *PS: Political Science & Politics*, *46*(01), 23–27.

Horton, J. (2011). Geoengineering and the myth of unilateralism: Pressures and prospects for international cooperation. *Stanford Journal of Law, Science & Policy*, *4*, 56–69.

Horton, J. B., & Keith, D. (2016). Solar geoengineering and obligations to the global poor. In: C. J. Preston (Ed.), *Climate justice and geoengineering: Ethics and policy in the atmospheric Anthropocene* (pp. 79–92). London; New York: Rowman & Littlefield International, Ltd.

Hourdequin, M. (2018). Climate change, climate engineering, and the 'global poor': what does justice require? Ethics, Policy and Environment. Forthcoming.

Jamieson, D. (1996). Ethics and intentional climate change. *Climatic Change*, *33*(3), 323–336.

Jamieson, D. (2014). *Reason in a dark time: Why the struggle against climate change failed–and what it means for our future*. Oxford; New York: Oxford University Press.

Jinnah, S. (2018). Governing geoengineering research: the role of an advisory commission in orchestrating experimental governance in California. Ethics, Policy and Environment. Forthcoming.

Korsgaard, C. M. (2003). Realism and constructivism in twentieth-century moral philosophy. *Journal of Philosophical Research*, *28*, 99–122.

LaFrance, A. (2016, April 26). Genetically modified mosquitoes: What could possibly go wrong? *The Atlantic*. Retrieved from https://www.theatlantic.com/technology/archive/2016/04/genetically-modified-mosquitoes-zika/479793/

Maskin, E., & Tirole, J. (2004). The politician and the judge: Accountability in government. *American Economic Review*, *94*(4), 1034–1054.

Morrow, D. (2017b). International governance of climate engineering: A survey of reports on climate engineering, 2009-2015. *SSRN Electronic Journal*. doi:10.2139/ssrn.2982392

Morrow, D. R., Kopp, R., & Oppenheimer, M. (2013). Political legitimacy in decisions about experiments in solar radiation management. In W. C. G. Burns & A. Strauss (Eds.), *Climate change geoengineering: Philosophical perspectives, legal issues, and governance frameworks* (pp. 146–167). Cambridge: Cambridge University Press.

Morrow, D. R. (2017a). Fairness in allocating the global emissions budget. *Environmental Values*, *26*(6), 669–691.

O'Neill, O. (2006). Transparency and the ethics of communication. In: C. Hood & D. Heald (Eds.), *Transparency: The key to better governance?* (pp. 74–90). Oxford; New York: Oxford University Press.

Peter, F. (2014, Winter). Political legitimacy. In. E. N. Zalta (Ed.), *The stanford encyclopedia of philosophy*. Retrieved from http://plato.stanford.edu/archives/win2014/entries/legitimacy/

Preston, C. J. (2018). Recognitional justice, climate engineering, and the care approach. Ethics, Policy and Environment. Forthcoming.

Rawls, J. (1999). *A theory of justice*. Cambridge, MA: Harvard University Press.

Rawls, J. (2001). *Justice as fairness: A restatement*. Cambridge, Mass: Harvard University Press.

Rawls, J. (2005). *Political liberalism*. New York, NY: Columbia University Press.

Rayner, S., Heyward, C., Kruger, T., Pidgeon, N., Redgwell, C., & Savulescu, J. (2013). The Oxford principles. *Climatic Change*, *121*(3), 499–512.

Simmons, A. J. (2001). *Justification and legitimacy; essays on rights and obligations*. Cambridge, Mass: Cambridge University Press.

Smith, P. T. (2018). Legitimacy and non-domination in solar radiation management research. Ethics, Policy and Environment. Forthcoming.

Weber, M. (1997). *The theory of social and economic organization*. (T. Parsons, Trans.). New York, NY: Free Press.

Wong, P.-H. (2016). Consenting to geoengineering. *Philosophy & Technology*, *29*(2), 173–188.

Legitimacy and Non-Domination in Solar Radiation Management Research

Patrick Taylor Smith

Introduction

The environmental impacts of anthropogenic climate change, from an increase in global temperatures melting polar ice caps to the generation of extreme weather events, appear to be happening even more quickly than anticipated (Graester et al, 2018). Yet, many are worried that global action towards emissions mitigation has stalled or will be inadequate. As a consequence, there has been growing interest in geoengineering strategies that may slow or reduce the negative consequences of climate change despite their apparent greater risk than responses that focus on mitigation. Research budgets are growing, extensive experimentation in the form of computer modeling is commonplace, and outdoor experimentation is under way (Dykema, James, Anderson, & Weisenstein, 2014). In light of this new interest, many groups – some activist, some governmental, and some scientific – have proposed a variety of governance schemes for the research and possible deployment of risky geoengineering strategies. The purpose of this paper is to explore what I take to be a significant limitation of the proposed regimes: their failure to grapple with the significant power imbalances between the scientific and policy-making communities of the developed world doing the research and those of the developing world that will suffer the worst consequences of both climate change and climate engineering. This paper will argue that the value of non-domination, including and especially how it relates to global inequality, should guide our thinking about how to legitimately engage in research into solar radiation management.

It would be unfair to expect a contextually limited governance regime to resolve all issues of global inequality and domination. Yet, I argue that three common features of proposed geoengineering regimes exacerbate inequalities in vulnerability, power, and capability in problematic and avoidable ways. These features are a reliance on the indoor/outdoor distinction, scientific self-governance, and information-sharing. In each case, the regimes underestimate the effects of SRM research on the political relations between the developed and developing world. These regimes treat research and deployment as fundamentally distinct activities that ought to be regulated according to distinct normative logics. However, when we focus on the way that research affects power relations, we can see that this picture is false and that underlying political values that govern deployment should be applied to research.

Towards a Better Taxonomy of Geoengineering Strategies

In what follows, I argue that some geoengineering strategies offer specific governance challenges due to the greater power they offer potential deployers over the environment and, subsequently, over the choices of others. An implication of this idea is that various taxonomies of geoengineering strategies – those based on risk or on mechanism – will fail to capture all of the governance issues related to these strategies.

Climate (or 'geo-') engineering is typically defined (The Royal Society, 2009) as 'intentional, large-scale intervention into the climate system.' Despite broad agreement on the definition, geoengineering is a vague concept, composed of strategies and interventions that vary considerably in their normative and public policy import. Thus, there is no singularly appropriate governance regime for geoengineering deployment or research. Some (The Royal Society, 2009) have suggested that a good proxy of normative relevance is the distinction between carbon dioxide removal (CDR) and solar radiation management (SRM). The former refers to strategies that purport to scrub greenhouse gases from the atmosphere, including and especially carbon capture and storage, while the latter refers to interventions to increase the reflectivity, or 'albedo,' of the planet. If more of the sun's energy is reflected back into space, then the planet will cool even if we assume a constant – or, within limits, even an increasing – amount of greenhouse gas in the atmosphere. The Royal Society argued that CDR broadly corresponded to the 'safe' interventions and SRM corresponded to 'dangerous' interventions. Yet, this distinction is misleading. First, even CDR strategies that are themselves safe will nonetheless require land-use decisions that, when compared to agricultural production, generate risks themselves.[1] What's more, there are CDR strategies that intrinsically generate significant environmental risks. Some of these CDR strategies – such as ocean iron fertilization – have been the subject of controversial experimentation (Strong, Chisholm, Miller, & Cullen, 2009). Second, some SRM strategies – such as whitening roads or roofs – are perfectly harmless if not especially effective.

The CDR/SRM distinction is, at best, an imperfect proxy for determining which geoengineering strategies require significant governance in order to be researched or deployed (Gardiner, 2011, 344ff). A significantly better distinction would rely upon risk. We could separate geoengineering strategies into those that exceed a certain likelihood of generating negative environmental impacts and, through those impacts, undermining human interests. On this view, the most controversial yet potentially effective solar radiation management strategy – the injection of reflective aerosols particles (RAIs) into the atmosphere[2] – would need to be governed because, for example, it risks disruption of welfare-bearing precipitation systems.[3] Yet, another intervention that relied upon a similar mechanism, such as making roofs more reflective, would not be a governance priority because it would not generate substantial environmental risks. It is important that these judgments do not assume that any particular intervention is unjustified. An intervention may be a governance priority in virtue of generating significant risks and yet be justified all things considered by the benefits its produces. So, this distinction would suggest that some interventions are more difficult to justify because they likely impose substantial probabilities of negative effects. On this view, governance plays an instrumental role on broadly consequentialist grounds since the less risky strategies need not be governed since things cannot go particularly wrong even if they are used unwisely. A consequence of this instrumentalist conception is that research governance is only relevant insofar as it allows us to manage the risks of deployment. Research into a risky strategy – such as ocean fertilization – might need to be governed in order to prevent 'technological lock-in' where the institutional incentives towards deployment will become more difficult to resist as greater resources are devoted to research and development (Cairns, 2014). Outside of the effects on deployment, the only relevant governance considerations for research are the potential environmental consequences of the experiments themselves. This will be relevant when we discuss the current governance assumptions about the 'harmless' nature of 'indoor' experimentation and modelling.

There are some important limitations of the risk conception. First, focusing on risk can lead to a problem of distribution. That is, if we aggregate risks across populations, then a large group of people suffering a minimal risk might outweigh a small group of people being subject to a catastrophic risk. This is especially true if we understand risks in economic terms, where the economic value of richer populations can create even more severe distributional inequalities. Second, risk evaluations of RAIs must implicitly make reference to a baseline. RAIs may not be especially risky when compared to a business as usual scenario of a greater than 3 degree increase in global average temperatures over the preindustrial baseline, but this is primarily because the consequences of such warming would be extremely severe. Yet, in comparison to a baseline where the developed world – those who are considering RAI deployment – satisfy their moral obligations through mitigation and adaptation, RAIs look very risky. Finally, and most importantly, the governance regime will be required to make decisions that generate winners and losers, distributing benefits and burdens. Yet, the risk-based regime is not inherently concerned with the procedural legitimacy or the relational equality of the agents and stakeholders involved in the regime. It seems clear that, at the least, the deployment of RAIs involves or constitutes an exercise of power, and there are normative considerations that ought to inform our evaluation of that exercise beyond whether it distributes risk in an optimal fashion.

As a result of these worries, I want to propose a different geoengineering distinction that emphasizes relational elements and human agency. There is a difference between capability-enhancing geoengineering strategies and passive, cost-reducing interventions. That is, some geoengineering strategies provide particular agents with more power, a greater ability to shape the environment and choices of others, and some do not. This feature of an intervention can be separated from its environmental risks. A geoengineering intervention can be risky yet do little to improve an agent's ability to manipulate nature to suit their interests while another geoengineering intervention can be less risky in terms of its environmental consequences while grounding an enhanced ability to manipulate nature. Another way to put this is to distinguish between two sets of risks: environmental consequences of deployment and the possibility of human abuse. So, a geoengineering strategy that *merely* blocked or prevented some negative consequence of climate change, was widely available, subject to veto from stakeholders, did little to develop follow-on capacities to change the environment but nonetheless generated negative environmental impacts would represent a different governance challenge from a less immediately risky intervention that could be deployed by an elite few in ways that disproportionately served their interests and served as a further foundation for manipulations of the environment.[4] So, let's call interventions that exacerbate existing or generate new capabilities to exercise greater control over others 'dominating' while interventions that have sufficiently high likelihood of generating bad environmental consequences 'risky.' In 'pure risk' cases, the instrumental governance paradigm seems much more plausible: if those risks do not exacerbate current governance inequalities, then it seems correct to suggest that the governance objective is to ensure that the intervention is deployed wisely. Yet, the instrumental paradigm is inadequate when faced with dominating interventions that look relatively safe, or at least present themselves as clearly passing any cost-benefit analysis, but nonetheless exacerbate asymmetries in power and capacity between privileged and non-privileged agents.

This, I submit, is a reason that many are more skeptical or wary of geoengineering interventions than a strict understanding of the risks would entail.[5] For example, the possibility of routine weather manipulation represented by cloud seeding or brightening generates concerns about who will be able to control these fundamental preconditions of our day-to-day lives. Yet, it is clear that reflective aerosol injections are both risky and dominating, sharing the features that demand robust governance. First, RAIs have the potential to cause negative climate effects, especially with regards to precipitation (Tilmes et al., 2013). Second, research and deployment of RAIs are currently justified – if they are justified – on maximizing, consequentialist grounds in ways that ignore other considerations (Barrett, 2008; Crutzen, 2006). RAIs will be bad, but climate change is worse, or so the argument goes (Gardiner, 2010). This generates strong pressure to adopt an instrumentalist conception of governance. Third, RAIs also have the greatest dominating potential of existing geoengineering strategies. RAIs can be deployed by a small group of people, require considerable engineering skill, and lead to improved capabilities to manipulate the environment. While RAIs, perhaps, can be used carefully in order to moderately shave off the worst aspects of climate change (Keith, 2013), many people could potentially benefit – while failing to pay the full costs – from more extensive RAI interventions. While it is true *for the moment*, that SRM interventions into the atmosphere are fairly crude such that it is not currently possible to carefully control temperature and precipitation, the very ideas of 'peak-shaving' and engaging in RAI *safely* demand that scientists develop more fine-grained control over RAI and its effects.[6] Thus, scientists are aiming to generate RAI and other SRM techniques that will increasingly give us control over important components of the atmospheric system which could be potentially be used for well or ill. So, the governance imperatives for RAIs are strong due to both their riskiness and their potential for domination. Unfortunately, we shall see that most governance initiatives – especially those that govern geoengineering research – are much more interested in former, failing to take the latter sufficiently seriously.

It is worth pausing here to discuss the role of power and domination in the context of legitimacy.[7] I argue that some geoengineering strategies look especially problematic, in part, because they exacerbate or lead to new opportunities for domination. I want to suggest, following the republican tradition, that domination represents an intrinsic relational wrong and a serious injustice. That is, if I dominate someone, our relationship is a source of political injustice (e.g. is wrongful) even if I benefit them or treat them well. Domination occurs when a person is subject to the superior arbitrary power of another. This definition involves two key components: superior power and arbitrariness. One agent has superior power over another when the superior is in a position to asymmetrically alter the choice situation – what choices are available to the agent and their relative costs – of the inferior position. In other words, superior power gives its possessor a degree of asymmetric control over its subjects. Superior power can be generated in a variety of ways: legal entitlements, physical strength, or technological capability. These different elements are all ways of providing control over the choices of another. It is important to see that one can dominate even if one never exercises the superior power; the mere fact that one's choices are subject to the decisions of the superior is sufficient. However, superior power, by itself, does not always generate domination because superior power can be controlled and made accountable to those subject to it. If superior power is not contestable, accountable, or controlled, then superior power can be exercised unilaterally and without the permission of those in the inferior position. A legitimate state, on this view, does an adequate job of maintaining the freedom of everyone by instantiating a political order that either prevents people from having superior power or ensuring that all exercises of superior power are, in some sense, structured such that they are accountable to those who are less powerful. That is, preventing domination should be understood as a criteria of legitimacy for a political system and dominated people have a serious complaint against political systems that permit it.

With this in mind, we can understand the idea that some geoengineering strategies grant those who develop them greater direct control over the choice situation of others while some do not. This focus on 'dominating' geoengineering responses also explains the intuition that geoengineering is *especially* in need of justification when compared to other strategies such as mitigation or adaptation. The reason is that most mitigation strategies – such as renewable energy – do not imbue their users with additional influence over the environment such that they can exercise greater power over others. Of course, the possession of the intellectual property, resources, and manufacturing capacities for renewable energy production can give individuals social or economic power and thus indirectly lead to domination. But that is a feature of a social system that distributes benefits and entitlements in certain ways. Renewable energy, painting roads and roofs white, or carbon capture do not grant the possessor the ability to directly intervene into the choice situations of others, but some forms of geoengineering – including and especially RAIs – allow the possessors of that technology to asymmetrically structure the options that are available to other agents. Those technologies require more stringent governance.

Three Features of Geoengineering Research Governance

The problematic nature of geoengineering research governance is not a consequence of malice. Rather, since most research governance is by scientists and for scientists, the governance community not implausibly assumes that most of the people subject to its potential rules are acting in good faith. Those who will be subject to the power the research community generates are not especially salient and the relevant policy choices will likely be made by others. Yet, this lack of focus on power dynamics has led to three gaps in the research governance proposals. In what follows, I rely upon a general paradigm of research governance that brings together common features of the proposed regimes.

Not every regime contains every feature but all contain at least one and most more than one.[8] These features are: dramatically increased regulatory requirements for outdoor research as compared to indoor research, information-sharing as the primary mechanism of generating research equity, and dependence on scientific self-governance.

The Salience of the Indoor/Outdoor Distinction

One of the core research distinctions in the context of RAIs is between that which occurs 'indoors' and that which occurs 'outdoors.' More specifically, indoor research is a hodgepodge of experimental and scientific techniques that can occur entirely within a lab and generate no intrinsic environmental risks. These 'experiments' can be physically robust, such as testing the effect of various geoengineering interventions within a sealed chamber designed to mimic outdoor environmental conditions or occur entirely within a computer. It is telling that even the most stringent scientists who argue that there must be governance in place before experimentation engage in computer modelling and other indoor experimentation; indoor experimentation is viewed as harmless and thus need not be governed. Outdoor investigations, on the other hand, are deeply controversial and it is routine among the scientific community to suggest that there should be a moratorium on such experiments until a proper governance regime can be put in place. Popular condemnation has forced outdoor experiments in ocean iron fertilization to cease and planned experiments in RAIs to be cancelled. The justification for this distinction falls squarely within the risk paradigm: outdoor experiments risk negative environmental or health consequences and also create lock-in incentives towards deployment.[9] So, drawing a sharp distinction between outdoor and indoor experiments is about managing risk.

Yet, it is hard to sustain this distinction. First, it is unclear why *small* outdoor experiments carry meaningful environmental risks, especially in comparison to other human activities that are considered routine. Of course, large-scale experiments that are *de facto* deployments carry environmental risks, but that is to recapitulate the distinction between research and deployment, not between indoor and outdoor research. Second, the logic behind lock-in is the idea that committing resources towards the development of a technology motivates people to deploy the technology so that the resources are not 'wasted.' Yet, indoor experiments also require resources – labs, work-hours, and the like – so it cannot be the case that outdoor experiments uniquely activate lock-in (McKinnon, 2018). It is perhaps true that moving to outdoor experiments makes lock-in reasons for governance *stronger*, but it seems that if indoor experimentation makes outdoor experimentation more likely, then lock-in considerations would also demand governance for indoor experimentation.

But the strongest reason to reject the idea that there is a sharp distinction between and indoor and out experimentation and to govern indoor experiments – including computer modelling – is that the indoor/outdoor distinction is not especially relevant to domination. To see this, consider the following scenario:

> ISLAND: Imagine that two groups find themselves washed ashore on a desert island. After a period of time, they manage to build sustainable, if temporary societies, and eventually come into contact. Each group has equal amounts of technical skill and engineering resources. However, one group – the Morlocks – has discovered the plans for materials and know-how to construct a dam, which can with difficulty be built with their existing

technology base, while the other group – the Eloi – has not and will be vulnerable to flooding if the dam is built. However, the Eloi could manage the flooding if they could build a similar dam or levee.

The groups are equal in terms of material resources, population, and technological skill. Yet, one group has discovered *information* that would allow them to create a technology that would lead a significant improvement in their capabilities to structure the choice situation of the other. Do the Morlocks have power over the Eloi

Does the asymmetric availability of information on how to create and use powerful technologies grant power to agents even if they have not yet created them? The answer seems to be, 'yes.' If so, this means that coming to possess additional information represents a potential threat of domination. Imagine a debate between the Morlocks and Eloi over the distribution of a valuable resource. The Morlocks could, of course, threaten to develop the dam unless the resources are distributed in ways favorable to them. The Eloi would need to take that threat very seriously and would likely need to make concessions in order to prevent it from happening. So, the mere discovery of information, even without any production or deployment, can result in a power imbalance if that information changes the ability of one agent to control what happens to another.

Let's adapt the scenario somewhat. Imagine that the Morlocks do not merely discover the plans for the dam and the advanced materials to build it, but that they have some unique set of resources or skills that, through great effort, allow them to create both the plans and a workable design. That research program, aimed at simply creating more knowledge, changes the power dynamics between the two communities. Again, a good indication of the consequences, in terms of power dynamics, of this research program is that Eloi would likely be willing to take steps – through negotiation or otherwise – to prevent the research program from disrupting the equality between the communities. In other words, the research program is, from the Eloi perspective – a threat that must be contained and that reflects greater possibilities for domination. What's more, the example nicely illustrates another feature of domination: the irrelevance of intention. It does not matter whether the scientists are developing the dam – or intend to actually build it from the discovered plans in the first scenario – in order to change the inter-group power dynamics. Perhaps the Morlocks are developing the dam for peaceful purposes – hydroelectric power or flood management – and plan to scrupulously avoid even the appearance of an explicit threat. This might make the change in power dynamics more palatable, but the peaceful program would *still* make the Morlocks more powerful because acquiring that knowledge allows them to develop the dam and then use it as they see fit. The Eloi are subject to the whims of the Morlocks, and one might think that that fact by itself changes the relationship between the two groups.

This is salient for RAI research. Like the dam, RAI can give one side an outsized ability to control the environment: a more sophisticated dam can determine water levels and a more sophisticated RAI program will give individuals the power to ratchet temperature up and down. RAI programs will research how to effectively deploy the reflective particles, build particles that are more durable and less costly, and model the consequences of various deployment suites. In other words, these programs – regardless of intention – allow those who control the relevant decision-making nexus to develop new and better capabilities to manipulate the climate in the way they prefer. Again, perhaps the people developing RAIs prefer to use them in ways that benefit mankind, but this is irrelevant to the question of domination. What is relevant is that even indoor experiments develop these capabilities, decrease the cost of deployment, or help policymakers deploy RAIs more effectively. That is, they change the power relations, exacerbating the inequalities between people who have and can use the information to manipulate the environment. So, as a consequence, the distinction between indoor and outdoor begins to dissolve; it *may* be true that outdoor experiments lead to greater marginal increases in capability, but it is clear that even environmentally harmless computer modelling can increase one's potential ability to use RAIs effectively, assuming the modelling tracks some relevant feature of deployment and one has some capability to translate the.[10] I hasten to note that this does not mean that we should refrain from RAI research. I am only suggesting that insofar as we are concerned with domination, we should be interested in governing both indoor and outdoor research.

Information-Sharing as the Primary Mechanism of Research Equity

Geoengineering researchers are aware of the sharp inequalities between researchers in the developed and developing world. They correctly perceive that these inequalities can generate worries about the legitimacy of the decision-making process. Many have subsequently argued[11] that some inclusive political process needs to govern eventual deployment decisions, but the primary mechanism for maintaining the legitimacy of the research process is to ensure that scientists around the world have access to its results. This sort of transparency, again, is consistent with a risk paradigm. Having an open and transparent discussion where individuals can test and replicate various results helps ensure that people of good will can avoid serious mistakes that would lead to imprudent interventions. It will surface potential problems and decrease the chances of groupthink. Yet, transparency is an inadequate compensation for existing inequalities in resources and capability. In the face of severe inequality, any objections generated through transparency still depend upon the good will of the powerful for uptake.

To illustrate, let us return to ISLAND. In the first scenario, transparency would solve the potential domination of the dam construction. This is because we assumed no differences in capability or resources such that the only difference is one of knowledge. So, if the Morlocks wish to develop the dam for peaceful purposes and avoid domination, they can give the Eloi a copy of the plans, making their plans completely transparent and then the Eloi could counter potential flooding by building a dam of their own. Yet, in the second scenario, the reason the Morlocks were able to develop the weapon is that they had a unique set of resources or capabilities. Similarly, imagine that the Morlocks were uniquely positioned to benefit in an outsized fashion from the technology – or to suffer considerably lower costs than the Eloi for production – in ways that were not fully reducible to their knowledge or skill; perhaps the production of material for the dam required access to raw materials that were far more accessible to the Morlocks or the Eloi were positioned on the high ground such that they enjoyed a massive advantage in dealing with the subsequent flooding. In each of these cases, the mere availability of the plans would not equalize the position between the two groups. Even if the Eloi were to come upon the plans for the dam, they would not be in a position to exploit that knowledge as effectively as the Morlocks.

Yet, even if we assume a less extreme version of the example where the Morlocks are only in a more favorable position to exploit the relevant knowledge, information-sharing will at best decrease the power differential between the groups without eliminating it. The fact that the Morlocks can research or deploy the technology more effectively or efficiently will grant them superior power over the Eloi even if the availability of the relevant information makes it possible for the Eloi to deploy the relevant technology. This is an unsurprising result: if two agents begin with different capabilities, those capabilities can intersect with new information such that the combination can result in greater, equivalent, or only marginally reduced asymmetries in power. This need not only apply to technology. Without the resources and institutional support to actually participate in the international legal system, mere knowledge of, for example, human rights law will not necessarily lead to an improvement in the political position of a marginalized group. Or, at the very least, the introduction of a new legal regime will not sufficiently benefit the marginalized group such they improve their power relative to the other agents in the system.

The relevance to RAIs is two-fold. First, while many regimes argue for widespread availability of scientific research in the name of open-source transparency, they fail to require researchers to give up their intellectual property on the underlying invention. Yet, scientific transparency is quite different from legal entitlement. Much of the future work on geoengineering will involve the design of reflective particles, delivery systems, or monitoring mechanisms that can nonetheless be proprietary even if information about the consequences of those mechanisms is readily available. If researchers can use domestic and international legal systems in order to place significant obstacles on the ability of marginalized people to generate new capabilities, then information-sharing will have greatly diminished effects. The second, and more important, connection is that RAI research is and will be disproportionately performed by agents with asymmetric and superior resources and capabilities. Developed nations, where the vast majority of RAI research is done, have superior resources, engineering skill and educational infrastructure. RAI research will concern a variety of issues that will increase effectiveness, decrease costs, and grant greater ability to direct or minimize negative impacts. RAI research, therefore, could lead to the development of usable particles and delivery methods, increasing the power of the deploying agent to manipulate the environment. Yet, those advantages will accrue, or accrue at a greater level, to those who have the relevant institutional support. The developing world insofar as they are less advantageously positioned will be less capable of deploying RAIs. Thus, research into a new technology (where only one party is positioned to deploy optimally) threatens to exacerbate existing power inequalities by providing one party with capabilities that cannot be matched by the agents with antecedently inferior resources and power. In a context where the capacity to exploit information is differentiated, open and transparent information-sharing may not prevent power asymmetries from becoming more severe.

Scientific Self-Governance

The third problematic element of RAI governance is that political and legal interventions occur too late in the process. The relevant scientific research regimes are almost universally committed to scientific self-regulation. So, they believe that fellow scientists and researchers should be the primary agents when it comes to creating and then applying rules to regulate

RAI research. These regimes suggest some limits to this scientific self-governance. In the event of large-scale outdoor research, the various political regimes designed to regulate deployment – should they exist – or legal regimes governing pollution could potentially be invoked in order to include other agents. And there are obvious advantages to having scientists play the primary role in collectively determining what research is appropriate. The greatest of these advantages is epistemic. Scientists have essential knowledge to evaluate the potential risks and rewards of any particular research regime; others agents will almost certainly need to rely – to at least some extent – upon the expert testimony of scientists.[12] Furthermore, the agents who will be most directly constrained by the regime will be the scientists themselves, a structure that invites all the objections of self-regulation.[13] After all, it will be scientists that will be named and shamed, sanctioned, or otherwise constrained in their research pursuits and experiments. Careers and lives will be affected by these regimes, and researchers have a legitimate claim that their interests be represented substantively and procedurally. What's more, excessive political interventions into the scientific process might undermine its epistemological status and one might expect scientists to be more sensitive to those worries than outside agents.

There are significant worries about the effectiveness of scientific self-governance. Scientists are strongly incentivized to continue with their research and experimentation, so it is difficult to shut down promising research avenues. Furthermore, scientific institutions lack effective and timely enforcement mechanisms. What's more, what enforcement mechanisms that do exist may not be applicable to researchers in the private sectors.

However, even if these issues could be resolved, there would be a deeper problem. Self-constraint is an inadequate response to the possibilities of domination. Imagine a climate governance regime made up of rich and powerful high-emitting individuals who have fairly intense economic relations with poor low-emitting individuals who will experience negative impacts from carbon emissions. A set of social norms and incentives amongst high-emitting individuals causes them to check each other and to take steps to mitigate their emissions and serve the interests of the low-emitting class but without any participation or contribution by the low-emitters themselves. In other words, high-emitters are constrained but only by other high-emitters. Does this reflect a set of sufficient political protections for low-emitters such that the exercise of power is legitimately non-arbitrary? Obviously, we should be skeptical that a scheme that relied upon the good will and interests of the powerful agents to serve the interests of the powerless is going to fail. Deploying the mechanisms of social sanction against a fellow member of your class is unpleasant, expensive, difficult, and unsure. And of course, one would not wish to establish precedents that would undermine the economic usefulness of low-emitters and so the protection of their interests would always be, in this sense, indirect and limited. Some high-emitters may feel genuinely altruistic towards low-emitters, but any system that that relied upon those feelings and routinely set the economic and social interests of the powerful against the interests of the powerless would likely be unstable.[14]

But we might think that the system was problematic even if it was relatively successful and a stable equilibrium that served the interests of the serfs existed. There seems to be something unjust about the fact that these serfs would need to rely upon the charity of powerful aristocrats for the furtherance of their interests beyond their economic usefulness. In other words, the serfs seem dependent upon the will of the aristocrats for the protections of their rights; they cannot participate in their own protection or contest the decisions of their masters. Perhaps those masters are benevolent, but it is unjust to be so subject to others no matter how benevolent they might be. In the literature on domination,[15] this is understood as the idea that true political equality requires that no person be subject to the arbitrary will of another. Arbitrariness is then understood as being subject to the decisions of another where there are no effective avenues of control or accountability requiring that the powerful take the interests of weaker stakeholders into account.

The consequences of this view for RAI research are profound. If this type of normative claim is correct, then those who are subject to the exacerbated power imbalance created by RAI research have a claim to avenues of accountability, contestation, and control when it comes to the decisions of research scientists in the developed world. In fact, not only is scientific self-governance inadequate if those collective decision-making bodies are dominated – as they must be, given their membership – by the developed world, but even additional political or legal governance by national or regional scientific bodies would be insufficient (Pettit, 2010; Laborde, 2010). After all, those political and legal agents are designed to represent the interests of their constituents. Adding political regulations to RAI research may correct for *some* of the limitations of self-governance, but the fundamental normative weakness of a scientific regime whereby rich and powerful members of the global and scientific elite decide, for themselves, to what extent they will accept the criticisms and serve the interests of the developing world will remain illegitimate. And this is no criticism of the motivations or sincerity of the agents developing their self-governing regime; the political structures in place simply do not make room for legitimate research decisions if they are done unilaterally regardless of why the decisions are made. The developing world needs to engage with RAI research that offers mechanisms of contestation beyond mere persuasion. Legitimating avenues of contestation and accountability cannot be mere charity.

To sum up, proposals for RAI research governance have attempted to avoid being too political. They accept that wide swathes of RAI research – including essentially all research currently being done – need not be regulated at all. Only research that has generated public controversy or generates significant environmental risks is normatively relevant. Second, insofar as these regimes are concerned with power imbalances, they argue that these imbalances can be made good through transparent information-sharing. Finally, when it comes to RAI research, the primary agents in developing and implementing these regimes will be the scientists themselves. In each case, these proposals do not take the power dynamics or political implications of their research sufficiently seriously. In fact, it seems likely that many participants in these regimes are not even aware of the ways their actions intersect with global power inequalities. As a consequence, these regimes are unlikely to prevent these inequalities from worsening as research creates more and better means for manipulating the climate.

Two Strategies for Non-Domination and SRM Research: Power Accountability and Power Equalization

It is important to see that RAI research is not necessarily impermissible because it exacerbates current power imbalances. The question is, rather, whether we can prevent RAI-domination by adopting different rules, regulations, or institutional structures. In this section, I wish to propose two ways in which RAI research governance might be able to avoid my criticisms. Yet, I want to argue that each of these proposals have significant weaknesses and that they work at cross-purposes. As a result, the decision to engage in either strategy will be controversial, generating costs and benefits. Unfortunately, the political structures to make that initial strategic decision in a non-dominating fashion does not yet exist. After all, the strategies are themselves supposed to be what resolve domination. This raises a pressing question of global justice: how should powerful agents decide amongst options for reducing non-domination when those agents are, at the moment, in a dominating position. Powerful agents must make a choice about how to respond to the potential of domination in RAI research, but even the choice to create accountability mechanisms for the less powerful will be made without those accountability mechanisms in place; how can the decision of which legitimate procedures should be put in place be itself legitimate? In a sense, the question concerns how one should behave when all available actions are, in some important sense, politically illegitimate.

Domination occurs when two conditions are met. An agent dominates another when the former has superior power over the latter and that power can be exercised arbitrarily. This implies two anti-domination strategies (Pettit, 1997, Chapter 6). The first is *power equalization* while the second is *power accountability*. When one agent becomes more powerful, non-domination demands that other agents gain equivalent power or that the power be checked and balanced such that the superior power is held accountable to those who are less powerful. We can understand the differences between these strategies by considering the power relations between the United States and the Soviet Union in the beginning of the Cold War. The success of the Manhattan Project and the subsequent development of atomic weapons initially gave the United States a decisive, qualitative superiority over potential rivals. There were two strategies for dealing with this newfound power asymmetry. One possible avenue – advocated by Albert Einstein, Bertrand Russell, and others who distrusted the ability and the political will of competitive nation states to act responsibly with the new weapons – was controlling, directing, and checking the political effects of nuclear weapons though a centralized and democratic global regime. They argued that nuclear weapons ought to be centrally controlled by the United Nations, which was in turn structured by an international rule of law and a global democracy. The United Nations, on this view, would have control over the most powerful weapons ever devised. The power differential between the UN and member states could be extreme, but the exercise of that power could be made non-arbitrary and accountable. We might, following Pettit, call this the modern conception of how to regulate superior power. Developed in a political milieu where states became more powerful domestically, Enlightenment era domination theorists argued that by dividing state power along functional lines, creating independent judiciaries, and introducing democratic politics, the absolutist authority and increasing institutional power of the state could be made consistent with non-domination of its citizenry.

There is, however, an alternative conception that has its origins in the political constitutions of Greece and Rome. In those societies, non-domination was not generated through institutional regulation but by having the relevant groups within society have equivalent power. For example, both the plebeian and patrician classes of ancient Rome had legally equal[16] legislative entities that could issue binding decrees on the entire polity. In the nuclear context, we can see this as the strategy of the Soviet Union in its development of equivalent nuclear weapons and delivery systems as the United States. The Soviet Union never achieved complete parity with the United States; its deterrence triad was never entirely secure. In response, the Soviet Union developed a nuclear deterrent *and* pursued conventional superiority in Europe using tactics and organizations that were designed for nuclear survivability. So, the Soviet Union pursued a position of power parity with the United States through a combination of developing equivalent military technology and creating asymmetric counter-measures that substitute and compensate for the shrinking differences in capability. It is important to see that the non-dominating effects are contextual and relational. The Soviet Union was able to produce rough parity in geopolitical position viz. the United States in this specific policy domain. Despite the more general economic and diplomatic superiority of the United States, it was unable to exploit its advantage in nuclear weapons. Furthermore, the nuclear parity between the two superpowers did not 'reset' the power relations to before the technological development; the new parity created entirely new power dynamics and risks. It is, however, clear that the Soviet development of nuclear weapons changed the power dynamics between the superpowers, substantially equalizing them. Of course, this equalization is relational; Soviet technological developments may very well have led to greater domination elsewhere or may have exacerbated domestic injustice. The claim I am making is quite limited: the dominating potential of nuclear weapons in the relationship between the United States and Soviet Union was dramatically reduced by their mutual possession of the relevant capabilities.

When discussing RAI research, analogous policy responses suggest themselves. The power accountability strategy would involve the creation of a new or the restructuring of an existing body to create a centralized agency that will make it possible for stakeholders in the developing and developed world to come to together to legislate and administer a set of checks and balances on RAI research and deployment. We could attach these processes to the existing UNFCCC structure or create a separate management regime. The key features of the regime would be the following. First, it would require a legislative component where various stakeholders could come together, in a position of deliberative equality, to formulate research rules. Second, the regime would need a fact-finding and enforcement component where compliance and sanctions could be monitored in the event of a violation. The regime would need to have real teeth, being able to impose significant penalties on those found in violation. Finally, the regime would need a judicial component that could make impartial and independent judgments concerning violations. What I am describing would combine elements of the International Criminal Court (ICC) with its sophisticated checks designed to insure judicial independence, with the World Trade Organization's Dispute Settlement Board (DSB) with its linkage between rule violation and sanction. From the ICC, we would add include various mechanisms to include broad-bases geographical support – a certain number of judges would need to come from each area – and safeguards, such as multiple appellate levels, to prevent any one judge from acting unilaterally. From the DSB, judges would be empowered to issue sanctions in the event of a breach. Unlike the DSB which only empowers the state subject to unfair trade practices to issue sanctions, we would need omnilateral, rather than bilateral, enforcement such that all members of the regime would be required to enforce the relevant sanctions. Also, a separate ombudsperson whose objective was to maintain the integrity of the regime and could be empowered to bring cases before the executive would be essential, especially if paired with democratic processes for selecting the members of the office (Caney, 2006). In other words, the kind of robust regime that many imagine for the RAI *deployment* should also be developed for RAI *research*. Or perhaps more accurately, we should imagine a robust governance regime that treated research and deployment as a continuous whole, understanding that decisions about research and deployment are intertwined. If this regime was created, then stakeholders from the developing world would have a set of mechanisms by which they could ensure that their interests were represented. Non-domination does not require that one always get one's way, but it does require that one's basic rights and one's position in the deliberative process are ensured; this regime would be a significant step in the right direction.

The accountable power path seems to be ideal. It better captures the idea that climate change is a global problem that should be resolved collectively. Furthermore, it seems substantially less risky. There is a reason why relying upon the unilateral judgments of actors to constrain each other is unstable. Actors make mistakes, and the states that control the relevant capacities might come to be ruled by unjust governments. It can be difficult to reliably send or accurately receive the relevant policy signals, and multiplying the actors with the relevant capabilities increases the risk that misperception will lead to escalation (Sagan & Waltz, 1995). What's more, the legislative, judicial, and executive functions of the regime help – in Lockean terms – to solve problems in the state of nature (Ripstein, 2009, Chapter 6). Nations will overvalue their own interests and undervalue others; collective decision-making amongst stakeholders leads to more impartial judgments. Further, creating a set of public rules that are enforced and administered will make it easier to adapt to new developments in technology or in existing climate risks. What's more, if the danger of climate change passes, it will be easier for a centralized authority to coordinate the reduction or elimination of these policy tools. Insofar as people finds this strategy attractive, one of the main contributions of this paper, then, is that it shows that the non-instrumental reasons to develop this accountability regime in one context – deployment – are also non-instrumental reasons to develop this regime for research.

With those advantages, it would seem that very little could be said for power equalization. Yet, this is deceptive. The reason for power equalization in the context of RAI research is feasibility. Those who called for a global centralization of nuclear weapons failed in the face of the obstacles presented by geopolitics and state sovereignty. The political architecture that I described in the previous section would be one of the most politically robust global governance regimes ever devised. It is immensely difficult to create regimes that can level sanctions in ways independent of the geopolitical interests of the constituent states and stakeholders. Once we include impartial adjudicative institutions – which are expensive and difficult to maintain – and executive institutions free of regulatory capture, we can see the extent of the task. What's more, we have consistently failed to see the creation of similarly robust regimes in the context of climate governance, despite the greater urgency and salience of the issue. A key reason for the potential infeasibility of the accountable power avenue is the intersecting number of affected stakeholders. The regime is sufficiently powerful and sufficiently implicates state sovereignty that it would invoke powerful interests in self-determination and security while the eventual benefits are, for the moment, considered distant and tenuous.

While the power accountability strategy is attractive but perhaps unfeasible, the power equalization strategy – at least in the context of solar radiation management – is feasible but less attractive. This can be seen, for example, in discussions of nuclear proliferation, where arguments that a wider distribution of nuclear weapons would be beneficial are met with considerable skepticism (Waltz, 1981). Nonetheless, this strategy is available in the context of RAI research. Let's return to the desert island. In the first scenario, domination could be avoided simply by giving the Eloi the *information* needed to build the damn. Yet, in our second scenario, the Morlocks had access to skills or resources that the Eloi did not. As a consequence, mere information sharing was insufficient. So, let us suppose that the Morlocks wish to prevent or resolve the domination inherent to their position. Sharing the plans of the dam would be insufficient. They could provide a sufficient amount of material to build their own levees to the Eloi, but this would be problematic on a number of levels. First, they would need to provide enough to make the Eloi *self-sufficient* and that would require complicated and difficult policy judgments. Second, this would be an insufficient response if there was some expectation that the Morlocks would be able to build on the existing plans to develop superior mechanisms of environmental control. In that case, the Eloi would not be able to keep up. So, the clear solution is to provide some of the actual material but to *also* help the Eloi build their native engineering capacity such that they would be able to manufacture the material and engage in their own research program. Then, if the Morlocks developed new capacities, these now equalized engineering skills or resource access would help the Eloi keep pace. This would be especially true if regular technology transfers were part of Morlock policy.

We could adopt a similar policy with regards to RAI research. Rather than merely making information available to the developing world, the scientific community could make a robust effort to engage in scientific and engineering capability building in the developing world. Not only would this increase global ability to understand and critique existing research, it would also allow the developing world to decide for themselves if their interests would be served by RAI deployment rather than relying on the good will of scientists in the developed world. Capability building, thus, would need to go beyond educational programs but would need to include improvements in the ability of the developing world to engage – on more equal terms – with the actual research and development of reflective particles, delivery mechanisms, and monitoring tools. This, combined with open source transparency and an equitable intellectual property regime, would allow the developing world to be a RAI partner, granting them similar capabilities to manipulate the climate. Like the Soviet development of nuclear weapons, this would balance the power relations within the context of solar geoengineering. Obviously, the broader inequalities that structure our current pursuit of climate justice in particular and global justice in general would still exist, but RAI researchers would be contributing to an increase in power for developing countries.[17]

Building these capabilities would also have further knock-on effects that would decrease opportunities for domination. First, developing nations may have different priorities when it comes to RAI research. For example, perhaps they are more interested in developing particles or interventions that would *inoculate* the atmosphere from risky interventions, preventing RAIs from being deployed effectively. Perhaps there are particles that would bond with proposed reflective aersols, causing them to fall to earth more quickly and dramatically increasing the costs of the intervention. Or more optimistically, perhaps the fact that the negative consequences of RAIs will fall predominantly on the developing world[18] will lead to research that focuses on risk reduction rather than cost reduction. In other words, capability building amongst the developing world may generate research avenues, reduce risk, and generate politically and epistemically useful veto points that will make RAI science better, politically more legitimate, and the engineering more innovative and responsive.

Finally, building these capabilities among the developing world will create interesting game theoretic possibilities in the context of termination shock (Ross & Matthews, 2009).[19] Termination shock is one of the more severe risks of RAIs. The worry is that if we use RAI to reduce global temperatures without mitigating our greenhouse gas emissions, we will need to engage in geoengineering to an increasingly greater extent in order to compensate for the greenhouse effect of our carbon emissions. Yet, if we have to cease RAIs for some reason – a new understanding of the risks, political disruption, and so on – then the climate could undergo rapid warming such that no society could effectively adapt, generating catastrophic consequences. The widespread distribution of the ability to engage in RAI research and eventually, deployment, has two interesting effects on the dynamics of termination shock. First, if many groups are positioned to engage in RAI, it provides redundancy in the event that termination shock is caused by a political or economic disruption amongst the deploying agents. The second effect is based on the important insight that the termination shock scenario, like nuclear war, is unacceptably costly to everyone. Since the developed world does not wish to risk termination shock, the possibility that the *developing* world may engage in unilateral RAI could become an important bargaining chip in climate negotiations. RAI has the potential to reduce the overall climate risks the developing world faces while, at the same time, creating a future threat where parts of the world that are relatively inoculated from the immediate consequences of climate change would need to take seriously. So, a more equitable distribution of geoengineering capability could prevent disaster while at the same time allowing less powerful agents to use the threat of the disaster to equalize their bargaining position in future negotiations.

The two primary worries concerning power equalization are abuse and escalation (Gardiner, 2013). First, we might worry that the broader dissemination of the technology will encourage bad actors to use it in their own interests. Second, we might worry that multiplying the actors, in a strategic environment, might generate problematic incentives to deploy. In order to generate this sort of collective action problem, several things would need to be true. Agents would need to have preferences as to *how* RAIs were deployed and there would need to be path dependent dynamics whereby an agent deploying in one way would preclude an agent deploying in the second way. If those two features of RAIs are true, then we could have escalatory dynamics. The first agent could feel the need to sub-optimally deploy in a 'use it or lose it' scenario where they anticipate the second agent might use RAIs, thus blocking their preferred deployment scenario. In this way, RAI deployment would exhibit the same pressures as nuclear deterrence, where the possibility that one side develops an asymmetric 'first strike' technology undermines the other side's ability to retaliate forces the threatened side to escalate before their technology becomes obsolete. Thus, we might get deployment that neither side wants because they feel pressured to escalate before their deployment suite becomes obsolete.

Though RAI research is still in its infancy, there is *some* evidence that these two features obtain but we cannot be certain. It looks like deploying to different regions, with different particles, or to differing extents and altitudes could generate different environmental and social impacts. It might matter to precipitation patterns whether, for example, particles are inserted into the atmosphere at the poles rather than the equator. Different types of particles might also affect the ozone layer in different ways. What's more, the benefits of any particular strategy could dissipate if another agent deploys differently. Suppose that one agent decreases precipitation risk by a polar insertion and compensates for the additional expense by using reflective particles that are cheaper to produce yet more harmful to the ozone (Tilmes, 2017). A second actor comes along using different particles that are themselves harmless to the ozone but inserts at the equator. If both agents so act, then proposed benefits of ozone safe particles will be erased *and* the two agents will generate greater risk to the precipitation system than either would accept. And since precipitation risks seem to directly correlate with increased RAI intervention, both agents intervening would likely push the risks beyond what either agent deemed acceptable prior to deployment.

There are two potential responses to the escalation and abuse worries, given these dynamics. First, the ability to unilaterally deploy without input from relevant stakeholders characterizes the status quo. So, it seems like abuse is a severe risk at the moment without any power equalization. Second, it is important to see that the widespread capability to deploy increases the potential costs of deployment to those who do it without an international consensus. Since other agents will be able to disrupt the benefits of first deployment, then optimal RAI usage will demand coordination and those who have traditionally had little say in climate policy will have some degree of influence. Again, the goal is not to resolve all global inequalities but to ensure that particular decisions to research and deploy RAIs are made within a more equal footing. The ability of the developing world to undermine the benefits the developed world may receive from engaging in solar radiation management helps equalize their positions, though it does so in an admittedly risky way. If a developed country wishes to use RAIs in a way that maximally favors its interests at the costs of imposing risk on developing nations, then developing nations are in a position to deploy their own RAI technologies in order to change the cost-benefit calculation of the initial deployer. Thus, in order for the deploying agent to be assured that they will receive a net positive set of benefits and costs, they will need to coordinate with other agents who possess the relevant capabilities. This gives other agents greater influence and power over the eventual policy outcome than they would have had absent their RAI capability. However, since other agents can exercise similar influence, there is little advantage is going first if others can intervene.

These responses only temper but do not eliminate the risks of equalization. However, given how unlikely the accountability strategy is, it is time to think explicitly about how to manage the problems that accompany power equalization. The analogy with nuclear weapons is instructive. Centralization, either through the United Nations or an American monopoly, was either infeasible, unstable, or dominating. This forced the international system to devise ways to manage the proliferation of nuclear weapons. Of course, the features and dynamics of the Cold War and of nuclear weapons are dramatically different from the current context of climate change and solar radiation management, so we should not assume that an effective set of institutions in one case will also be appropriate in the other. Rather, the key insight of that period was that people took the lack of centralization and the relevant failure of power accountability as a given and then designed ways to limit the risks of power equalization. This included alliances structured around 'nuclear umbrellas,' emergency redlines for communication, and treaty regimes *symmetrically* limiting the relevant actors. In the context of RAIs, the sooner we realize that the current structure of unilateral power is untenable and accountability regimes are not forthcoming, the sooner we can begin to create equalization regimes that limit that strategy's risks and encourage its upsides.

Thus, we have two fundamental strategies for reducing domination in the context of RAI research. Yet, these strategies invoke opposing dynamics. The accountable power avenue demands centralization and collective decision-making while the equal power avenue demands decentralization as we build up the capabilities in a more disparate set of agents that are not subject to collective authority. We cannot easily do both strategies, just as nuclear proliferation continues to make global regulation of nuclear weapons more difficult. However, while de-centralizing capability and centralizing authority often work at cross-purposes, there are sources of hope. Perhaps the more equitable political contestation the power equalization strategy makes possible will contribute to the generation of the power accountability strategy in the long run. The point of this analysis is not to discourage the pursuit of power accountability but that avoiding some forms of serious injustice require that we think hard about how to manage the power equalization strategy. Ultimately, this is how the world barely avoided catastrophe in the case of nuclear weapons and it should not be surprising if such muddling through is required again.

Notes

1. For justice-based discussions of land use, see Buck (2016) and Shrader-Frechette (2013).
2. Henceforth RAIs, for reflective aerosol injections. Some (Svoboda, Kellar, Goes, & Tuana, 2011) focus on *sulfate aerosols*, but sulfates may not be the particles that are actually used. Some (Weisenstein, Keith, & Dykema, 2015) are considering, for example, aluminum-based aerosols. Similarly, the focus on *stratospheric* aerosol injections assumes a deployment strategy that, while very likely, is not absolutely guaranteed. Thus, I use the broader category in this paper.
3. This is due to interaction effects between the cooling consequences of an albedo increase – reflecting more sunlight into space – interacting with the warming effects of the greenhouse gases that remain in the atmosphere. See Tilmes et al. (2013).
4. As an example of this distinction, consider bio-energy, carbon capture, and storage technologies (BECCS). These technologies generate considerable environmental risks due to their extensive land and water use, but they also have the potential to block many of the worst climate impacts. However, due to how costs are distributed and the relatively small effect of any individual act of BECCS, these technologies rely on broader acceptance by

stakeholders and need societal support. It is harder for a group of agents to coordinate in order to control the environment. Thus, managing the risks of BECCS is a different problem from managing the possibilities of human abuse and domination presented by RAI.

5. Surveys of marginalized groups about geoengineering consistently show worries that move beyond the strict risk profile of solar based geoengineering (Carr & Preston, 2017).
6. This kind of control could include a greater ability to manipulate precipitation levels, injection sites, particle types and could include potentially include interventions into extreme weather events. In other words, a suite of geoengineering technologies could, at some point in the future, give agents a significant ability to manipulate the environment. In the far future, we can even imagine swarm nanotechnology that can great localized RAI effects. While we are far from these capabilities, the logic of RAI research is in the direction of greater precision. I thank an anonymous reviewer for pressing me on this point.
7. Pettit (1997), Ripstein (2009), and Iris Marion Young (1990) are all examples of the republican political tradition.
8. Morrow, "International Governance of Geoengineering: A Survey of Reports on Geoengineering Governance, 2009–2015" *FCEA Working Paper Series: 001* (2017).
9. A typical but by no means uncommon of an example of 'research first' for outdoor experimentation but essentially no restrictions on indoor experimentation due to worries about lock-in is that of Andy Parker (2014).
10. Does this sort of reasoning also apply to merely thinking about new technologies that may generate a power imbalance? After all, if modelling can increase knowledge about how to manipulate the environment, so can merely taking the time to think about, imagine, speculate, and deliberate upon geoengineering. Can this kind of activity contribute to domination I believe the answer is a deeply qualified, 'Yes.' That is, it does not strike me as obvious that some thoughts could be, in and of themselves, unjust without the appropriate political order; this shows how much we rely upon good institutions in order to act ways that are consistent with everyone's freedom. However, two significant qualifiers need to be included. First, most technological development these days is sufficiently complicated that without sophisticated computer modelling, we cannot expect much progress. So, usually, the possibilities for domination made possible by mere thought are *de minimis*.- Second, any attempt to directly restrict freedom of thought – as opposed to incentivizing other research or banning more substantial research down the line – will almost certainly be dominating in its consequences, intended or otherwise. So, we should not infer from the fact that some thoughts could lead to new possibilities for domination that state is permitted to coercively restrict those thoughts.
11. A single but influential example is Rayner et al (2013) that emphasizes the role of information sharing in establishing the legitimacy of geoengineering research.
12. Of course, others have essential knowledge as well, but scientific self-regulation is committed to the idea .
13. See Roy Baumeister and Todd Heatherton (1996) for a summary and, as a specific example, consider the failures of banks to regulate their risk exposure to subprime mortgages in *The Financial Crisis Inquiry Report (2011)* especially Chapter Four.
14. This is, for example, one reason to think that principles of justice apply to the basic structure; principles that applied directly to individuals would be burdensome and excessively demanding. See Rawls (2001, Section 15 and 16) and Cohen (2001).
15. E.g. Pettit, (1997, pp 116–118) and see Stilz (2009, pp. 67–68).
16. Of course, this was not sufficient for power between the classes to be equal, but it is an example of decreasing domination by granting additional legal powers rather than making superior power accountable.
17. One could argue that this does not help very much because the developed world will simply prevent the developing world from deploying RAIs and so capability building will be irrelevant. First, various positive benefits to the developing world's research agenda, including the focus on what would best serve their interests and how to block RAI should that be necessary, would remain untouched. Second, the developed world would need to pay

economic and reputational costs to block RAI deployment, costs that could play a role in the negotiations I describe below. Third, a developing world coalition that aimed at RAI deployment would be difficult for the developed world to block. Capability building does not change large scale economic and military inequalities, but it would reduce their effects in this particular context. I thank an anonymous reviewer for pressing me on this point.

18. E.g. precipitation disruption will be concentrated in the developing world.
19. Some SRM researchers (Reynolds, Parker, & Irvine, 2016) have argued that termination shock is not necessarily a significant problem for SRM deployment, but I do not find the arguments especially convincing. First, the paper ignores some of the scenarios – such as additional knowledge about SRM effects – that might lead to rapid termination. Second, the paper tends to make overly optimistic assumptions about the motivations of the actors and the capabilities of international civil society in order to safeguard against termination shock.

Disclosure statement

No potential conflict of interest was reported by the author.

References

Barrett, S. (2008). The incredible economics of geoengineering. *Environmental and Resource Economics*, *39*(1), 45–54.

Baumeister, R., & Heatherton, T. (1996). Self-regulation failure: An overview. *Psychological Inquiry*, *7* (1), 1–15.

Buck, H. J. (2016). Rapid scale up of negative emissions technologies: Social barriers and social implications. *Climatic Change*, *139*(2), 155–167.

Cairns, R. (2014). Climate Engineering: Issues of path dependence and socio-technical lock-in. *Wiley Interdisciplinary Reviews: Climate Change*, *5*(5), 649–661.

Caney, S. (2006). Cosmopolitan justice and institutional design: An egalitarian liberal conception of global governance. *Social Theory and Practice*, *32*(4), 725–756.

Carr, W., & Preston, C. (2017). Skewed vulnerabilities and moral corruption in global perspectives on climate engineering. *Environmental Values*, *26*(6), 757–777.

Cohen, J. (2001). Taking people as they are. *Philosophy and Public Affairs*, *30*(4), 363–386.

Crutzen, P. (2006). Albedo enhancement by stratospheric sulfur injections: A contribution to resolve a policy dilemma? *Climatic Change*, *77*, 211–220.

Dykema, J., Keith, D., Anderson, J., & Weisenstein, D. (2014). Stratospheric controlled perturbation experiment (SCoPEx): A small-scale experiment to improve understanding of the risks of solar geoengineering. *Philosophical Transactions of the Royal Society A* 372 (2031).

Gardiner, S. (2010). *Is 'arming the future' with geoengineering really the lesser evil? Some doubts about the ethics of intentionally manipulating the climate system. Climate Ethics: Essential Readings*. Oxford, UK: Oxford University Press.

Gardiner, S. (2011). *A perfect moral storm*. Oxford, UK: Oxford University Press.

Gardiner, S. (2013). The desperation argument for geoengineering. *PS: Political Science and Politics*, *46*(1), 28–33.

Graeter, K. A., Osterberg, O. C., Ferris, D. G., Hawley, R. G., Marshall, H. P., Lewis, G., & Birkel, S. D. (2018). Ice core records of west greenland and climate forcing. *Geophysical Research Letters*, *45*, 3164–3172.

Keith, D. (2013). *A case for climate engineering*. Cambridge, MA: The MIT Press.

Laborde, C. (2010). Republicanism and Global Justice: A Sketch. *European Journal of Political Theory*, *9*(1), 48–69.

McKinnon, C. (2018). "Sleep-walking into walk-in: Avoiding wrongs to future people in the governance of solar radiation management research," Environmental Politics, Online: March 2018

Morrow, D. (2017) International governance of geoengineering: A survey of reports on geoengineering governance, 2009–2015. *FCEA Working Paper Series: 001*

Parker, A. (2014). "Governing solar geoengineering research as it leaves the laboratory,". *Philosophical Transactions of the Royal Society A, 372*, 20140173.

Pettit, P. (1997). *Republicanism: A theory of freedom and government*. Princeton, NJ: Princeton University Press.

Pettit, P. (2010). A republican law of peoples. *European Journal of Political Theory, 9*(1), 70–94.

Rawls, J. (2001). *Justice as fairness: a restatement*. Cambridge, MA: Harvard University Press.

Rayner, S, Heyward, C, Kruger, T, Pidgeon, N, Redgwell, C, & Savulescu, J. (2013). "The Oxford Principles,". *Climactic Change, 121*, 499–512.

Reynolds, J., Parker, A., & Irvine, P. (2016). Five solar geoengineering tropes that have outstayed their welcome. *Earth's Future, 4*(12), 562–568.

Ripstein, R. (2009). *Force and Freedom*. Cambridge, MA: Harvard University Press.

Ross, A., & Matthews, D. (2009). Climate engineering and the risk of rapid climate change. *Environmental Research Letters, 4*(4), 045103.

Sagan, S., & Waltz, K. (1995). *The spread of nuclear weapons*. New York, NY: WW Norton.

Shrader-Frechette, K. (2013). Biomass incineration: Scientifically and ethically indefensible. In: A. Maltais & C. McKinnon (Eds.), *The ethics of climate governance*. Lanham, MD: Lexington Press, 155–172.

Stilz., A. (2009). *Liberal loyalty*. Princeton, NJ: Princeton University Press.

Strong, A., Chisholm, S., Miller, C., & Cullen, J. (2009). Ocean fertilization: Time to move on. *Nature, 461*(7262), 347–348.

Svoboda, T., Keller, K., Goes, M. P., & Tuana, N. (2011). Sulfate aersol geoengineer: The question of justice. *Public Affairs Quarterly, 25*(3), 157–179.

Tilmes, S., Fasullo, J., Lamarque, J.-F., Marsh, D. R., Mills, M., Alterskjaer, K., & Watanabe, S. (2013). The hydrological impact of geoengineering in the geoengineering model intercomparison project (GeoMIP). *Journal of Geophysical Research, 118*(19), 11036–11058.

Tilmes, S., Richter, J. H., Mills, M., Kravitz, B., MacMartin, D., Vitt, F., & Lamarque, J.-F. (2017). Sensitivity of aerosol distribution and climate response to stratospheric SO_2 injection locations. *Journal of Geophysical Research: Atmospheres, 122*(23), 12591–12615.

Waltz, K. (1981). The spread of nuclear weapons: More may be better. *Adelphi Papers* 171, 1–28.

Weisenstein, D. K., Keith, D. W., & Dykema, J. A. (2015). Solar geoengineering using solid aerosol in the stratosphere. *Atmospheric Chemistry and Physics, 15*, 11835–11859.

Young, I. M. (1990). *Justice and the politics of difference*. Princeton, NJ: Princeton University Press.

Reports

The Royal Society. (2009). *Geoengineering the climate: science, governance, and uncertainty*. London, UK: The Royal Society

Financial Crisis Inquiry Commission. (2011). *The Financial Crisis Inquiry Report*

Toward Legitimate Governance of Solar Geoengineering Research: A Role for Sub-State Actors

Sikina Jinnah, Simon Nicholson and Jane Flegal

ABSTRACT

Two recently proposed solar radiation management (SRM) experiments in the United States have highlighted the need for governance mechanisms to guide SRM research. This paper draws on the literatures on legitimacy in global governance, responsible innovation, and experimental governance to argue that public engagement is a necessary (*but not sufficient*) condition for any legitimate SRM governance regime. We then build on the orchestration literature to argue that, in the absence of federal leadership, U.S. states, such as California, New York, and other existing leaders in climate governance more broadly have an important role to play in the near-term development of SRM research governance. Specifically, we propose that one or more U.S. states should establish a new interdisciplinary advisory commission to oversee and review the governance of SRM research in their states. Centrally, we propose that state-level advisory commissions on SRM research could help build legitimacy in SRM research decisions through the inclusion of, at minimum: meaningful public engagement early in the research design process; an iterative and reflexive mechanism for learning and improving both participatory governance mechanisms and broader SRM governance goals over time; as well as mechanisms for adaptation and diffusion of governance mechanisms across jurisdictions and scales.

Introduction

It is increasingly clear that, even if fully implemented, the first round of national pledges to address climate change will not in themselves be enough to meet the temperature targets under the 2015 Paris Agreement (Chen & Xin, 2017; Rogelj et al., 2016). This reality has led to increased scholarly attention, and more recently attention from policy makers and civil society more broadly, on the possibility of developing climate engineering technologies.

Climate engineering is an umbrella term encompassing two different categories of technology – carbon dioxide removal (CDR) technologies that might remove large amounts of carbon from the atmosphere and hold it in storage, and solar radiation management (SRM) or albedo modification technologies that might dampen temperatures by reflecting some amount of incoming solar radiation back into space. We focus here on SRM technologies, and in particular on the exploratory roles that sub-state (that is, sub-national) actors ought to play in the near-term governance of SRM research. Our focus is on the SRM governance roles that might be played by states like California, Washington, and New York in the United States, since this is the country from which the best-known and most advanced proposals for open-air experimentation related to SRM are emanating. The article concludes by advocating for and sketching out design considerations to guide development of one or more sub-state 'Advisory Commissions on SRM Research,' as a reasonable and manageable near-term step that a state like California, Washington, or New York might take.

Despite their early stages of investigation, a rich body of scholarship has already emerged surrounding SRM technologies.[1] Two notable strands of this literature explore the ethical and governance dimensions of SRM development and potential deployment. We briefly, here, outline and mine this SRM-focused literature for lessons about *legitimate* governance (our chief theoretical concern in the following pages is the legitimacy of governance efforts). In the next section we turn to an examination of a number of different strands of literature from political science, international relations, and science and technology studies, for additional insights about what it takes to craft legitimate forms of governance. The paper then translates the insights from this range of complementary literatures into a concrete and actionable proposal for near-term sub-state action.

As a starting point, we note that there already exists a particularly rich literature surrounding the ethical and normative dimensions of climate engineering. Scholars have unpacked, for example, the moral issues raised by climate engineering, the normative implications of technological lock in on future generations, the conditions under which climate engineering could be considered morally permissible, and the ethical frameworks embedded in key climate engineering reports (Gardiner, 2011; Jamieson, 1996; McKinnon, in press; Preston, 2013). Particularly relevant to our work here are ethical arguments that political legitimacy in decision-making are sufficient to guide SRM research (Morrow, Kopp, & Oppenheimer, 2013). Still others have questioned whether humans should be meddling with the climate system via large-scale technological interventions at all (Hamilton, 2013), raising questions about the legitimacy of the entire enterprise, and who should be entrusted with decisions surrounding the shaping of the 'Synthetic Age' (Preston, 2018).

Daniel Callies' contribution to this special issue pushes forward these conversations in proposing a set of normative criteria, which he argues any institution ought to have if it is to legitimately coordinate action around SRM (Callies, this issue). Callies argues that any such institution governing SRM, must at a minimum confer: (1) comparative benefit; (2) accountability; (3) transparency; (4) substantive justice; and (5) procedural justice. His institution-specific, and 'necessary but not sufficient' approach to developing legitimate institutions is worth underscoring here, as it aligns with our own view. That is, although we can identify some of the necessary ingredients, institutional legitimacy cannot be derived from a recipe. Rather, on the basis of an ingredient list policy actors can begin to develop institutions, but those undertaking such work must always be reflexive and willing to pivot to the needs of specific contexts and issues as institutions evolve.

Due to the inherent transboundary impacts of these technologies if ever deployed, scholars of global governance and international law have also been active in advancing insights into and proposals for the governance of SRM. Several have put forth specific governance proposals at the international and regional levels (e.g. Armeni & Redgwell, 2015; Hubert, 2017; Lloyd & Oppenheimer, 2014; Nicholson, Jinnah, & Gillespie, 2018; Parson & Ernst, 2013; Reynolds, Jorge, Contreras, & Sarnoff, 2018; Sugiyama et al., 2017), debated various framings and tradeoffs between climate engineering and traditional mitigation and adaptation measures (Horton, 2015; Reynolds, 2015), and argued that expert bodies are already playing an important role in 'de facto' governance across scales (Gupta & Moller, in press). In response to calls for more engagement by scholars of international relations (Horton & Reynolds, 2016), recent work has also explored, for example, how international relations theory might contribute to governance design (Jinnah, 2018) and the collective action implications of geoengineering (Sandler, 2018).

With a few notable exceptions (e.g. BPC, 2011; Craik, 2017; Craik, Blackstock, & Hubert, 2013; Winickoff & Brown, 2013; Mahajan, Tingley, & Wagner, 2018; US GAO, 2010; Stilgoe, Owen, & Macnaghten, 2013), the role of domestic politics and institutions for SRM governance has received comparatively little attention. This is surprising given the broader devolutionary trend in the climate governance literature with, for example, the increased attention to city-level governance (e.g. Gordon, 2013), and the fact that some federal government agencies and departments in a range of countries have begun to fund small amounts of SRM research, though the bulk of this research consists of indoor modeling studies and social scientific research.

We aim in this paper to fill this gap by outlining a particular role for sub-state actors in laying the initial governance foundation for SRM research. Although research governance may be necessary at the global scale as well, governance at the sub-state level is more tractable in the near-term given the nascent nature of formal climate engineering governance, its contentious character among many publics, and the proactive role that these actors are already taking on climate change issues, including as related to climate engineering. Moreover, privately-funded small-scale outdoor experiments are currently planned/being planned in U.S. states, often without triggering national or international oversight. Early attempts at governing research at sub-state scales can serve as sites of experimentation and contribute to collective social learning, if designed carefully, and with this goal in mind (Stilgoe, 2016).

Our other concern in this paper is with the legitimacy of governance efforts. Joining other governance scholars who are centrally interested in the intersection between issues of equity and governance in this space (e.g. Burns & Flegal, 2015; Flegal & Gupta, 2018; Svoboda, Buck, & Suarez, 2018), this paper explores how various strands of governance theory can inform the design of legitimate political institutions. To be clear, developing legitimate political institutions cannot be achieved by applying some predetermined formula. Nor are the prescriptions we suggest here alone sufficient to develop legitimate political institutions. However, in looking across several strands of governance theory, there are certain practices, centrally, participatory ones, that can *contribute to* the development of legitimate political institutions. We are not so naive to think that any participatory process could or would be successful in this regard. To the contrary, participatory processes done poorly can, in fact, undermine efforts to develop political institutions that enjoy high degrees of legitimacy. As such, this paper draws from governance theory to propose some initial steps towards development of one political institution that could, if designed carefully, begin to build norms and mechanisms for broad stakeholder engagement in SRM research governance.

Specifically, building on previous work, this paper makes a policy recommendation to advance a slightly revised version of one of Nicholson et al.'s SRM governance objectives. Namely, we are proposing here policy mechanisms for soliciting active and informed public and expert community engagement with a view towards improving the quality of research conducted, especially its responsiveness to societal goals and norms (2018).[2] This objective is rooted in the literature on the role of public engagement in climate policy, including – although not exclusively – for SRM (see e.g. Bellamy, Chilvers, & Vaughan, 2016; Corner, Parkhill, Pidgeon, & Vaughan, 2013; Macnaghten & Szerszynski, 2013; Pidgeon, Parkhill, Corner, & Vaughan, 2013; Winickoff, Flegal, & Asrat, 2015).

The rationales for public engagement are not always explicit, but have been described in three categories: instrumental, normative, and substantive (Fiorino, 1990). The instrumental rationale is principally about staving off public resistance, while the normative rationale holds that democratic ideals require that potentially affected parties have a say in scientific decision-making, in accordance with principles of transparency, informed consent, and political legitimacy (NASEM, 2016). The substantive rationale holds that experts do not have a monopoly on expertise and concerns relevant to research, and therefore broader participation can enhance the quality – and legitimacy – of science itself. To the extent that forms of public engagement can open up expert discourse, these efforts can help to facilitate 'honest brokering' (Pielke, 2007) of policy options and ensure that parties with potentially quite different worldviews have an opportunity to be heard and responded to by each other (Heyward & Rayner in Heazle & Kane, 2015). Under some circumstances, the interplay of divergent perspectives enabled by robust public engagement can lead to a kind of 'distributed technology assessment' and/or social learning (Rayner, 2004) in ways that may contribute to the perceived legitimacy of institutions; as we note above, however, in poor circumstances it can lead to social conflicts, misallocation of resources, and/or the scientization of politics (Rayner, 2004).

To summarize, the goal here, then, is to develop a concrete recommendation for how sub-state actors can promote meaningful public engagement in order to construct legitimate SRM governance. The remainder of the paper proceeds as follows. The next section provides a theoretical foundation for our recommendations on participatory SRM governance. Picking up on the brief introductions to these topics above, we detail how insights from the literatures on legitimacy in global governance, responsible innovation, and experimental governance support central elements of our subsequent policy recommendation that U.S. states should spearhead experimental governance initiatives that are rooted in meaningful public engagement. We then turn to the recommendations themselves. Centrally, we propose that U.S. states establish one or more expert advisory bodies to orchestrate a set of experimental research governance objectives that are deeply rooted in meaningful public engagement surrounding SRM research governance goals. Our conclusions and recommendations reinforce the rationale for such a Commission and point again to the unique leadership role that U.S. states can play in this increasingly salient arena.

Theoretical Foundations for Legitimate Governance of SRM Research

In this section, our focus is on the legitimacy of SRM governance efforts. We look across a number of different strands of governance literature to glean insights about the links between meaningful public engagement and the legitimate governance of something as complex and contentious as SRM research. Although we focus on governance of SRM research, we recognize that other science policy questions – including around mitigation efforts – might also benefit from greater attention to the construction of legitimacy and public engagement. Moreover, we recognize the somewhat artificial separation between research and deployment in the context of SRM. The need for governance of the former is rooted in the risks and uncertainties surrounding potentiality for the latter. We focus on near-term governance of research because activities in the early stages of technological exploration can help to define and enable socially acceptable research, constrain or halt potentially dangerous or undesirable research, begin establishing institutions to manage these technologies *if* they are ever deployed, and to shape the trajectories of technological development while these technologies are inchoate and therefore susceptible to social steering. Our goal here is to derive a set of design considerations from the broader governance literature that can guide sub-state actors in moving forward with SRM research governance.

To be clear, we do not mean 'legitimate' in the narrow sense that public engagement or the development of a governance architecture are a means to rubber-stamp a research effort. Rather, as we explain in some detail below, legitimate research requires, among other things, opportunities for public engagement and feedback that could, ultimately, hold a particular research pathway to be *illegitimate* or undesirable in some way. That is, in advocating for research governance, we are including the possibility that the mechanisms of governance might themselves result in a closing off of all SRM research or particular avenues for SRM research.

Therefore, we draw insights here from various strands of the governance literature to inform the development of SRM research governance. That is, we review those literatures to inform the design elements of our proposed governance framework that is rooted in principles of transparency, participation, and learning over time. We demonstrate through this review how this multidisciplinary governance literature supports our assertion that there are essential baseline conditions for any legitimate SRM research governance scheme. Specifically, we discuss below the literatures on legitimacy in global governance, responsible innovation, and experimental governance in order to theoretically frame our core policy recommendation in the subsequent section. We further draw from the orchestration literature to explain how sub-state actors, such as U.S. states, can coordinate governance in this space in the near-term while their official political positions on SRM are still undetermined. We argue for an experimental approach to SRM research governance that takes seriously public engagement, social learning, and is orchestrated, in the near-term, by U.S. states. As highlighted by the 2018 Intergovernmental Panel on Climate Change's (IPCC) special report on global warming of 1.5 degrees, much of the literature on SRM governance appears as 'commentaries, policy briefs, viewpoints and opinions' (2018, p. 52). As in our previous work (Nicholson et al., 2018; Chhetri et al., 2018), we seek here to delve into the governance literature to develop a theoretically-derived policy proposal for SRM research governance.

The global governance literature on *political* legitimacy is instructive in constructing a foundation for SRM research governance. Distilling from this literature a definition that is relevant to the SRM research governance context, we can say the following: political legitimacy refers to the acceptance of a governance institution by its stakeholders who are willing to defer to the decisions of that body in lieu of making their own decisions on specific matters (Bernstein & Cashore, 2007; Bodansky, 1999; Cashore, 2002; Esty, 2006). The conditions under which political actors would be willing to defer in this way are directly related to elements of process in institutional design, and are a central topic of analysis in the literature on political legitimacy.

For example, Bernstein and Cashore (2007) analysis of non-state market driven governance schemes offers some important insights into how elements of institutional design can contribute to (not guarantee) the development of politically legitimate institutions. Importantly, they demonstrate that political legitimacy is often something that institutions work towards rather than something that is inherent in their initial design. They highlight the importance of shared norms, such as transparency and democratic participation, in providing the foundation of political legitimacy of governance systems, but underscore that norms continue to evolve in politically legitimate systems. Specifically, they argue that 'what is "good" is often precisely the issue to be worked out within politically legitimate arenas' (p. 355). That is, built-in processes for learning within governance institutions is central to developing legitimacy over time. This can be done through, for example, forums/institutions for expert discussion, which allow for debate, critique, and learning, and building of shared databases of experience (Bernstein and Cashore, 2007, p. 362–3).

In keeping with a point that we underscored in the opening section above, Bernstein has further argued elsewhere that specific determinants of political legitimacy cannot be developed *a priori* from a checklist (2011). Rather, he argues that stakeholders themselves must identify the specific determinants of legitimacy through interactions over time (2011). Interestingly, Bernstein further notes that understanding the construction of legitimacy frameworks in this way further helps to explain why, in some cases, legitimacy determinants for some institutions can ultimately be dysfunctional for achieving the institution's core purpose (p. 43). This suggests that in designing an SRM research governance framework that strives for political legitimacy, compromises may have to be made between legitimacy and the speed with which SRM technological development proceeds, at least in the near-term. We come back to this point and discuss implications for our core recommendation in the concluding section.

The literature on science and technology studies (STS) also provides guidance in the quest to move toward legitimate experimentation and governance of SRM. First, STS research suggests that legitimacy is itself a contested concept, and that experts do not have a monopoly on its definition. Rather than orient research toward pre-defined normative goals, STS-inflected frameworks, such as responsible research and innovation (RRI), anticipatory governance, real-time technology assessment, and constructive technology assessment, seek to open up the question of what is good/desirable for broader democratic negotiation and debate, alongside questions such as, 'what is the purpose [of research]; who will be hurt; who benefits; and how can we know?' (Jasanoff, 2003, p. 240) As Jack Stilgoe points out, viewing physical science or engineering experiments themselves as socially constructed and determined, at least in part, 'challenges the attempt to hermetically seal [them] from public scrutiny' (Stilgoe, 2016). In its focus on anticipation, inclusion, reflexivity, and responsiveness, RRI underscores the importance of 'experimenting with experimentation,' considering who should be involved in the definition and conduct of experiments themselves.[3]

As with the global governance literature on political legitimacy, STS literature also underscores that participation through public engagement is an important, although insufficient, element in legitimate SRM research. The substantive rationale for public engagement described earlier (which holds that experts do not have a monopoly on the kinds of expertise and concerns relevant to research, and therefore broader participation can enhance the quality of science itself), although not explicitly discussed in terms of political legitimacy above, comports with Bernstein's (2011) argument that stakeholders themselves must define the determinants of political legitimacy in any given context, and Bernstein and Cashore's call for forums for presentation, discussion, and *critique* of expert's proposals/plans/etc. by broader stakeholders. Far from a panacea for the legitimacy of science and technology, public engagement has itself been critiqued in STS literature, both with regard to inputs (Lovbrand et al. point out that engagements are often only seen legitimate for those directly involved in them) (2011, p. 483); and outputs: engagements sometimes reinforce existing power structures, and do not meaningfully impact governance (Scharpf, 1999; Van Oudheusden, 2011).

While public participation can be pursued in light of concerns about the legitimacy of decision-making, it is important to underscore that the act of engagement does not necessarily confer legitimacy on decision-making in any straightforward way. A particularly important challenge, which we take up directly in our recommendation below, relates to making policy decisions responsive to the outputs of public engagement efforts. There is some risk that weak public engagements do not facilitate true deliberation, and instead serve to legitimate existing policies (Rayner, 2003). Furthermore, these mechanisms of engagement are most likely to be impactful when technologies are further 'upstream,' or before they are locked-in (Collingridge, 1980). This means that, while especially effective in cases of emerging technologies, such as SRM, public engagements should ideally be initiated in the early stages of research, including in the setting of goals and priorities for research itself.

The literature on experimental governance provides some further and highly complementary insights into elements of a legitimate governance scheme for SRM research. Experimental governance is a 'recursive process of provisional goal-setting and revision based on learning from comparison of alternative approaches to advancing these goals in different contexts' (Overdevest & Zeitlin, 2014, p. 25). Although the idea of experimenting with governance in this area may raise concerns for some, experimental governance need not take sweeping, irreversible governance decisions, and need not presume that all research should face additional governance from the top down (Stilgoe, 2016). Rather, experimental governance can proceed in an incremental way, with built-in processes for feedback and adjustment. As such, experimental governance has gained much attention in the global environmental governance literature, particularly surrounding issues related to complex environmental problem-solving that require innovation beyond tried and tested models, such as climate change (e.g. Castán Broto & Bulkeley, 2013a, 2013b; Hoffmann, 2011).[4] This is in part because, as with RRI described above, its reflexive and adaptive approach allows for suppleness, including in response to unexpected policy options as they arise (Heilmann, Shih, & Hofem, 2017). Experimental governance further provides opportunities for SRM governance to be more publicly-accountable, in that central to any experimental governance architectures are broad participatory processes for goal-setting (Overdevest & Zeitlin, 2014).

Central, therefore, to experimental governance are processes for learning and improvement over time on the basis of multiple experiments and broad participation (McFadgen & Huitema, 2018; Overdevest & Zeitlin, 2014). Importantly, this learning process could include pivoting to the possibility that SRM research should halt. As Castán Broto and Bulkeley have argued in their studies of urban climate politics, governance experiments allow for exploration of 'unchartered policy territories to either learn or open up new forms of intervention ... [Such experimental governance] interventions... try out new ideas and methods in the context of future uncertainties. They serve to understand how interventions work in practice, in new contexts where they are thought of as innovative' (Castán Broto & Bulkeley, 2013a, p. 1953, 2013b, p. 93). They go on to argue that such experiments can establish new forms of political space that blur public and private authority and, of particular relevance to SRM governance, have the potential to challenge dominant understandings of political response to climate change. This flexible and adaptive approach is particularly well suited to public engagement in SRM research because it allows for policy learning within the nascent and highly uncertain empirical terrain of SRM research and its impacts.

An experimental governance approach could further lead to the development of participatory models of governance design that could be adapted and diffused across jurisdictions. By building in processes for learning and information exchange, experimental governance can aid in diffusion by developing networks across jurisdictions (Hildén, Jordan, & Huitema, 2017). For example, in an experimental governance framework, local peer jurisdictions share information and take corrective measures to better meet centrally defined goals, which themselves are provisional and updated over time based on review of experience on-the-ground (Overdevest & Zeitlin, 2014; Sabel & Zeitlin, 2012). Such review is a critical element of the experimental governance process (Jordan, Huitema, Schoenefeld, Van Asselt, & Forster, 2018). Through this process, experimental governance initiatives can diffuse both horizontally across peer jurisdictions and vertically to other levels of governance (Hildén et al., 2017).

Finally, there is a question of agency here. Who should be responsible for initiating and stewarding public engagement, learning, and review processes surrounding experimental SRM research governance? What role should that actor(s) play? How should that role change over time? The literature on orchestration lends some insights here. Orchestration is a mode of governance in which one actor (the orchestrator) enlists one or more intermediary actors to govern a third actor or set of actors (the targets) in line with the orchestrator's goals (Abbott et al., 2015). The concept has gained much attention in the global governance literature, particularly surrounding transnational climate governance (Abbott & Hale, 2014; Chan, Brandi, & Bauer, 2016; Chan & Pauw, 2014; Gordon & Johnson, 2017; Hale & Rogers, 2014), and is increasingly recognized as a key mode of global governance (Bäckstrand & Kuyper, 2017). Although initially focused on the orchestrating role played by IGOs and states (Abbott & Snidal, 2010; Hale & Rogers, 2014), more recent works have broadened this to include non-state actors, networks, corporations, foundations, and non-governmental organizations (Chan & Pauw, 2014; Gordon & Johnson, 2017).

We contend here that orchestration can further offer analytical leverage in considering how sub-state actors, such as California or New York in the U.S. context, can engage in SRM research governance by creating intermediaries (e.g. the advisory commission detailed below) to launch and oversee experimental research initiatives. This approach is particularly useful in the case of solar climate engineering because the scientific uncertainties and complex ethical issues surrounding it has led to much political controversy. This controversy makes it incredibly risky for any political actors to take ownership of the issue for fear of electoral or other forms of political backlash. In charging other actors with constructing voluntary near-term governance structures, an orchestrating role can provide political cover for elected officials to catalyze governance at an early stage without taking a big political risk. Centrally therefore, we propose below that one or more U.S. states should create new interdisciplinary advisory commissions on SRM research to serve as a chief intermediary for this work.

U.S. states are well suited to this role because several of them possess the key characteristics necessary to orchestrate SRM research governance. Namely, in order to orchestrate, an actor must possess convening power, moral legitimacy,[5] financial support, technical expertise, the ability to serve as a focal point (Abbott et al., 2015; Hale & Rogers, 2014), and the availability of or capacity to create appropriate intermediaries and sufficient resources to enlist, support, and steer intermediaries (Abbott & Bernstein, 2015). In addition to having the political, financial, and logistical means to create a new advisory commission, states such as California and New York can also provide the material and/or ideational support to facilitate the intermediary's work towards specific governance goals. Material support can strengthen the intermediary's operational capacity, whereas ideational support can include guidance, formal approval, or political endorsement to enhance the intermediary's effectiveness and legitimacy vis-à-vis private, non-state or state targets (Abbott et al., 2015).

Sub-national political bodies in the U.S. have played this kind of orchestration role in relation to other arenas of scientific and technological development, particularly in instances when research or the implications of research have broader social implications. Illustrative examples of issues and mechanisms include investigation by the interstate Delaware River Basin Commission (among other bodies) into fracking (Davis & Hoffer, 2012); a range of state-level initiatives to examine and regulate research into and the use of products resulting from genetic engineering and biotechnology (Vito, 2018); and the establishment of broad-based commissions and councils empowered to provide independent assessment of issues stemming from scientific and technological developments affecting a given state (see, for instance, the California Council on Science and Technology).

While a substantial literature has explored how and why orchestration takes place, the normative dimensions (e.g. fair and just governance), power relationships, and the effectiveness of orchestration in attaining the intended goals remain underexplored. One exception is Bäckstrand and Kuyper (2017) recent proposal for a 'democratic values approach' to evaluating orchestration. Dovetailing with the literatures on political legitimacy in global governance, RRI, and experimental governance discussed above, this approach argues for participation, deliberation, accountability, and transparency as key metrics for evaluating, and in our case, designing and maintaining an experimental governance architecture for SRM research.

In summary, the global governance literature offers several insights into how sub-state actors might begin to build a legitimate governance framework for SRM research. Namely, this literature suggests that:

(1) Near-term trade-offs between legitimacy and speed of technological development may be necessary.
(2) Evolving and responsive governance processes that allow for institutional learning over time based on public input are critical.
(3) The public – and/or its representatives – should have early opportunities to critique and contribute to development and broad assessment of research proposals in order to shape innovation in socially acceptable directions, including in the setting of goals and priorities for research itself.
(4) Controlled experimental governance approaches to participatory engagement can be helpful in developing context-specific governance mechanisms for engagement surrounding SRM research over time.
(5) Information sharing across jurisdictions about experimental governance approaches will be crucial to learning and adapting governance institutions over time. Such information sharing should allow for the taking of corrective measures to better meet centrally defined governance goals, which themselves are provisional and updated over time based on review of experience on-the-ground.
(6) In the absence of clear political positions on SRM research from governments, sub-state actors can orchestrate participatory processes as a way to catalyze governance in this area.

Recommendation: One or More U.S. States Should Establish an Interdisciplinary Advisory Commission on SRM Research Governance

In this section we build from prior work by others to sketch a concrete proposal for an 'Advisory Commission on SRM Research.' We build this proposal on theoretically-derived insights from the governance literatures examined in the prior section. We apply those theoretically-derived insights to the particular empirical and political contexts surrounding proposed SRM experiments and emerging leadership in climate governance more broadly. Development of one or more advisory commissions would enable U.S. states such as California or New York to take an orchestrating role in SRM research governance, with an eye toward facilitating early public engagement on questions of research design and governance. Specifically, we suggest here that states should create a new state-level advisory commission to serve as an intermediary (i.e. overseen by but not necessarily representing the state) in catalyzing and shaping the development of experimental SRM research governance in the state.

Our intent here is to sketch a set of design characteristics and features of a legitimate, functioning, and workable sub-state commission. We do so by building on previous proposals for a similar federal-level body (e.g. BPC, 2011; Winickoff & Brown, 2013) and the six core insights from the literature delineated above.

As a starting point, we note that various commentators and experts have put forward substantive standards for the governance of research, including standards for the kinds of research that ought to move forward (or not) (e.g. Parson & Keith, 2013). However, as Winickoff and Brown argue, the articulation of these standards only addresses one piece of the problem, because it does not address procedural questions around the need for standards themselves to 'emerge from trusted institutions and a transparent process' (2013, p. 80). Addressing procedural questions is the first critical role that an advisory commission might play.

In terms of outlining design considerations, there are a variety of prior efforts on which to draw. Advisory commissions have been established in a variety of issue areas, such as bioethics and neuroethics, to provide information, analysis, and recommendations to policymakers. In their proposal for a federal-level advisory commission for geoengineering, Winickoff and Brown (2013) look to such prior efforts to outline several necessary characteristics to ensure the effectiveness and legitimacy of any advisory commission developed to look specifically at SRM. Specifically, Winickoff and Brown (2013) argue for a commission that is independent, transparent, deliberative, publicly engaged, and broadly framed. They further suggest that a Commission's membership be: interdisciplinary, including natural/physical and social scientists; politically diverse, including 'experience-based experts' from, for example civil society organizations and the private sector; and, critically in our view, also including representatives from potentially affected communities (p. 83). We endorse those broad and general characteristics here as well, noting that when it comes to 'affected communities,' lines will have to be drawn between those potentially impacted in the near-term by small-scale research (e.g. those who are close to experimental sites) and those who would be impacted should SRM technologies ever be deployed at scale (i.e. implying regional or global-scale impacts). The question of who draws those lines and how is also a matter of concern from the perspective of legitimacy; advisory commissions should consider these issues and provide advice.

Advisory commissions also typically establish explicit avenues for public input, either through the hosting of public meetings or the seeking of written commentary on aspects of the commission's work. Here, the global governance literatures reviewed above suggest a need for special care. This is because exercises designed to elicit public input too often become rote exchanges. Given the broad social and political ramifications of SRM and its status as an imaginary technology at this stage, experts – including ethical experts – do not have a monopoly on the kinds of expertise or concerns relevant to its governance. In order to maintain political legitimacy, then, a Commission on SRM Research must move beyond thinking in terms of public input and the provision of consensus policy recommendations, and instead work to generate meaningful public *engagement* to clarify value disputes and political disagreements, enabling compromise and honest brokering of policy choices. Indeed, meaningful engagement is a necessary (but not sufficient) design element to ensure any future SRM research governance regime enjoys some degree of political legitimacy. Moreover, advisory commissions should be designed to enable social learning around efforts to govern SRM research (flexibility, adaptability are features that might help enable this).

Research in other emerging technologies has shown that wider publics often raise issues related to innovation missed by experts.[6] Public engagements should also focus on the kinds of governance architectures that might be appropriate for various kinds of research. This addresses the gap in existing approaches to governing SRM experimentation, which seek to develop substantive standards and/or thresholds for research based on criteria developed in isolation of public involvement (e.g. technically-defined thresholds or 'allowed zones' for experimentation). Engagement work might also overlap with the anticipatory functions of the Commission, including via the involvement of lay publics in foresight exercises and scenario-building. Centrally, meaningful public engagement can take many forms, and should be integrated across phases of the research process.

The Commission should ideally operate at the level of research programs, rather than at the level of specific proposed experiments. This means that the public should be involved in *early* stage discussions about ethics and other social issues that may arise from SRM research. At the same time, public input could prove valuable for individual researchers and experimental efforts, by helping to sharpen research questions and revealing possible risks and concerns that might have been overlooked by principal investigators. Similarly, there are a variety of established methods that can be employed to this end, for example, focus groups, deliberative mapping, citizen panels, and scenario planning (see Stilgoe et al., 2013).

Critically, there must be a mechanism for outputs of such public engagement processes to feed back into governance design, ideally across scales and jurisdictions. Here, again, insights from both experimental governance models and RRI underscore the importance of reflexivity and learning. The Commission should play a central role in developing such processes, through for example, 'peer review' of public engagement work across jurisdictions (Overdevest & Zeitlin, 2014, p. 25), and asking localities to report back to the Commission on experience with engagement processes identified above, and by creating pathways for such experiences to feed directly back to policy makers through, for example, formal recommendations from the Commission. Public engagement outputs should also feed into less formal governance mechanisms such as refinement of existing codes of conduct and/or through ties to public funding decisions, which conditionally link responsiveness of project design to outputs of public engagement processes. In addition, the outputs of the engagement work should meaningfully affect the advice the Commission gives to researchers and research funders.

Finally, experimental governance models can provide for diffusion across jurisdictions. As suggested in the RRI literature, governance diffusion can indeed be dangerous if done in a 'one size fits all' manner. However, this tension can be managed if procedures for adapting governance models for specific localities are a baked in and expected part of the experimental governance process. States are particularly well suited to orchestrate diffusion due to existing institutional mechanisms, such as the Global Climate Leadership Memorandum of Understanding ('Under2MOU') and the C40, which provide policy frameworks to facilitate cooperation between sub-national jurisdictions across the globe, including as related to public outreach and transparency. Specifically, the Under2MOU can serve as an intermediary to organize communication and learning between the Commission and its peer institutions in other jurisdictions.

The Commission should adopt the procedures outlined above across its work program on SRM research governance. In addition, building on the Bipartisan Policy Center's Report on climate engineering governance, the Commission on SRM Research Governance should undertake, at a minimum, the following three core functions:

1. Identify key research questions and capacities

An important early task for the Commission would be an authoritative review of existing SRM research, from modeling to possible open-air experimentation, with an eye to existing or planned research efforts in the relevant jurisdiction. The goal should be the establishment of a detailed understanding of the existing research landscape and, to the degree possible, the establishment of a set of near- and longer-term research priorities.

SRM technologies, should they ever move from the laboratory into the world, will likely function as only one part of a portfolio of climate change response options. SRM research, this is to say, is best seen as science in the service of important social and environmental aims, rather than as science determined and defined ultimately by the proclivities of individual scientists or investors (Long, 2017). This view suggests the need to ensure that research is multidisciplinary, and/or that it is integrated with stakeholders/users, so that research is responsive to evolving social needs and norms. The Commission will need to build from existing research reviews, most notably those carried out by the National Academies of Sciences (see National Research Council, 2015a, 2015b), direct inputs from relevant scientific communities, and broad and deep public engagement to establish research needs.

In assessing research questions and determining priorities, the Commission will need to pay attention to the trade-offs between legitimacy and efficiency of technological development noted above. That is, meaningful public participation will necessarily take time. However, if done well this added time will not only add legitimacy to the process of policy development, but will also yield stronger policy outcomes in future. As such, the trade-off is a short term one, with public participation yielding efficiency benefits in the longer-term by way of enhanced problem solving. A forward-reaching research agenda that appears intent on bringing SRM technologies quickly into the world would likely generate public backlash and could lock-in undesirable research pathways. On the other hand, too restrictive a research agenda could make the effort appear legitimate in the eyes of certain stakeholders, but could unduly shackle the activities of researchers.

This is all to say that it will be an additional and important early step to have a full accounting of the state's existing SRM-relevant research capacity and efforts and to establish mechanisms for information sharing across jurisdictions. The Commission will need to identify individuals and research programs that are already undertaking research that is relevant to an understanding of climate engineering and its impacts, along with identification of individuals and research programs that are in a strong position to contribute to the development of such knowledge. This should include identification of key individuals, programs, and perspectives in other jurisdictions, to increase flows of information and learning.

2. Advise state governments on social and ethical issues that may arise from research

Another function of the Commission will be to offer advice on how best to navigate the social and ethical issues that arise from SRM research and potential deployment scenarios, potentially building on substantive ethical frameworks put forward in the literature (Rayner et al., 2013; Bipartisan Policy Center, 2011; Asilomar 2, 2010). This suggests two things. First, the makeup of the Commission must take explicit account of the need for a range of different kinds of expertise and viewpoints. In addition to those with a strong grasp of relevant aspects of physical science, the Commission should have membership that includes ethicists and religious scholars, representatives from non-governmental organizations, along with members versed in social scientific research and policy engagement, and perhaps lay membership. Second, here, too, there is explicit need not just for public input but for public engagement, to spur deep social learning and deliberation on SRM's ethical dimensions. The learning and diffusion mechanisms described above are easily transferrable to this function as well.

Importantly, in the early stages of SRM governance, this function of the Commission should be seen as advisory and evaluative rather than regulatory. The SRM conversation and research into possible technologies is at too early a stage to make big decisions on formal control over research efforts. The evolving and responsive governance frameworks argued for above will be fostered by information gathering and sharing and the daylighting of activities, coupled with careful consideration of the social and ethical dimensions of present-day and possible future research decisions.

3. Recommend criteria for research oversight and the apportionment of state funding for CE research

Research on SRM is not simply normal scientific investigation, since the technologies to which such research could lead have enormous, world-shaping implications, and are characterized by deep uncertainty and high decision stakes. By this, we mean two things. First, decisions about SRM research taken or supported by sub-state actors could ripple into the global climate change response conversation. If a major US state, for instance, were to signal support for rapid development of SRM deployment capacities, this could force a reevaluation of international commitments to climate change mitigation activities. The political signals sent by research prioritization and funding decisions is, then, an important consideration. Second, if particular SRM technologies were to be shown through a research program to be viable or, by contrast, to be too risky to support, this too would shape understandings of how best to respond to climate change.

Given all of this, any decisions about funding are immensely important. In addition to establishing research priorities, it might be supposed that a Commission could make recommendations about directing public research dollars, or a coordinating and daylighting role for private research dollars by contributing to ongoing efforts to establish clearinghouse and other transparency mechanisms (see e.g. Turkaly, Nicholson, Livingston, & Thompson, 2017). A key additional feature of this work will be the determination and inculcation of norms and rules that would be relevant to publicly funded research. This suggests that any Commission should, working with other relevant bodies, be tasked with determining appropriate forms of research oversight, either through voluntary means (e.g. requesting that researchers adhere to a voluntary established code of conduct) or formal means (e.g. requiring particular forms of reporting and impact assessment of research related to SRM, or requiring that researchers adhere to a code of conduct). This final point relates directly to the insights from the orchestration literature unpacked above. Sub-state actors are in a prime position to develop the guidelines and guardrails for SRM research, and to disseminate rules and norms across jurisdictions.

Conclusions

Governance questions surrounding climate engineering are becoming increasingly salient. This is driven by the sense that traditional forms of mitigation will be insufficient to avoid breaching the 1.5 C threshold of atmospheric warming set out in the Paris Agreement, and by a number of proposed small-scale field experiments in the U.S. In parallel, sub- and non-state actors have demonstrated leadership in climate change response through a bevy of creative and largely experimental climate governance response measures. In light of these developments, we urge states to play a role in the experimental governance of SRM research.

Nevertheless, engagement from U.S. states on research governance is unlikely to be straightforward. Why should states take the lead on this issue when they can demonstrate climate leadership in other less controversial ways? One central reason they may wish to do so is because field experiments are already proposed and may soon be underway in various states. These states can either reactively address SRM experiments and public responses after the fact, or they can do so proactively in an effort to steer the responsible development of SRM research and its governance. There is a non-negligible risk, of course, that efforts to develop research governance at the state level will be unable to contain political backlash or opposition to technology development and associated governance institutions. This could be viewed as a kind of informal technology assessment (Rayner, 2004).

Another key consideration if U.S. states consider developing advisory commissions for SRM research governance is the potential for competition between parallel commissions in various states. To the extent that advisory commissions across states make different recommendations about research governance needs, there is some risk that researchers simply propose experiments in the most permissive states. This issue has been raised in the context of experimentation with driverless vehicles in the U.S., for example (Stilgoe, 2017). Our recommendations here do not preclude governments and political decision-making bodies at a range of scales from governing SRM research, and we do not imply that advisory commissions at sub-national levels ought to substitute for national or even international approaches to governing SRM research. Rather, our more modest ambition is to advise state decision-makers in jurisdictions where experiments have been proposed, in order to enable collective learning and harmonization of research governance approaches over time – precisely the things that might help avoid regulatory competition, at least within a given political culture.

We have suggested above the development of an Advisory Commission on SRM Research as a valuable first step. The design of such a Commission deserves careful consideration. Drawing from several important strands of governance literature, we proposed that sub-state actors, such as California or New York, adopt an experimental governance model to inform and guide the Commission's work. By this, we mean that the Commission's governance recommendations and activities should, by design, emerge in an iterative and reflexive fashion. A key piece of this will be the development of a Commission that has not just the seeking of public input but rather the fashioning of ongoing and deep public *engagement* as one of its chief guiding principles. Such engagement should involve publics early in the research process, in everything from the construction and framing of research questions, to the dissemination of results and reflection on the meaning of those results for the future of SRM research in the state.

Notes

1. Given their centrality to the Paris Agreement's net zero emissions mandate, the literature surrounding carbon dioxide removal is also quickly growing (e.g. Burns & Nicholson, 2017). Given the divergent political implications and thus governance frameworks demanded by CDR technologies we do not address them directly here. Nonetheless, the core governance proposal detailed below could be adapted to include CDR technologies as well.
2. Nicholson et al. (2018) derive, from a review of governance literatures, four objectives to guide the development of SRM governance: *Objective #1*: Guard against potential risks and harms; *Objective #2*: Enable appropriate research and development of scientific knowledge; *Objective #3*: Legitimize any future research or policy-making through active and informed public and expert community engagement and the development of democratic systems of control; and *Objective #4*: Ensure that SRM is considered only as a part of a broader portfolio of responses to climate change.
3. For an illustrative table of potential techniques and approaches that correspond to these dimensions, see (Stilgoe et al., 2013).
4. Experimental governance is closely related to what Bryan Norton and others have termed adaptive management, which also focuses on experimentalism, multi-scalar analysis, and place sensitivity (Norton, 2005, p. 92). However, whereas adaptive management is by Norton's definition used to make decisions impacting the environment specifically, experimental governance is largely used to discuss governance decisions that impact social systems specifically.
5. Moral legitimacy refers to the recognized authority by its polity to govern in a particular space.
6. For example, a 2007 study on public views of nanotechnology found that the public was significantly concerned about the possibility that the technologies might eliminate jobs, while this was of little concern to nanoscientists. (Scheufele et al., 2007).

Acknowledgments

We are grateful for comments on earlier versions of this manuscript from Steve Gardiner, Catriona McKinnon, Augustine Fragniére, Ted Parson, and an anonymous reviewer. We are also grateful to Zach Dove for this research assistance. This paper was in part supported by a fellowship from the Andrew Carnegie Corporation of New York. All content remains the responsibility of the authors.

Disclosure statement

No potential conflict of interest was reported by the authors.

ORCID

Sikina Jinnah http://orcid.org/0000-0003-4528-3000

References

Abbott, K. W., & Bernstein, S. (2015). The high-level political forum on sustainable development: Orchestration by default and design. *Global Policy*, *6*(3), 222–233.

Abbott, K. W., Genschel, P., Snidal, D., & Zangl, B. (2015). *International organizations as orchestrators*. Cambridge, UK: Cambridge University Press.

Abbott, K. W., & Hale, T. (2014). Orchestrating global solutions networks: A guide for organizational entrepreneurs. *Innovations: Technology, Governance, Globalization, 9*(1–2), 195–212.

Abbott, K. W., & Snidal, D. (2010). International regulation without international government: Improving IO performance through orchestration. *The Review of International Organizations, 5* (3), 315–344.

Armeni, C., & Redgwell, C. (2015). *International legal and regulatory issues of climate geoengineering governance: rethinking the approach* (Climate Geoengineering Governance Working Paper Series #21). Retrieved from http://www.geoengineering-governance-research.org/perch/resources/workingpaper21armeniredgwelltheinternationalcontextrevise-.pdf

Asilomar 2. (2010). Full text of "Asilomar 2" Statement. *Science | AAAS*. Retrieved from http://www.sciencemag.org/news/2010/03/full-text-asilomar-2-statement

Bäckstrand, K., & Kuyper, J. (2017). The democratic legitimacy of orchestration: The UNFCCC, non-state actors, and transnational climate governance. *Environmental Politics, 26*(4), 764–788.

Bellamy, R., Chilvers, J., & Vaughan, N. E. (2016). Deliberative mapping of options for tackling climate change: Citizens and specialists 'open up' appraisal of geoengineering. *Public Understanding of Science, 25*(3), 269–286. doi:10.1177/0963662514548628

Bernstein, S. (2011). Legitimacy in intergovernmental and non-state global governance. *Review of International Political Economy, 18*(1), 17–51.

Bernstein, S., & Cashore, B. (2007). Can non-state global governance be legitimate? An analytical framework". *Regulation and Governance, 1*, 347–371.

Bipartisan Policy Center's Task Force on Climate Remediation Research. (2011). *Geoengineering: A National Strategic Plan for research on the potential effectiveness, feasibility, and consequences of climate remediation technologies*. Washington, DC.

Bodansky, D. (1999). The legitimacy of international governance: A coming challenge for international environmental law? *American Journal of International Law, 93*, 596–624.

Burns, W. C. G., & Flegal, J. A. (2015). Climate geoengineering and the role of public deliberation: A comment on the US National Academy of Sciences' recommendations on public participation. *Climate Law, 5*, 252–294. Print.

Burns, W. C. G., & Nicholson, S. (2017). Bioenergy and carbon capture with storage (BECCS): The prospects and challenges of an emerging climate policy response. *Journal of Environmental Studies and Sciences, 7*(4), 527–534.

Cashore, B. (2002). Legitimacy and the privatization of environmental governance: How non-state market-driven (NSMD) governance systems gain rule-making authority. *Governance, 15*, 502–529.

Castán Broto, V., & Bulkeley, H. (2013a). A survey of urban climate change experiments in 100 cities. *Global Environmental Change, 23*, 92–102.

Castán Broto, V., & Bulkeley, H. (2013b). Maintaining climate change experiments: Urban political ecology and the everyday reconfiguration of urban infrastructure. *International Journal of Urban and Regional Research, 37*(6), 1934–1948.

Chan, M., & Pauw, W. P. (2014). *A global framework for climate action (gfca)-orchestrating non-state and subnational initiatives for more effective global climate governance* (German Development Institute Discussion Paper 34).

Chan, S., Brandi, C., & Bauer, S. (2016). Aligning transnational climate action with international climate governance: The road from Paris. *Review of European, Comparative & International Environmental Law, 25*(2), 238–247.

Chen, Y., & Xin, Y. (2017). Implications of geoengineering under the 1.5 °C target: Analysis and policy suggestions. *Advances in Climate Change Research, 8*, 123–129.

Chhetri, N., Chong, D., Conca, K., Falk, R., Gillespie, A., Gupta, A., ... Nicholson, S. (2018). *Governing solar radiation management*. Washington, DC: Forum for Climate Engineering Assessment, American University.

Collingridge, D. (1980). *The social control of technology*. New York, NY: St. Martin's Press.

Corner, A., Parkhill, K., Pidgeon, N., & Vaughan, N. E. (2013). Messing with nature? Exploring public perceptions of geoengineering in the UK. *Global Environmental Change*, *23*(5), 938–947. doi:10.1016/j.gloenvcha.2013.06.002

Craik, A. N. (2017). *Developing a national strategy for climate engineering research in Canada. Center for International Governance of Innovation* (Working Paper No. 153).

Craik, N., Blackstock, J., & Hubert, A. M. (2013). Regulating geoengineering research through domestic environmental protection frameworks: Reflections on the recent Canadian ocean fertilization case. *Carbon & Climate Law Review*, *7*(2), 117–124.

Davis, C., & Hoffer, K. (2012). Federalizing energy? Agenda change and the politics of fracking. *Policy Sciences*, *45*(3), 221–241.

Esty, D. (2006). Good governance at the supranational scale: Globalizing administrative law. *Yale Law Journal*, *115*, 1490–1561.

Fiorino, D. J. (1990). Citizen participation and environmental risk: A survey of institutional mechanisms. *Science, Technology & Human Values*, Print. *15*(2), 226–243.

Flegal, J. A., & Gupta, A. (2018). Evoking equity as a rationale for solar geoengineering research? Scrutinizing emerging expert visions of equity. *International Environmental Agreements: Politics, Law and Economics*, *18*(1), 45–61.

Gardiner, S. (2011). Some early ethics of geoengineering the climate: A Commentary on the values of the royal society report. *Environmental Values*, *20*(2), 163–188.

Gordon, D. (2013). Between local innovation and global impact: Cities, networks, and the governance of climate change. *Canadian Foreign Policy Journal*, *19*(3), 288–307.

Gordon, D. J., & Johnson, C. A. (2017). The orchestration of global urban climate governance: Conducting power in the post-Paris climate regime. *Environmental Politics*, *26*(4), 694–714.

Gupta, A., & Moller, I. (in press). De facto governance: How authoritative assessments construct climate engineering as an object of governance. *Environmental Politics*.

Hale, T., & Rogers, C. (2014). Orchestration and transnational climate governance. *The Review of International Organizations*, *9*(1), 59–82.

Hamilton, C. (2013). *Earth masters: The dawn of the age of climate engineering*. New Haven, CT: Yale University Press.

Heazle, M., & Kane, J. (2015). *Policy legitimacy, science and political authority: Knowledge and action in liberal democracies*. London, UK: Routledge.

Heilmann, S., Shih, L., & Hofem, A. (2017). National planning and local technology zones: Experimental governance in china's torch programme. *The China Quarterly*, *26*, 896–919.

Hildén, M., Jordan, A., & Huitema, D. (2017). Special issue on experimentation for climate change solutions editorial: The search for climate change and sustainability solutions - The promise and the pitfalls of experimentation. *Journal of Cleaner Production*, *169*, 1–7.

Hoffmann, M. J. (2011). *Climate governance at the crossroads: Experimenting with a global response*. New York, NY: Oxford University Press.

Horton, J. B. (2015). The emergency framing of solar geoengineering: Time for a different approach. *The Anthropocene Review*, *2*(2), 147–151.

Horton, J. B., & Reynolds, J. L. (2016). The international politics of climate engineering: A review and prospectus for international relations. *International Studies Review*, *18*(3), 438–461.

Hubert, A.-M. (2017). *Code of conduct for responsible geoengineering research*. Retrieved from http://www.ucalgary.ca/grgproject/files/grgproject/revised-code-of-conduct-for-geoengineering-research-2017-hubert.pdf

Intergovernmental Panel on Climate Change. (2018, October 6). Global warming of 1.5 °C. *UNEP and WMO*. Retrieved from http://www.ipcc.ch/report/sr15/

Jamieson, D. (1996). Ethics and intentional climate change. *Climatic Change*, *33*(3), 323–336.

Jasanoff, S. (2003). Technologies of humility: Citizen participation in governing science. *Minerva*, *41* (3), 223–244.

Jinnah, S. (2018). Why govern climate engineering?: A Preliminary framework for demand-based governance. *International Studies Review.*, *20*(2), 272–282.

Jordan, A., Huitema, D., Schoenefeld, J., Van Asselt, H., & Forster, J. (2018). Governing climate change polycentrically. In A. Jordan, D. Huitema, H. Van Asselt, & J. Forster (Eds.), *Governing climate change: polycentricity in action?* (pp. 3–26). Cambridge: Cambridge University Press.

Lloyd, I. D., & Oppenheimer, M. (2014). On the design of an international governance framework for geoengineering. *Global Environmental Politics.*, *14*(2), 45–63.

Long, J. C. S. (2017). Coordinated action against climate change: A New world symphony. *Issues in Science and Technology*, *33*, 3.

Lövbrand, E., Pielke, R., & Beck, S. (2011). A democracy paradox in studies of science and technology. *Science, Technology & Human Values*, *36*(4), 474–496.

Macnaghten, P., & Szerszynski, B. (2013). Living the global social experiment: An analysis of public discourse on solar radiation management and its implications for governance. *Global Environmental Change*, *23*(2), 465–474.

Mahajan, A., Tingley, D., & Wagner, G. (2018). Fast, cheap, and imperfect? US public opinion about solar geoengineering. *Environmental Politics*, 1–21.

McFadgen, B., & Huitema, D. (2018). Experimentation at the interface of science and policy: A multi-case analysis of how policy experiments influence political decision-makers. *Policy Sciences*, *51*(2):161–187.

McKinnon, C. (in press). Sleepwalking into Lock-In: Avoiding wrongs to future people in the governance of solar radiation management research. *Environmental Politics.*

Morrow, D. R., Kopp, R. E., & Oppenheimer, M. (2013). Political legitimacy in decisions about experiments in solar radiation management. In C. G. William, Burns, & A. Strauss (Eds.), *Climate change geoengineering: Philosophical perspectives, legal issues, and governance frameworks* (pp. 146–167). Cambridge, UK: Cambridge University Press.

National Academies of Sciences, Engineering, and Medicine. (2016). *Gene drives on the horizon: advancing science, navigating uncertainty, and aligning research with public values*. Washington, DC: The National Academies Press. doi:10.17226/23405

National Research Council. (2015a). *Climate intervention: Carbon dioxide removal and reliable sequestration*. Committee on Geoengineering Climate: Technical Evaluation and Discussion of Impacts. Washington, DC: National Academies Press.

National Research Council. (2015b). *Climate intervention: Reflecting sunlight to cool earth*. Committee on Geoengineering Climate: Technical Evaluation and Discussion of Impacts. Washington, DC: National Academies Press.

Nicholson, S., Jinnah, S., & Gillespie, A. (2018). Solar radiation management: A proposal for immediate polycentric governance. *Climate Policy*, *18*(1), 322–334.

Norton, B. G. (2005). *Sustainability: A philosophy of adaptive ecosystem management*. Chicago, IL: University of Chicago Press.

Overdevest, C., & Zeitlin, J. (2014). Assembling an experimentalist regime: Transnational governance interactions in the forest sector. *Regulation & Governance*, *8*(1), 22–48.

Parson, E. A., & Ernst, L. N. (2013). International governance of climate engineering. *Theoretical Inquiries in Law*, *14*(1), 307–337.

Parson, E. A., & Keith, D. W. (2013). End the deadlock on governance of geoengineering research. *Science*, *339*, 1278–1279.

Pidgeon, N., Parkhill, K., Corner, A., & Vaughan, N. (2013). Deliberating stratospheric aerosols for climate geoengineering and the spice project. *Nature Climate Change*, *3*(5), 451. doi:10.1038/nclimate1807

Pielke, J. R. A. (2007). *The honest broker: Making sense of science in policy and politics*. Cambridge, UK: Cambridge University Press.

Preston, C. J. (2013). Ethics and geoengineering: Reviewing the moral issues raised by solar radiation management and carbon dioxide removal. *Wiley Interdisciplinary Reviews: Climate Change*, *4*(1), 23–37.

Preston, C. J. (2018). *The synthetic age*. Cambridge, MA: MIT Press.

Rayner, S. (2003). Democracy in the age of assessment: Reflections on the roles of expertise and democracy in public-sector decision making. *Science and Public Policy*, *30*(3), 163–170.

Rayner, S. (2004). The novelty trap: Why does institutional learning about new technologies seem so difficult? *Industry and Higher Education*, *18*(6), 349–355.

Rayner, S., Heyward, C., Kruger, T., Pidgeon, N., Redgwell, C., & Savulescu, J. (2013). The Oxford principles. *Climatic Change*, *121*(3), 499–512.

Reynolds, J. (2015). A critical examination of the climate engineering moral hazard and risk compensation concern. *Anthropocene Review*, *2*(2), 174–191.

Reynolds, J. L., Jorge, L., Contreras, J., & Sarnoff, D. (2018). Intellectual property policies for solar geoengineering. *Wiley Interdisciplinary Reviews: Climate Change*, *9*, e512.

Rogelj, J., Den Elzen, M., Höhne, N., Fransen, T., Fekete, H., Winkler, H., ... Meinshausen, M. (2016). Paris agreement climate proposals need a boost to keep warming well below 2°C. *Nature*, *534* (7609), 631–639.

Sabel, C. F., & Zeitlin, J. (2012). Experimental governance In D. Levi-Faur ed., *The Oxford handbook of governance* (pp. 169–183). Oxford, UK: Oxford University Press.

Sandler, T. (2018). Collective action and geoengineering. *The Review of International Organizations.*, *13*(1), 105–125.

Scharpf, F. W. (1999). *Governing in Europe: Effective and democratic?* Oxford: Oxford University Press.

Scheufele, D. A., Corley, E. A., Dunwoody, S., Shih, T.-J., Hillback, E., & Guston, D. H. (2007). Scientists worry about some risks more than the public. *Nature Nanotechnology*, *2*(12), 732–734.

Stilgoe, J. (2016). Geoengineering as collective experimentation. *Science and Engineering Ethics*, *22* (3), 851–869.

Stilgoe, J. (2017, April 7). Self-driving cars will only work when we accept autonomy is a myth. *The guardian*, sec. Science. Retrieved from https://www.theguardian.com/science/political-science /2017/apr/07/autonomous-vehicles-will-only-work-when-they-stop-pretending-to-be-autonomous

Stilgoe, J., Owen, R., & Macnaghten, P. (2013). Developing a framework for responsible innovation. *Research Policy*, *42*(9), 1568–1580.

Sugiyama, M., Asayama, S., Ishii, A., Kosugi, T., John, C., Moore, J. L., ... Xia, L. (2017). The Asia-Pacific's role in the emerging solar geoengineering debate. *Climatic Change*, *143*(1–2), 1–12.

Svoboda, T., Buck, H. J., & Suarez, P. (2018). Climate engineering and human rights. *Environmental Politics*, 1–20.

Turkaly, C., Nicholson, S., Livingston, D., & Thompson, M. (2017). *Climate engineering clearinghouse meeting report*. Washington, DC. Retrieved from http://ceassessment.org/fcea-reports/climate-engineering-clearinghouse/

US GAO. (2010). Climate change: A coordinated strategy could focus federal geoengineering research and inform governance efforts. GAO-10-903. Retrieved from http://www.gao.gov/products/GAO-10-903

Van Oudheusden, M. (2011). Questioning 'participation': A critical appraisal of its conceptualization in a flemish participatory technology assessment. *Science and Engineering Ethics*, *17*(4), 673–690.

Vito, C. (2018). State biotechnology oversight: The juncture of technology, law, and public policy. *Maine Law Review*, *45*(2), 329–383.

Winickoff, D. E., & Brown, M. B. (2013). Time for a government advisory committee for geoengineering research. *Issues in Science and Technology*, *29*, 79–85.

Winickoff, D. E., Flegal, J. A., & Asrat, A. (2015). Engaging the global south on climate engineering research. *Nature Climate Change*, *5*(7), 627.

Fighting risk with risk: solar radiation management, regulatory drift, and minimal justice

Jonathan Wolff

ABSTRACT
Solar radiation management (SRM) has been proposed as a means of mitigating climate change. Although SRM poses new risks, it is sometimes proposed as the 'lesser evil'. I consider how research and implementation of SRM could be regulated, drawing on what I call a 'precautionary checklist', which includes consideration of the longer term political implications of technical change. Particular attention is given to the moral hazard of 'regulatory drift', in which strong initial regulation softens through complacency, deliberate deregulation ('regulatory gift') and the limited constituency of people with the skills to regulate ('thin markets'). I propose the strengthening of civil society groups to keep regulators in check.

Introduction

It will not be necessary for me to repeat here in detail the accounts of how the earth's climate is in the process of changing, or the projections for the coming decades given differing scenarios of human action. My focus is on the attempt to reduce the effects of climate change by using new forms of technology that reduce the effect of the sun's rays rather than by cutting carbon emissions. Solar radiation management (SRM) is one example, but what I say here will apply to any attempt to reduce the perceived effects of climate change by means of technical innovation, using techniques that have not previously been tried out on any significant scale. I draw on some general considerations about the regulation of risky new technologies, but they are adapted to the circumstances of the proposal to use untested innovations to help reduce the urgent threat of climate change. My strategy here will be to explore how issues of power and politics can become intertwined with the more technical aspects of the regulation of new technology, thereby introducing new risks. In outline, this is nothing new, but I hope to offer some conceptual assistance that will help us appreciate those difficulties, and to examine ways of protecting us against what I call regulatory drift (and its dangers), using SRM as an illustrative case.

Although SRM is new, the introduction of new technologies is routine. We are aware that any new technology runs the risk of unwanted dangers. Concerned citizens were worried that microwave cookers or mobile phones could cause cancer. Due diligence would normally require thorough testing and investigation, to the degree that it is possible. And it is already the case that detailed thought has been given to the regulation of research into geoengineering (e.g. Hubert, 2017). But no amount of testing can really prepare us for the effects of full-scale implementation, and some SRM technologies, so it is said, cannot be piloted in any representative way (Robock, Bunzl, Kravitz, & Stenchikov, 2010). In any case, sometimes a situation calls for swift action. If we were simply in the position of incrementally adding a further technology to the stock we already have then there is time for careful analysis. But if we are in a situation where a new technology promises to deal with a serious present problem, then delaying its introduction could have significant consequences. Delay poses risks, even though introduction does too. Hence, my title 'fighting risk with risk', which, of course, refers to attempting to reduce the risks of climate change with risky new technologies such as SRM. It is a familiar situation regarding, for example, the introduction of new pharmaceuticals in a medical emergency, when following ordinary protocols about safety testing could lead to years of delay and deaths on a massive scale. It has been argued that we are in a parallel situation with respect to SRM (Parson & Keith, 2013).

How, though, should we understand SRM? It encompasses many possible techniques. To take one helpful description from the literature:

> SRM is a form of geoengineering that aims to reflect solar radiation, thereby reducing the amount of solar energy in the atmosphere.... Terrestrial forms of SRM include land use changes such as increasing reflective vegetation, increasing the number of white roofs and surfaces, and desert reflectors. More technologically involved forms of SRM include stratospheric sulfate injections, cloud brightening via tropospheric cloud seeding, and space-based reflective surfaces. All forms of SRM involve intentionally manipulating our climate on a global scale. (Hartzell-Nichols, 2012, p. 163)

The forms mentioned here run from the apparently innocuous, such as increasing the number of white roofs, to the highly speculative and risky. Of these, the most notable is stratospheric sulfate injections (SSI), which has risks, both known and unknown, aimed to mitigate the known risks of climate change (Crutzen, 2006). SSI will be the main technology discussed here.

The idea of 'fighting risk with risk' may seem a dramatic way of describing a commonplace state of affairs. Asbestos was introduced to reduce the risk of fire, and CFCs to reduce the risk of the explosion of chemical refrigerators. I have chosen these examples for the obvious reason that the new technologies, in different ways, turned out to be highly problematic. When known risks increase or receive a great deal of publicity, and political leaders become anxious, it can seem defensible to rush into new possibilities that we might not have approved under calmer circumstances. And the problem is compounded if we are in a situation which is both harmful and deteriorating sharply. We can come to believe that we simply have no time to waste and must get on with whatever can help us. The more doom-laded the predictions about the effects of climate change, both in terms of speed and extent, the more reasonable it appears to be to reach for whatever technologies are available to reduce or remove such threats. Hence, in a seminal paper, Crutzen writes,

> If sizeable reductions in greenhouse gas emissions will not happen and temperatures rise rapidly, then climatic engineering ... is the only option available to rapidly reduce temperature rises and counteract other climatic effects. (Crutzen, 2006, p. 216)

This 'lesser evil' framing does not go unchallenged, of course, for it could lock us into pursuing options that themselves structure the situation in adverse ways, by, for example, putting other reasonable options out of reach (Gardiner, 2010, p. 2013). Still, such arguments cannot be dismissed out of hand, and it seems prudent to consider what forms of regulation are appropriate for technologies such as SSI, should we become convinced they need to be taken very seriously. In what follows, I will draw on some background issues about risk, regulation, and justice that I have discussed elsewhere, and then consider how geoengineering should be considered in the light both of the theory of risk and regulation, and relatedly, of justice. I will not be proposing a new set of principles for regulating geoengineering, or advancing a judgement about whether it is appropriate to engage in geoengineering now. I will, however, set out considerations that explain why strong independent oversight is both needed and not enough, and must be supplemented by external scrutiny by civil society groups.

SRM and the precautionary checklist

Our topic is, in effect, the regulation of very risky forms of SRM, where, by regulation, I include the possibility of comprehensive prohibition. Although my primary concern here is deployment of SRM technology on a significant scale, especially when fortunes are to be made from it, much of what I say will apply to research. In my work on regulation, I have assumed that the point of regulation is to reduce the likelihood of adverse events, which I summarise by the slogan that regulation should be relative to risk (Wolff, in press). This view could be contested, for example, by arguing that at least one point of regulation is to bring out the potential in new technology, for example, by having common standards that harmonise combinations of technologies made by different manufacturers. But certainly at least one function of regulation is to reduce risk, and in the present case of SRM, the issue is one of regulation to reduce the chance that SRM itself brings in a new range of adverse effects. The issue, of course, is not one of eliminating all risks, but how to strike the right balance between reaping the potentially considerable advantages of new technology against the new risks of introducing it. This may mean permitting the technology in some manner, but prohibiting it in its most dangerous forms.

Before proceeding at all, we need some reason to believe the technology has a chance of succeeding. At the moment the best evidence comes from the natural experiment of volcanic eruptions which have appeared to have a cooling effect on the climate. (Crutzen, 2006; Robock, 2008). In what follows I shall assume that a good, potential, case has been made for SSI. Nevertheless, even with the best science, there is a good reason for taking a precautionary approach, as everything has risks and some risks will remain unknown. It is sometimes said that in cases like this we need to apply the 'precautionary principle'. Yet it is increasingly recognised that, strictly, there is no single precautionary principle that can be used to settle the contours of regulation in practice (Gardiner, 2006a; Manson, 2002; Munthe, 2011). Instead, we need to apply what can be called a 'precautionary approach' (Hartzell-Nichols, 2017; Hubert, 2017). With this in mind, a number of regulatory frameworks have been proposed to govern research and implementation of geoengineering (Gardiner & Fragnière, 2018; Hubert, 2017). Elsewhere, I have outlined what I have called a precautionary checklist that can be used to inform the assessment of any risky new technologies, although it is not proposed as a complete account for any particular case (Wolff, 2020, see also Myers, 2005). It proceeds through the following steps:

(1) Are the costs and risks of the new technology tolerable?
(2) Does it have significant benefits?
(3) Do these benefits solve important problems?
(4) Could these problems be solved in some other, less risky, way?
(5) What are the possible longer term economic consequences of introducing the technology?
(6) What are the possible longer term political consequences of introducing the technology?

In different circumstances, other questions will also be relevant, and there is no limit, in principle, to the issues that could become salient. The rationale for this list is, roughly, questions 1–4 are a formulation of 'common sense' risk management, while 5 and 6 are not, but are often neglected with potentially disastrous results. Considering longer term factors in advance, though fraught with guess-work, can save considerable difficulties later. But the absence of a question from this list should not be taken as a reason to believe it will not be relevant in some circumstances.

Although I call it a checklist, it is not intended as a series of yes/no questions in the form, say, advocated by Atul Gawande, useful though that may be for other contexts (Gawande, 2009). Here the questions, and especially the first four, are to be considered holistically: what might be an unacceptable risk in one context could be acceptable in another, if the problems are much more serious.

Basic cost-benefit analysis will focus on the first two questions, naturally looking at the risks and benefits. However, formal cost-benefit analysis, if applied in a mechanical manner, is likely to be of very little help in the case of geoengineering as the risks of solar radiation management are unknown and therefore unquantifiable. The situation is one of uncertainty rather than risk, and any assignment either of the values of possible losses or their probabilities can, at best, be expert estimates that could differ by orders of magnitude and in any case are very likely to leave out salient factors. For comparison, the possible effects on the ozone layer, I presume, were not part of the discussion when the decision to authorise CFCs for domestic use was made. Hence, other decision approaches are needed.

Where a new technology has very low risk, then there is little reason for halting its development, even if the problems it solves are relatively trivial. The matter becomes more serious as the dangers become more grave in terms of the nature of the hazards introduced by the new technology, their probability, and their likely extent if realised. At some point, risks will be so high that unless we really are on the verge of catastrophe the technology should not be introduced. Even here it is hard to make exception-free rules. But the main point is that the first four conditions need to be taken together. What counts as an intolerable risk in the circumstances will depend on its benefits, whether those benefits solve important problems, and whether those problems can be solved in some other way. Compare principle 8 of the 'Tollgate Principles':

> Climate policies that include geoengineering schemes should be socially and ecologically preferable to other available climate policies, and focus on protecting basic ethical interests and concerns (e.g. human rights, capabilities, fundamental ecological values). (Gardiner & Fragnière, 2018, pp. 167–168)

Although this is a much broader principle, it implicitly asks the same question: is this technology really necessary? If not, and it has new possible risks, why do it? (See also (Hartzell-Nichols, 2012; Hubert, 2017)). There may be an answer in terms of feasibility; that there are alternatives but very little chance of their adoption. If so, then there is something to debate here, to which I will return in the next section. But generally, this 'necessity' question is a protection against technophilia (the love of technology for its own sake, which can be a serious problem within certain communities) or a protection against vested interests (again to which we will return).

This central question – is the technology necessary? – has been widely emphasised. Even Crutzen, often regarded as a defender of SRM, writes:

> Finally, I repeat: the very best would be if emissions of the greenhouse gases could be reduced so much that the stratospheric sulfur release experiment would not need to take place. (2006, p. 217)

Yet, caustically he adds (and thereby ends his paper):

> Currently, this looks like a pious wish. (2006, p. 217)

Putting this important remark aside for the moment (I will return to it in the next section) the argument here is that SRM is unnecessary, for it poses new risks in attempting to solve a problem that we already know how to solve in less risky ways. This is the key issue, which I will take up shortly. Before doing so I want to complete the brief commentary on the last two questions on the 'precautionary checklist'.

The questions are:

(1) What are the possible longer term economic consequences of introducing the technology?
(2) What are the possible longer term political consequences of introducing the technology?

While the first four questions of the checklist take us through familiar questions about costs, benefits and alternatives, this more distinctive part of the precautionary checklist looks at longer term effects, and brings political questions to the fore. Although not explicit, one particularly sensitive issue, straddling economic and political questions, involves the issue of the nature of the regulator. In effect, it asks how the regulator is to be regulated. Of course, this question threatens a regress; one we attempt to solve in politics by setting up systems of checks and balances, most notably through a separation of powers. Probably similar structures are possible for regulators although it is likely to lead to very slow-moving bureaucratic, and possibly over-cautious procedures. An alternative, which we will explore towards the end of this paper, is to build a set of strong norms into regulation, and to encourage various forms of challenge to the regulator if those norms are violated or ignored. But what will count as effective regulation, and how is it to be achieved?

Is SRM really necessary?

To recap, SRM is a form of technology that, if all goes well, will mitigate some of the effects of climate change, but in its most ambitious forms it seems very risky. Critics have suggested that, as we know how to mitigate climate change in less risky ways, the introduction of SRM is difficult to justify. SRM is, so it is said, both highly risky and unnecessary, or, at least quite a long way down the list of options (Gardiner, 2010). Therefore, there is no obvious case to allow it to proceed. In this section, I want to look at this argument in more detail.

First, in the previous section, I mentioned the principle that regulation should be relative to risk. If this is right, then before going any further it is incumbent on us to come to an appreciation of the possible risks of SRM. What, therefore, might happen if we start to spray very large quantities of sulfates into the stratosphere in the hope that they will deflect a sufficient proportion of the heat from the sun so as to mitigate temperature increases at the surface of the earth? What could go wrong? To start with, what does it involve? In the words of Crutzen, placing sunlight reflecting aerosol in the stratosphere 'can be achieved by burning S_2 [sulfide] or H_2S [hydrogen sulphide], carried into the stratosphere on balloons and by artillery guns to produce SO_2 [sulfar dioxide]' (2006, p. 212). (Incidentally, Crutzen suggests that sulfur dioxide is naturally produced by burning fossil fuels, and so, somewhat ironically, successful attempts to reduce pollution may have contributed to the increase in global temperatures.) Other forms of SRM, involving the spraying of seawater into the sky to brighten clouds and increase their reflectivity are also under discussion.

At the most basic level, SRM could simply be a costly failure. We could spend a great deal of time and effort to achieve no effect whatsoever because of political or technical obstacles meaning that there is no deployment of the technology. This generates at least two types of problems. The first is the waste of resources it would involve, which may be relatively minor in the great scheme of things. But the second, and more important, is that the belief that we are on the verge of a successful intervention could make us complacent about the issues, and disincentivise us from taking other measures, which is one way in which 'moral hazard' is introduced into the debate (Preston, 2013; Robock, 2008). However, the bare possibility of failure is not enough, on its own, to suggest that SRM should not go ahead, but rather we would need to keep it constantly under review, and to appreciate that other efforts should not be stopped.

A second type of risk is that the technology could have problematic side effects. These could range from minor to very serious. Perhaps imperfect implementation (such as leakage of materials) will pose health hazards to some populations, such as increased susceptibility to non-life-threatening chest infections or allergies. Crutzen remarks 'locally, the stratospheric albedo modification scheme, even when conducted at remote tropical island sites or from ships, would be a messy operation' (Crutzen, 2006, p. 213). Concerns have been expressed that SRM could also reduce rainfall, leading to worse overall results than there would have been even under climate change (Robock, 2008), especially if the monsoon pattern is disrupted. Whether this would actually happen has been questioned, based on climate models, but it is clearly something that needs to be considered. (Irvine et al., 2019). Other possible problems could be increased acidification of the oceans, the disappearance of blue skies (albeit compensated for by colourful sunrises and sunsets), or the misuse of the technology (Robock, 2008).

A third, related, risk is that adverse effects, such as those of loss of rainfall, would fall especially on populations that are already vulnerable. Once more, some models suggest that particular fear may be exaggerated (Irvine et al., 2019), but these models are themselves controversial, and the risks again cannot be ignored, for this is only one of the indefinite number of possible problems. Such differential effects bring in questions of justice as well as collective harm. After all, there may be other risks that fall asymmetrically on different populations, just as we see with the risks of climate change itself, where those who live in low-lying areas, such as Bangladesh, are especially vulnerable to rising sea levels, for example. One important form of asymmetry, of course, is between those living now and those living in the future (Gardiner, 2006; McKinnon, 2018).

The worst case of any risk is that we cross some sort of tipping point, leading to irreversible, cumulative, and highly harmful changes. In advance, this will be hard to predict, but it is a major concern with SSI. Given that SSI intervenes in a climate system that it not well-understood, it cannot be known whether it would disrupt some critical climate variables. In the literature, one often discussed risk is the possibility that while the technology is effective, it needs to be discontinued abruptly, for example, because of war, sabotage, or natural disaster (McKinnon, 2018). Its termination could induce a major shock that could possibly put in train some longer term effects (Preston, 2013).

Given these concerns, it is very natural to return to the suggestion introduced in the last section that SRM is unnecessary. The risks are grave and there is an easy alternative in reducing emissions. Therefore, there is no good reason to push ahead with SRM. But still, can we really ignore Crutzen's remark that a significant global effort to reduce emissions 'looks like a pious wish'? (2006, p. 217). The fact that we know how something can be done does not entail that we know what steps are needed to make it happen (although in this case we probably do). Still less, though, does it mean that taking those steps is achievable, and will actually be achieved, from where we are now. Hence, the 'necessity principle' – highly risky technologies should only be attempted when there are no lower risk ways of achieving the same ends – is ambiguous. It could mean 'no conceivable' alternative ways or 'no easily achievable' alternative ways or anything in between. The danger of the 'no conceivable' interpretation is that if a route is conceivable, but very unlikely to be taken, then we can end up taking no action at all to mitigate a threat even when a risky technology is available. At the least, and this seems to be Crutzen's 2006 position, it seems we need stand ready to deploy the new technology (in the safest possible form) should it turn out that it is too late to take any safer option. And little has changed in terms of climate change mitigation in the decade and more since he wrote that should make us change that opinion (cf McKinnon, 2018). But this too is fraught with danger. Knowing that we have the technology near-ready can make us more blasé about failing to reduce emissions. This is a dilemma for SRM that many have noted. If we do not develop the technology we might have an inadequate response to runaway climate change, but if we do develop it then we are more likely to need to use it, leading to termination risk threats, other unknown risks, and increasing acidification of the oceans, and so forth.

The risk triangle and regulating the regulator

If we need to 'stand ready' to deploy SRM if we do not manage, collectively, to reduce emissions to any significant degree, what does that, in fact, mean? I interpret it as having two strands. Research is needed to develop the technology in safe and effective forms. And in addition, we need to set out a strong regulatory framework to govern any possible large-scale deployment in a suitably precautionary fashion, should it ever be needed. But what should that regulatory framework look like?

To provide a conceptual framework to assist in the task, I want to draw on work I have done elsewhere that builds on an insight from Hermannson and Hansson that we can call 'the risk triangle' (2007). Consider any decision to introduce a change or technology that imposes some risk on some people. As Hermannson and Hansson acutely pointed out, there are three key roles in this scenario, those of the agent taking the decision; the agent or entity that will benefit if all goes well; and the agent or entity that will suffer if things go badly.

Extending this insight, we can see that it is possible that one agent can occupy all three roles, thereby making it an entirely personal decision, which we can call 'individualism'. At the other extreme, three different agents or entities can occupy the different roles. This will be the case, for example, if a government is asked to adjudicate on a development of some sort that will benefit one group if it goes well, but have costs for a different group if it goes badly (perhaps through pollution effects of some kind). I have called these three-party scenarios 'adjudication' as an independent party is judging a situation which affects two other parties in different ways.

As we see in the table below, if there are three roles in any decision about risk, and we assume each role is occupied by only one agent, then there are only five possible risk structures. So far we have looked to at two: individualism and adjudication. To complete the picture there are three different ways in which one agent or entity can occupy two roles and another agent the third role. The most troubling is the case where one agent makes the decision and would gain any benefit, while any potential losses fall on others. In economic theory, this is the case of 'moral hazard'[1] or 'negative externalities'. It is closely related to what Gardiner has called 'buck passing' (2016) and can be a highly dangerous situation as it can encourage reckless risk taking. If the benefits, but not the costs, of a decision accrue to the decision maker, why be cautious? Arguably this is the central problem of our age.[2]

		Decision maker	Benefits go to	Costs go to
1	Individualism	A	A	A
2	Paternalism	A	B	B
3	Moral Hazard	A	A	B
4	Moral Sacrifice	A	B	A
5	Adjudication	A	B	C

I will return to this highly problematic case of moral hazard shortly but it is also worth pausing to mention moral sacrifice. Here the idea is that the agent taking the risk will suffer the consequences if it goes wrong, but others will benefit if it goes right. In such cases who would take the risk? But consider the case where any potential loss would be small but the benefit very great; there would be 'positive externalities' to use the economists' phrase. Very often it would be beneficial collectively for the risk to be taken, but the relevant agent has no incentive to do so as their share in any benefits will be zero or small. We might think, for example, of the possible decision whether or not for a company to research a new green technology that would be impossible to patent. In this case, the developers of the technology are guaranteed a loss if they go ahead, but humanity could benefit significantly if they do. In these cases, governments need to work out incentives for people and companies to take beneficial risks, otherwise much collective value will be forgone.

Moral sacrifice cases will also, then, be very relevant to issues of climate change. But it is moral hazard that primarily concerns us in this paper. Pure moral hazard cases are rare. Normally any decision maker will, in part, be overseen by their government or regulatory agencies, and therefore there is a type of shared decision-making. Furthermore, taking reckless risks that go badly threatens reputational damage, so even if there is no direct loss the decision maker risks indirect loss. Other people may well gain spill over benefits; hence, there are likely to be others that share the benefits. But still we can recognise there are many situations in which the incentives encourage people to take risks that they would not have done if more of the costs were to fall on them. These would be cases where an individual has a dominant role in decision-making, will collect significant benefits if things go well, and risks only a comparatively minor loss if things go badly. And, of course, it is exactly the situation many people believe we are in with respect to climate change generally, where the decisions makers and citizens in the current generation are dumping the costs of their recklessness on the future.

But to return to SSI, let me outline what I see as the main danger in the regulation of any complex new technology. When a new technology is proposed, politicians may well have considerable doubts about it, and insist on rigorous initial oversight into its introduction. Campaigners and protest groups are likely to point out where costs and risks could fall, while those who promote the technology will argue for its benefits and downplay the risks. When the promoters are also those who would benefit, whether financially or through prizes, honours, or enhanced reputation, we need to be aware of their vested interests. It is also true that opponents can have vested interests. For example, they could have an interest in existing technologies that will be displaced, or have other privileges that will be eroded, but instead of being honest about how they will be affected look for ways of attacking new developments. Hence, the regulators cannot take any argument at face value, but, ideally, would carry out their own independent assessment. Ultimately the regulator will need to make a decision, and this exemplifies the structure in the risk triangle that I have called adjudication. Note that in the case of adjudication, it is not those who may lose that are party to the decision but rather the government or its agents are taking a decision that is intended to take the interests of those who may lose fully into account. The degree to which this actually happens will vary from case to case. For example, when the costs are far in the future, as they will be in intergenerational cases, even governments may discount them to zero (Gardiner, 2006b).

But let us consider what happens next. Suppose the regulator has given approval to the technology, perhaps with safeguards of various sorts. The months and the years tick over and if there is no major problem the issues will fade from public concern. The technology develops over time, and further approvals may need to be given, but if nothing goes wrong, regulatory resources may be moved away from the area and press interest will decline. A type of regulatory complacency could set in before long, with approvals being nodded through. Even worse it could be seen as an area ripe for deregulation, leading to what Jude Browne has called 'regulatory gifting' in which government allows industry to self-regulate, perhaps as part of an electoral promise to 'cut red tape', and often on the agreement that other benefits will be provided in return, yet these benefits may never in fact materialise (Browne, 2018).

Another threat to regulation, which in the present case is highly likely, is that after the initial regulators are moved on to other tasks, others will need to be drafted in to join a regulatory panel. If the area is a very technical one, as in this case, the market of potential regulators will be 'thin' as Ramanna has put it (Ramanna, 2015), and those who have some history in the industry may be the only plausible candidates. Ramanna's examples concern the regulation of financial standards, given the complexity of contemporary auditing and financial products, but the same concerns affect any emerging complex area. After all who, other than those who have worked full time for an extended period in a complex industry, will have the knowledge and interest to be a competent regulator? Even academics in the field may have little practical experience, and in any case are likely to divide their time between several different activities. Whatever their explicit motivations, and however well-meaning they are, it is very likely that regulators recruited from industry will show a degree of sympathy for its concerns, whether or not they still have any official role there. In other words, thin markets will have a tendency to lead to regulatory capture, accidental or otherwise. In the case of geoengineering, it has already been observed that a 'geoclique' of experts appear on panel after panel (Kintish, 2011). Little by little, by means of the three mechanisms of regulatory complacency, regulatory gifting, and regulatory capture due to thin markets, we may well see what we can call regulatory drift to the point where there is de facto near self-regulation. This, of course, runs a great risk of moral hazard. In other words the danger is that once the immediate public concern has passed, the structure of adjudication, in which a neutral regulator judges between two cases, we gradually slide into to a situation in which the decision makers become aligned with those who will benefit most, who are those who have a financial interest in the implementation and further use and development of the technology. There is no certainty that this will lead to adverse outcomes, but the probability will increase if this transformation of the regulation takes place. Costs and risks are most likely to pile up on those who have least representation at the table; the poorest and most vulnerable. And this is on top of any concerns of moral corruption in the framing of the problem, leading to other forms of regulatory change, such as ignoring the most serious long-term issues (Gardner, 2016). As often noted, the most serious risks can be those that do not even make it on to the risk register.

What, then, need we worry about if the area becomes regulated by those sympathetic to the industry? Here are some possibilities. First, we have been warned that SSI is likely to be a 'messy business'. Suppose it is much messier than expected or that the costs fall mostly on the most vulnerable. A partially captured regulator may be very slow to take these problems seriously. Second, a number of economic assessments of costs and benefits will no doubt be undertaken. Because of the significant uncertainties any responsible analysis should provide a reasonably wide range of valuations, with an error bar. But typically to make a decision a point value must be chosen. From experience elsewhere, it seems that choosing a particular value is often a political decision masquerading as an economic one. For example, it is typical for those supporting a project to choose high values from within a range to demonstrate benefits, but low values when calculating costs (Wolff & Orr, 2009). More generally, it has been found that right-wing, middle age, white, wealthy, men tend to trust government and technology more than other groups (Slovic, 2000). In many areas, this is the demographic likely to flow from industry to regulator, although in the area of SRM this may be less of a danger.

Regulating the regulators

To recap the argument to this point, I have argued that, as in many other areas of life, one of the great risks of geoengineering is likely to be the three mechanisms of regulatory drift: complacency; gifting; and, via 'thin markets' capture by those who have primary sympathies in favour of geoengineering. Even if those drafted in to regulate try to be objective and impartial it is likely they will see the benefits of the technology in sharper focus than its risks, and this can be highly problematic, especially for those who are most vulnerable, including future generations. There are ways of setting up the regulatory body to reduce the chances of this happening, but they require resources and constant vigilance, and as priorities change, may well get neglected. I propose a different approach in which regulation explicitly includes principles of justice, and civil society organisations are given the standing by which the regulator can be held to account through social pressure, and, if necessary, in law, if those principles are violated.

Here there is an obvious line of scepticism. Principles of justice are disputed. Who is to decide which is the most appropriate? Furthermore, no actual society lives up to any well-known theory of justice, so how can we expect regulators to enforce a higher standard with respect to SRM than is available elsewhere in society? But in my view, such concerns are overstated.

To gain some focus on these questions let us first reflect on the situation we are often thought to be in, and consider parallels. Roughly speaking, if we allow 'business as usual' we will see short-term economic growth, short-term climate shocks (storms, heatwaves), and then in the coming decades there will be more severe consequences such as disruption of water supplies from glaciers and monsoons, the rendering of some currently arable lands too arid to farm, and subsequent movement of peoples, disputes and potentially wars over land and water rights, and so on, which not only directly jeopardises basic rights but will hit economic growth fairly significantly. Successful mitigation may slow economic growth somewhat, in the short term, and it may never recover to past rates in all countries, but the worst consequences outlined above may be avoided.

Broadly, almost all climate scientists suggest we know what to do to avoid the worst consequences, which is to reduce emissions. But to do so will lead to short-term loss for some, and unless we are exceptionally creative, frustration of aspirations for many of the poorer people in the world. For these two very distinct reasons it will be hard to produce mass motivation. Accordingly, so its advocates argue, geoengineering could, in more modest versions, provide breathing space while we design the political and social institutions needed to make the changes, or more ambitiously, provide a solution which means no other change of behaviour is required. It will, nevertheless, have significant economic costs, which will have to be found from somewhere, and expose populations to the risks outlined above. As seen this framing can be challenged for ignoring other possibilities, but let us continue considering the scenario as laid out.

The bare structure of the situation is that a change is proposed that will have different effects on different people, and also distribute different risks to different groups. How should the costs, benefits, and risks, of such changes be allocated? Like it or not, this is a question of justice, which can be hidden but not avoided. Hence, it is as well to face it head-on. What does justice require in the face of change? To some degree, this is an under-theorised topic in political philosophy. If discussed at all it is assimilated to more general issues of distributive justice, of which the most prominent theory is John Rawls's Difference Principle, which in essence says that priority should be given to the worst off who should be made as well off as possible (Rawls, 1999). However, there may be a broader philosophical consensus around a version of Parfit's modification, which suggests that the worst off should be given highest priority, but there are possible circumstances in which they can legitimately suffer a small loss for the sake of a larger gain for others who are a little better off (Parfit, 1998). The thrust of both principles, though, is that justice primarily requires us to consider those who are worst off in society. But as I mentioned, these are set out as general theories of justice, as distinct to theories of justice in changing circumstances.

Welfare economics has done more to theorise the justice of change. We are very familiar with the idea of Pareto Optimality, which, in the current context suggests that no one at all should lose from a change. Applying it to the situation of risks, it would seem to imply that no one should be subject to increased risk as a result of change. Yet it is precisely because it is assumed that this is not possible that we have the questions we do. The main alternative is to adopt a version of the Kaldor–Hicks principle which suggests that a change is only permissible, or to be approved, if there is a potential Pareto improvement, which is to say that it is possible that the winners could compensate the losers, even if it doesn't actually happen. Yet, first, it is unclear why this should count as a justification, unless a fairly crude form of consequentialism is assumed, and second because of overall losses, in the immediate future such compensation is not possible. Both criticisms come together with the observation that those most badly affected will die before compensation can be offered. But in any event, Kaldor–Hicks in unmodified form is not immediately relevant to sharing the costs of decline.

Nevertheless, Rawls, Pareto, and Kaldor–Hicks can provide inspiration for the formulation of a pair of principles that I think represent an idea of natural justice in the face of change, and, I claim, although cannot demonstrate, are widely held at least in outline form. The first, which I have appealed to at various times in my writings, can be referred to as the Minimal Equity Principle (MEP):

> MEP: When a change generates a surplus, those who are already towards the bottom of the distribution should not lose as a result of that change.

I first formulated this principle in relation to the European Union, when discussions were underway for the entry of Poland and other accession states (Wolff, 1996). There was, in some quarters, a belief that Poland would benefit from joining the EU but small Polish farmers, who were believed to be living not far above the poverty line, would be likely to lose their livelihoods. Hence, it seemed essential that there should be some sort of safety net, and if there was a general boost to the Polish economy then there should be the resources available to finance such provision. In sum, though, it simply seemed very unfair that the poor should (avoidably) be made worse off for the sake of those who were already richer.

Note, though, that this is not the Difference Principle. It does not say that the worst off should be made as well off as possible. It is not even Parfit's prioritarianism, as it does not call for the worst off to be made better off, even if this would be highly desirable. In order to have very wide appeal, it needs to be less committed than that, and hence it only calls for the worst off not to be further damaged. It gives a form of priority to the worst off, but only in a weak sense. To be clear, I do not personally believe that this is all that justice requires, but I do believe that it should be possible to get a very wide consensus that this is one component of justice in that any change that violates this principle will have been handled unjustly.

It is also worth saying that this is also not Pareto Optimality. It allows the wealthy to be negatively affected. There may be other reasons for ruling out such changes, but if so they are not covered by the MEP. This is the sense in which it gives priority to the worst off. It is a type of lop-sided Pareto Principle. And for completeness, it is also not Kaldor–Hicks in that potential compensation is not enough; the MEP calls for actual compensation.

Thinking about actual compensation, however, brings out a central vagueness in the principle. What is the metric of well-being by which we can both use to know who is towards the bottom, and also tell whether anyone is being made worse off? As soon as it is specified in concrete terms we will get dragged into interminable debates about the currency of justice and question of commensurability and non-compensable harms. And it is also likely that any consensus on the principle depends on keeping it general. I prefer, however, to regard it not so much as vague but as context dependent. In each case, it is quite likely that a particular metric will be salient, but it could be that in different cases different metrics will apply. In some cases, we are going to be worried about salaries, in others air quality, and in others mere subsistence. For this reason, I will not specify it further.

A different, and highly pertinent, objection is that however appealing the principle is in general, like the Pareto Principle and Kaldor Hicks, it simply does not apply in the current circumstances. The principle begins 'if a change generates a surplus'. Well, that would be a nice problem to have. As things are, we are talking about managing a decline rather than sharing a new benefit. True, it may be that there is a future benefit to share, and there is a counter-factual benefit: if all goes well we would be better off than we would have been. But it is far from guaranteed that geoengineering will generate a net benefit compared to our current situation. Therefore, the question of justice we face is arguably one in which we need to consider how to distribute the costs (and benefits) of managing decline, rather than sharing a surplus.

Now, this may or may not be so, and perhaps adopting SRM would buy time to allow some form of positive transformation of the economy to cheap, efficient, renewables that would be available to all. But it could be that the more pessimistic predictions are correct, and the MEP does not provide a way of thinking about how to deal with allocating the burdens of decline. An obvious amendment would be what we could call the Baseline Minimum Equity Principle; that if the change leads to a decline in then those already at the bottom should not suffer as a result of that decline. In other words, all losses should fall on those best placed to bear them (for discussion see Shue, 1993; Moellendorf, 2018). I certainly would not want to reject this principle, especially when losses are relatively modest, but there may be cases where the losses are very significant and will need to be shared even by the worst off. To cope with these it seems necessary to generate a new principle, and one plausible version is what we could call the Negative Minimal Equity Principle (NMEP) which, like the MEP functions more like a constraint on action than a target:

> NMEP: When society is in a situation of decline, those who are already towards the bottom should not suffer disproportionately (relative to their well-being) compared to those who are better off.

For purposes of understanding the principle, imagine it is possible to put a number on each individual's well-being. When there is an overall decline, this principle calls for the burden to fall more heavily, in percentage terms as well as absolute quantities, on the well-off than the poor. One thing we have seen lately in welfare reform in the UK lately, so it has been claimed, is that those on lowest incomes have seen, proportionately, the greatest negative changes (Duffy, 2017). This is ruled as unjust by the NMEP. It is intended to block what we might call 'bad trickle-down': the dumping, proportionately, of the greatest costs on the least advantaged. One of the concerns about unchecked climate change is that it will have the greatest impact on those who are least able to cope with such shocks. It is equally critical that attempts to mitigate climate change do not have similar effects. What is the justification for MEP and NMEP? They are perhaps the most minimal possible interpretation of the idea that we are all to be regarded as equals. In particular, they express the thought that it is wrong for the rich to benefit at the expense of the poor.

This paper has been a long route to two connected thoughts. The first is that, like any complex new technology, initial active scrutiny, and strong regulation, faces the possibility of what I called regulatory drift. I identified three mechanisms of drift: complacency (neglect); gifting (deliberate de-regulation); and de facto regulatory capture by those with connections with the industry who will be among the few with the expertise to take on the role. Note this is not intended as a conspiracy theory or deliberate corruption. I called it drift for a reason. Others, no doubt, will suggest that I should have gone further.

The second danger follows in part from weak regulation, although it is not dependent on it, for strong regulation could have similar consequences; that the burden of the costs and side-effects of SRM fall disproportionately on those who are already the most vulnerable. To avoid such outcomes there should, in addition, be explicit adherence to the Negative Minimum Equity Principle as itself a principle of regulation. And it will be the critical (in more than one sense) role of civil society groups and international organisations to monitor and hold the regulators and SRM businesses to account. Regulation needs to be buttressed by a network of formal and informal pressure to keep regulators true to their task. Consider, for comparison, the area of animal rights where activists rarely give regulators a moment's rest. This is not to say that the activists get what they want – far from it – but they certainly get more of what they want than they would have done had they not been so active. And the collective power of the climate justice lobby could achieve a great deal.

Conclusion

I've argued that if we accept the suggestion that SRM is a critical stopgap that can provide time to allow a transition to a low emission economy, then we need to ensure protection from two related dangers: first regulatory drift that overly favours those who personally benefit from the technology, and second costs that disproportionately fall on the poor. I've suggested that principles of justice need to be incorporated into practices of regulation and that civil society groups should be encouraged and empowered to hold regulators to account. Obviously, there are many remaining questions. What mechanisms of accountability are available? What resources would be needed to pursue such claims? What if civil society organisations lose interest? I have not attempted to address such issues here. But I have, in effect, considered a different form of scepticism; that there can be no agreement on principles of justice. At one level I do not disagree with this criticism, but at another I think it is exaggerated. I believe that there will be wide acceptance of what I have called the Minimal Equity Principle, and the Negative Minimal Equity Principle, which set out constraints of justice but not a full theory. If either of these constraints are violated – and whether they are violated can often be contested – then regulators will have failed in their duty of justice. And for this they should be held to account, thereby mitigating one of the problems of regulatory drift. With this, and other safeguards, in place, we can stand ready to fight risk with risk, if it turns out that we really do have no practical alternatives. But at the same time, if, contrary to current scepticism, we begin to make real progress in curbing emissions, the case for SRM begins to evaporate.

Notes

1. The term 'moral hazard' is typically used in climate research to refer to the possibility that allowing SRM will distract us from taking appropriate measures to reduce carbon emissions (Hale, 2012). Here I mean something moral general: a structure that incentivises one agent to dump costs or risks on another.
2. For more systematic discussion, including that of 'paternalism' and 'moral sacrifice' see Wolff (2010) and 2010.

Acknowledgments

An early version of this paper was presented at Reading University, and I'd like to thank the participants for their comments. I'm particularly grateful to Stephen Gardiner, Catriona McKinnon and an anonymous referee for exceptionally helpful and generous written comments on an earlier draft of this paper.

Disclosure statement

No potential conflict of interest was reported by the author.

References

Browne, J. (2018). The regulatory gift: Politics, regulation and governance. *Regulation & Governance*. doi:10.1111/rego.12194

Crutzen, P. (2006). Albedo enhancement by stratospheric sulfur injections: a contribution to resolve a policy dilemma?. *Climatic Change, 77*(3-4), 211–219. doi:10.1007/s10584-006-9101-y

Duffy, S. (2017). *The politics of poverty*. Centre for Welfare Reform. Retrieved from https://www.centreforwelfarereform.org/library/categories/work/the-politics-of-poverty.html

Gardiner, S. (2006a). A core precautionary principle. *Journal of Political Philosophy, 14* (1), 33–60.

Gardiner, S. (2006b). A perfect moral storm. *Environmental Values, 15*(3), 397–413.

Gardiner, S. (2010). Is 'arming the future' with geoengineering really the lesser evil? Some doubts about the ethics of intentionally manipulating the climate system. In S. Gardiner and D. Jamieson, *Climate ethics: Essential readings Climate ethics: Essential readings*(pp. 284–312). Oxford: Oxford University Press.

Gardiner, S., & Fragnière, A. (2018). The tollgate principles for the governance of geoengineering: Moving beyond the Oxford principles to an ethically more robust approach. *Ethics, Policy & Environment, 21*, 143–174.

Gawande, A. (2009). *The checklist manifesto*. New York, NY: Henry Holt.

Hale, B. (2012). The world that would have been: Moral hazard arguments against geoengineering'. In C. Preston (Ed.), *Reflecting sunlight: The ethics of solar radiation management* (pp. 113–131). Lanham, MD: Rowman and Littlefield.

Hartzell-Nichols, L. (2012). Precaution and solar radiation management. *Ethics, Policy & Environment, 15*(2), 158–171.

Hartzell-Nichols, L. (2017). *A climate of risk*. London: Routledge.

Hermansson, H., & Hansson, S. O. (2007). A three-party model tool for ethical risk analysis'. *Risk Management, 9*(3), 129–144.

Hubert, A.-M. (2017). *Code of conduct for responsible geoengineering research*. Retrieved from https://www.ucalgary.ca/grgproject/files/grgproject/revised-code-of-conduct-for-geoengineering-research-2017-hubert.pdf

Irvine, P., Emanuel, K., He, J., Horowitz, L. W., Vecchi, G., & Keith, D. (2019). Halving warming with idealized solar geoengineering moderates key climate hazards. *Nature Climate Change, 9*(4), 295–299.

Kintish, E. (2011). *Hack the planet*. Hoboken: N.J Wiley.

Manson, N. A. (2002). Formulating the precautionary principle. *Environmental Ethics, 24* (3), 263–274.

McKinnon, C. (2018). Sleepwalking into lock-in? Avoiding wrongs to future people in the governance of solar radiation management research. *Environmental Politics, 28* (3), 441–459.

Moellendorf, D. (2018). Justice and climate change. In J. Wolff (Ed.), *Readings in moral philosophy* (pp. 457–467). New York: Norton.

Munthe, C. (2011). *The price of precaution and the ethics of risk*. Dordrecht: Springer.

Myers, N. J. (2005). A checklist for precautionary decisions. In N. J. Myers (Ed.), *Precautionary tools for reshaping environmental policy* (pp. 93–106). Cambridge, MA: M.I.T. Press.

Parfit, D. (1998). Equality and priority. In A. Mason (Ed.), *Ideals of equality* (pp. 1–20). Oxford: Basil Blackwell.

Parson, E. A., & Keith, D. W. (2013). End the deadlock on governance of geoengineering research. *Science, 339*(6125), 1278–1279.

Preston, C. (2013). Ethics and geoengineering: Reviewing the moral issues raised by solar radiation management and carbon dioxide removal. *Wiley Interdisciplinary Reviews: Climate Change, 4*, 23–37.

Ramanna, K. (2015). Thin political markets: The soft underbelly of capitalism. *California Management Review, 57*(2), 5–19.

Rawls, J. (1999). *A theory of justice* (Revised ed.). Oxford: Oxford University Press.

Robock, A. (2008). 20 reasons why geoengineering may be a bad idea. *Bulletin of the Atomic Scientists, 64*(2), 14–18.

Robock, A., Bunzl, M., Kravitz, B., & Stenchikov, G. L. (2010). A test for geoengineering? *Science, 327*(5965), 530–531.

Shue, H. (1993). Subsistence emissions and luxury emissions. *Law and Policy, 15*(1), 39–60.

Slovic, P. (2000). *The perception of risk*. London: Earthscan.

Wolff, J. (1996). Integration, justice, and exclusion. In U. Bernitz & P. Hallstrom (Eds.), *Principles of justice and the European Union* (pp. 15–26). Stockholm: Juristforlaget.

Wolff, J. (2010). Five types of risky situation. *Law, Innovation and Technology, 2*(2), 151–163.

Wolff, J. (2020). *Ethics and public policy* (2nd ed.). London: Routledge.

Wolff, J. (in press). Rules for regulators. In J.-S. Gordon (Ed.), *Smart technologies and human rights*. Leiden: Brill.

Wolff, J., & Orr, S. (2009). *Cross-sector weighting and valuing of QALYs and VPFs: A report for the Inter-Departmental Group for the valuation of life and health*. Retrieved from https://jonathanwolff.files.wordpress.com/2019/09/igvlh-final.pdf

The Panglossian politics of the geoclique

Catriona McKinnon

ABSTRACT

Solar radiation management (SRM) – a form of geoengineering – creates a risk of 'termination shock'. If SRM was to be stopped abruptly then temperatures could rise very rapidly with catastrophic impacts. Two prominent geoengineering researchers have recently argued that the risk of termination shock could be minimised through the adoption of 'relatively simple' policies. This paper shows their arguments to be premised on heroically optimistic assumptions about the prospects for global cooperation and sustained trust in an SRM deployment scenario. The paper argues that worst-case scenarios are the right place to start in thinking about the governance of SRM.

Introduction

Solar radiation (SRM) techniques propose to mask global warming by reflecting sunlight away from the Earth. The most developed SRM research programmes focus on one of the following methods.[1] Marine Cloud Brightening (MCB) involves increasing the reflectivity of marine clouds by spraying very fine droplets of sea water into them on a continuous basis. This could require fleets of ships continuously at sea with nozzles pointed upwards, spraying continuously. A research group at the University of Washington is working towards the outdoor experiments to test parts of the technology.[2] Stratospheric aerosol injection (SAI) involves spraying tiny reflective particles – sulfur, or nano-engineered particles – into the stratosphere in order to reflect solar radiation away. This could be done by large tethered balloons or by drones continuously circling the Earth. A research group at Harvard will undertake a small-scale outdoor experiment in the next couple of years, probably over New Mexico.[3]

In 2006 Paul Crutzen published a paper that made solar radiation management (SRM) a respectable topic for enquiry by the scientific community (Crutzen, 2006). Since then, there has been an increase in the interest paid to SRM by natural scientists, the media, and scholars concerned to assess not only the social and economic costs and benefits of the technologies but also the ethical worries and governance challenges SRM throws up (Callies, 2019; Chhetri et al., 2018; Gardiner, 2010; Gardiner & Fragnière, 2018; Heyward, 2015; Hourdequin, 2019; McKinnon, 2018; Morrow, 2014; Preston, 2012, 2016; Smith, 2014; Whyte, 2019). One of the most abiding of these worries relates the possibility of a potentially catastrophic 'termination shock' if SRM was to be stopped abruptly. If deployment was to be suspended by, for example, war, sabotage, or natural disaster then global average temperatures would be likely to rise very quickly to pre-deployment levels (Jones et al., 2013). Some studies suggest a similar, although less pronounced, effect on precipitation (Keller, Feng, & Oschlies, 2014; Zhang, Moore, Huisingh, & Zhao, 2015). These effects would be damaging to people in the future both as a result of the impacts of the temperature rises (and other changes) previously masked by deployment, but also because of the speed at which these impacts would occur (Goes, Tuana, & Keller, 2011; Reynolds, Parker, & Irvine, 2016; Svoboda, Keller, Goes, & Tuana, 2011). In a termination shock, climate impacts that would have taken decades or longer to materialise as a result of cumulative emissions would happen much more quickly, causing great damage to people affected by the shock (Baum, Maher, & Haqq-Mistra, 2013; Keith & MacMartin, 2015; MacMartin, Caldeira, & Keith, 2014).

In a recent issue of *Earth's Future*, Andy Parker and Peter Irvine argue that a handful of 'relatively simple policies' could minimise the risk of termination shock as a result of the abrupt cessation of SRM deployment (Parker & Irvine, 2018). They argue that these risks have been overstated because it is likely that there would be a time lag of a few months between abrupt cessation of SRM and the rapid temperature rises of a termination shock. The lag could allow for the resumption of deployment and so avert the shock.

Parker's and Irvine's paper matters not only because of what they argue but also because of who they are. Irvine and Parker both have notable pedigrees in the SRM research community. Irvine is a member of the *The Keith Group* at Harvard, led by David Keith, who has been 'the face' of SRM scientific research to date.[4] Parker was a Policy Advisor for the Royal Society's influential Report 'Geoengineering the Climate' (The Royal Society, 2009) and is Project Director for the Solar Radiation Management Governance Initiative.[5] Irvine and Parker have both co-authored with world leaders in the SRM research community. They are key rising stars in what Eli Kintisch has called the 'Geoclique' (Kintisch, 2010), and they have form in shaping the debate as it has developed in recent years.

In their paper, Parker and Irvine offer a number of reflections on the risks of termination shock under different scenarios. For example, they argue that SRM with a cooling effect of less than 0.1°C would not cause a significant shock (Parker & Irvine, 2018) and that termination shock under SAI could be avoided if deployment was phased out at a rate of 50 years per degree Celsius of cooling (Parker & Irvine, 2018). They also argue that the instantaneous cessation of SRM would not cause instantaneous termination shock. Instead, there would be several years of (what they call) a 'buffer period' before any appreciable global temperature rise. This warming could be avoided if deployment was to be resumed within a few weeks (for Marine Cloud Brightening) or a few months (for SAI) (Parker & Irvine, 2018).

What I shall focus on here are Parker's and Irvine's suggestions of how SRM deployment could be governed to minimise the risks of termination shock. They argue that global deployment of SRM could be designed and coordinated in ways that would significantly reduce the risk and that catastrophic risk anyway only attaches to the largest scale deployments. If they are right, then worries about termination shock provide greater impetus to focus on questions of how to govern any deployment, rather than reasons to turn away from SRM as a response to worsening climate change. Given that we might be on the cusp of an era in which SRM research enters the mainstream of climate policy responses, policymakers' perceptions of the landscape of risk created by SRM are important. An argument by two rising stars in the Geoclique that termination shock can be dealt with via 'relatively simple policies' (Parker & Irvine, 2018) could mean that the dangers of termination shock are not properly factored into decision-making around SRM research programmes, and any ultimate deployment. If – as I shall claim – their arguments are premised on a Panglossian view of world politics that we have good reasons to reject, their suggestions could encourage humanity to sleepwalk into a future dangerous deployment scenario (McKinnon, 2018).

I shall start by focusing on their claims that the risk of termination shock as a result of forced cessation of SRM at regional scales can be lessened – or even minimised – by a geographical distribution of backup deployment infrastructure, combined with adequate defences for that infrastructure (Parker & Irvine, 2018). I shall argue that there are serious tensions in their arguments which are all related to their background conception of political decision-making as never wantonly irrational, reckless, or morally corrupt. We know enough about the short-sighted, self-serving, and wilfully morally myopic, reasoning of political leaders in the face of global environmental catastrophe to design governance of SRM that avoids the Panglossian conception of political decision-making assumed by Parker and Irvine. They tell us that, 'it is not possible to reach useful policy conclusions based on the analysis of worst-case scenarios' (Parker & Irvine, 2018, p. 10) *Pace* Parker and Irvine, I shall argue that analysis of worst-case scenarios is exactly the right place to start for reaching policy conclusions about whether we should stimulate research into SRM now, and whether we should deploy this form of climate engineering in the future.

'Relatively simple' safeguards against forced termination shock: geographical dispersion and backup hardware

Parker and Irvine make three related suggestions of measures that could be taken to reduce the risk of termination as a result of forced disruption of deployment caused by destruction of, or damage to, the delivery infrastructure for SRM. The cases they have in mind are generated by political opposition – for example, terrorist attack – or regional collapse as a result of conflict, economic breakdown, or local natural disaster.[6] They claim that these measures should be 'relatively simple' to achieve. I shall argue that this is true only given heroically optimistic and unsupported assumptions about the prospects for future political cooperation between states. Let me start with two interrelated suggestions they make:

> *Geographical Dispersion*: Geographical distribution of deployment infrastructure would enable continuation of deployment in the event of any part of the distributed infrastructure going down. (5–6)
>
> *Backup Infrastructure*: 'Backup delivery hardware' kept by one state, or a group of states, would enable swift redeployment were the existing deployment infrastructure to be irreparably damaged. (5–6)

Taking *Geographical Dispersion* first, their claim is that the infrastructure for delivery of SRM should be distributed across many states and geographical locations so as to minimise the amount of disruption to deployment that could be caused by forced termination at any one place in the infrastructure. Far from being 'relatively simple', this proposal would require an unprecedented degree of cooperation and trust between all states in possession of delivery hardware. Furthermore, it is very likely that private companies will be involved in producing and maintaining deployment infrastructure. Interstate and corporate cooperation under Geographical Dispersion would have to be sustained across many decades, possibly across centuries, depending on (a) how quickly emissions could be brought down, (b) how quickly the atmospheric concentration of greenhouse gases could be reduced, and (c) how far disruption to deployment – by, for example, climate surprises – could be avoided.[7] This degree of sustained trust and cooperation between states and private companies is not impossible but it is certainly far from 'relatively simple' to achieve.

Furthermore – and more concerning – if efficiency and affordability require that different parts of the delivery infrastructure performing different functions are to be located in different parts of the world, it is not obvious that geographical distribution would properly address the danger of termination shock. If drones are manufactured in China, nozzles are made in Russia, sulfur compounds are made in Iran, and monitoring systems are made in Canada, then the geographical dispersion of the overall infrastructure across these and other sites would do nothing at all to limit the danger of termination shock as a result of an earthquake in China or terrorist attack in Canada.

Backup Infrastructure addresses circumstances in which deployment infrastructure is irreparably damaged.

> If any capable party, anywhere around the world, kept backup SRM delivery hardware, it could be redeployed to maintain the SRM cooling before temperatures started to rise rapidly … [i]f spare deployment capacity were maintained or numerous nations were capable of deployment, then the SRM system would be resilient against catastrophes that were confined to one country or region. (Parker & Irvine, 2018)

Their thought here is that if geographical dispersion of deployment capacity fails to prevent abrupt cessation of deployment, backup delivery hardware could be activated in what they call the 'buffer period' between abrupt cessation of deployment and the start of a termination shock. There are actually two ways in which this could happen (although Parker and Irvine do not make this distinction). The first way involves activation of the appropriate part of a 'shadow' deployment infrastructure that is also geographically dispersed, whereas the second way involves maintenance of backup hardware by one actor unilaterally, or by bilateral or minilateral cooperation between actors.

Taking the 'shadow infrastructure' version first, imagine that Germany has back up nozzles, the US maintains a fleet of backup drones, alternative sulfur compounds can be sourced in Australia, and replacement monitoring systems are kept up to date in Brazil. If one essential part of the original geographically dispersed infrastructure was to go down, the corresponding part of the shadow infrastructure would be activated. If drone manufacturing plants in China were destroyed, the US could step into the breach. Let me offer three reflections on the 'shadow infrastructure' rote to maintaining backup delivery infrastructure.

First, a 'shadow infrastructure' would amplify the problems of sustained trust and cooperation between states and private companies already mentioned: now we must not only assume cooperation with respect to primary deployment infrastructure but also with respect to the shadow infrastructure. And the problems of cooperation are given an additional dimension of complexity. Not only would states maintaining the shadow infrastructure have to cooperate with one another, but each state would also have to cooperate with whatever party is running the part of the deployment infrastructure that they are shadowing, so as to ensure their shadow components genuinely mirror the original. Without this, there would not be a guarantee that any part of the shadow infrastructure could replace its damaged counterpart in the event of forced termination.

Second, the shadow infrastructure suggestion would at least double the cost of deployment, which undermines one of the touted advantages of SRM as a tool to buy more time for mitigation, viz. that it is cheap.[8]

Third, reliance on a shadow infrastructure as a last line of defence against potentially catastrophic termination shock should commit us to maintain a shadow-shadow infrastructure. Given what is at stake, would we want to place all our hopes in just one layer of shadowing? If not, the problems of cooperation and cost already mentioned would balloon in ways that should alarm us.

Parker and Irvine do not envisage shadow infrastructure as the way in which to secure backup delivery hardware. Instead, as they state, they intend backup infrastructure to be maintained unilaterally, minilaterally or bilaterally by one or more capable parties, presumably including both states and private companies. Under this scenario, if deployment using geographically dispersed infrastructure was to be abruptly stopped, a capable party or club of parties could step into the breach and redeploy in the buffer period. If there were other parties – states or private companies – also in possession of deployment infrastructure through unilateral, bilateral or minilateral cooperation, then that would serve as backup to the now deployed backup, thus solving the shadow-shadow infrastructure problems mentioned above.

Parker and Irvine's suggestion that unilaterally, bilaterally or minilaterally maintained backup capability would be a good safeguard against the risk of termination shock in the event of a geographically dispersed delivery infrastructure going down rests, again, on a Panglossian vision of international politics. In fact, their proposal stands in serious tension with *Geographical Dispersion* because it would reintroduce the potentially catastrophic danger of abrupt cessation for which geographical dispersion was offered as a remedy. This is the case because it would incentivise a 'backup deployment infrastructure arms race' between states, private companies, and a mixture of these capable parties. Any party that is unilaterally (bilaterally, minilaterally) in possession of backup deployment hardware is in a position of immense global power, and all parties would know this. It has often been noted that unilateral, bilateral or minilateral control over SRM deployment technology itself would create a severe imbalance of power across states. That this imbalance could be echoed in governance structures enabling deployment has not been noticed, and would increase the risks of dangerous imbalances of power in a deployment scenario.

Finally, and worst, when Panglossian assumptions about the reasons for which states and private companies act are abandoned, it is apparent that in a backup deployment arms race, there are powerful incentives for parties to deliberately damage existing geographically dispersed infrastructure and so become unilaterally (bilaterally, minilaterally) positioned with their hands on the global thermostat. And if there is more than one party (or club) with backup deployment hardware, that incentive remains even if the first party to unilaterally use their backup deployment hardware manages to avert termination shock by redeploying. If other parties have backup capability, there is the world to be gained by bringing down the backup infrastructure of competitors who were the quickest to redeploy after the failure of a geographically dispersed deployment programme.

Parker and Irvine present *Backup Infrastructure* as the inverse of 'mutually assured destruction', whereby freedom from nuclear strikes is supposed to be enhanced by nuclear proliferation and the balance of power it creates. Their 'unilateral (bilateral, minilateral) backup' scenario is supposed to be a case of 'mutually assured survival': freedom from termination shock is enhanced by the proliferation of SRM delivery hardware and the global safety net this provides. But once we abandon heroically optimistic assumptions about the reasons for which states and corporations act, we can see that their proposal could easily create a greater risk of termination shock than the original scenario of geographically dispersed deployment infrastructure: the cure is worse than the disease.[9]

'Relatively simple' safeguards against forced termination shock: adequate defences

The third suggestion Parker and Irvine make for guarding against force termination shock is this:

Adequate Defences: Delivery equipment should be protected with adequate defences, 'such as those that guard power plants or military bases' (p. 6).

Adequate Defences is supposed to address elective drivers of termination shock such as deliberate attacks on infrastructure. However, there are immediately noticeable differences between the defences in place for military bases and power plants, and those that would be needed to provide adequate protection for SRM delivery infrastructure. Military bases and nuclear plants have a fixed geographical location, whereas SRM and MCB delivery infrastructure would have many mobile components, such as the drones delivering the sulfur particles to the stratosphere or the ships spraying saline into marine clouds. In addition to providing adequate defences for these moving parts, all the manufacturing and production lines, distribution and supply chains, and product design and improvement intelligence would require adequate defences. An additional layer of complexity is introduced once we recognise that many if not all of these functions would be very likely to be contracted out to private companies that would also require adequate defences for their operations across multiple sites.

Furthermore, we could only be sure that defences are adequate and reliable if they are regularly inspected by some body with the expertise, authority and legitimacy to do this. This, again, reveals Parker's and Irvine's extremely optimistic assumptions about cooperation, trust and transparency between states. If we avoid Panglossianism by assuming that levels of conflict and mistrust between states in the future will not significantly lessen, or disappear, then nothing about Parker and Irvine's recommendation on adequate defences for SRM infrastructure is 'relatively simple'.

One way to make the *Adequate Defences* challenge manageable would be to give just one state, or a small group of states, control over provision and inspection of adequate defences for delivery infrastructure, perhaps by sanctioning that state, or club of states', monopoly over production. Here, again, we see the fundamental tension in Parker's and Irvine's proposals. The conditions under which they would be effective are precise conditions in which they would fail to minimise the danger of termination shock. Taking seriously states' desires for geopolitical dominance, their corrupt political leadership, and the profit motive of private corporations makes Parker's and Irvine's proposals a manifesto for unilateral or minilateral production, defence and deployment of SRM which increases the risk of termination shock.

The bigger picture: worst-case scenarios are where we should start

Parker and Irvine are right that the mere possibility of catastrophic harm as a result of termination shock does not provide policy guidance (Parker & Irvine, 2018) and that consideration must be given to the likelihood of termination shock in thinking about how to govern the development and potential deployment of this technology. However, they misinterpret the significance of the imperative to think more about the likelihood of termination shock, and they misconstrue the phenomena to which it applies. This explains in part, I think, why they proceed using Panglossian political assumptions. They understand the requirement to think about the likelihood of termination shock as applying only to the physical processes that could cause a termination shock was deployment to be interrupted. This is why, at the beginning of their paper, they focus on defending the claims that termination shock is scalar i.e. not all 'shocks' would cause substantial warming, termination shock could be avoided if SAI was phased out at a rate of 50 years per degree Celsius of cooling, and instantaneous cessation of SRM would not cause instantaneous termination shock. What they massively underestimate is the scale of the political, social and economic challenges to the proposals they make to address the non-physical causes of termination shock. This leads them to ignore entirely the question of whether, and under what conditions, their governance proposals could make termination shock more likely. It is also fit to encourage complacency about the difficulty of SRM governance which, in my view, creates a real danger that we will sleepwalk into being locked-in to deployment, and/or to research into the most dangerous form of SRM (McKinnon, 2018).

My argument is not that we know that future social, political and economic conditions will be such that Parker's and Irvine's proposals could make termination shock more likely. Although we live in a deeply unjust world, these injustices – created by the action of states, corrupt political leaders, and the profit-seeking behaviour of firms in a capitalist global economy – are not unalterable facts. I do not believe humanity is stuck with its present unjust hand. A world of cosmopolitan justice is a live option (Caney, 2005). Capitalism does not have to be the future (Cohen, 2009). Human beings – including political leaders – can and should be expected to act ethically and justly (Estlund, 2011). We are deeply uncertain about the prospects for a more just future in which states and other actors resist the incentives that would be created by Parker's and Irvine's proposals, and which would increase the likelihood of termination shock. The fact – if it is one – that assessing the likelihood of termination shock as a result of physical processes is more nuanced than is often assumed provides only partial information for an overall assessment of the likelihood of termination shock. Understanding the full picture requires assessing how likely it is that the actors in charge of any future deployment will be willing and able to sustain transparent cooperation and maintain trust over a period of at least decades, perhaps longer. We are deeply uncertain about the prospects for this cooperation, and Parker and Irvine do not acknowledge this.

The core of the problem with Parker's and Irvine's approach is that they assume the incentives on states collectively to maintain SRM and avoid termination shock will generate the unprecedented levels of sustained interstate and cross-corporation trust, cooperation, and transparency necessary for their proposals to deliver on their promise of minimizing the risk of termination shock. They claim that they do not assume perfect rationality on the part of actors involved in deployment and that their minimal assumption is only that actors will avoid wanton irrationality in the face of disruption to deployment fit to cause a termination shock. One important thing to note here is that there has been a very strong – overwhelming – rationale for aggressive mitigation for the last few decades, and yet this has not moved states to act as they should and avert the climate crisis. Putting that to one side, if Parker and Irvine are right to assume no wanton irrationality from actors involved in SRM deployment, their 'relatively simple' proposals incentivize powerful states and corporations to minilateral implementation of the measures they contain. In a world of states with opposed interests and nefarious ambitions, the most instrumentally rational choice for a powerful state or corporation committed to minimizing the risk of termination shock is to band together with other powerful actors to produce, protect, and deploy SRM in the ways Parker and Irvine suggest. Actors that avoid wanton irrationality will seek the most efficient and reliable means to their ends. If those actors are themselves ethically questionable, or are hemmed in by structures of interaction that crowd out ethically required action, then the avoidance of wanton irrationality by these actors will not magically transform the ends of their action into those that serve the interests of the whole human community over time.

Where does this leave us? Underlying my criticisms of Parker and Irvine is the view that worst-case scenarios are the right place to start in our political thinking about SRM. This is not because we know that worst-case scenarios with respect to future actors in charge of deployment are more likely than best-case scenarios. As I have been emphasizing, we do not know this. Rather, starting with worst-case scenarios is required in thinking about termination shock because there is a powerful argument for taking a precautionary approach to such cases. When uncertainty about the consequences of a course of action is extensive and deep, when the costs of assuming more than we are entitled to give this uncertainty are catastrophic, and when these costs will fall on people separate to those making decisions under these assumptions, we are morally required not to make political decisions using these assumptions. Applied to the present case: we are not permitted to assume that states and private companies will cooperate with one another in terms of trust in order to build and sustain a system of governance for SRM deployment that realizes geographical dispersion, backup infrastructure, and adequate defences.

The bigger picture behind my objections to Parker's and Irvine's proposals is that they encourage political decision-making that violates a moral requirement, given that deployment governed by the proposals has features that bring it within the scope of this precautionary approach. These features are described by Henry Shue as follows.[10]

> *Massive loss*: the magnitude of the possible losses [in the case] is massive;
>
> *Threshold likelihood*: the likelihood of the losses [in the case] is significant, even if no precise probability can be specified [for these losses], because the mechanism by which the losses would occur is well understood, and the conditions for the functioning of the mechanism are accumulating. (Shue, 2010)[11]

Everyone agrees that the magnitude of the losses involved in a termination shock would be massive, so let me focus instead on *threshold likelihood*. I have claimed that we are deeply uncertain of the political, social and economic future. It is possible that states and other actors in control of deployment could evolve in ways that make sustained trust and cooperation feasible, but it is also possible that the unjust status quo could continue, or even worsen. If we do not know the probabilities of these futures, how can we know whether they are above any threshold of probability? The right approach here is to make a distinction between exactitude in our judgements of the probability of uncertain outcomes and accuracy in our judgements about whether these outcomes are above some threshold of probability: judgements of the latter type can be well founded when judgements of the former type are not. Shue defends this method (Shue, 2010, 2015): 'one can reasonably, and indeed ought to, ignore entirely questions of probability beyond a certain minimal level of likelihood'.

Following Shue, the relevant question is not 'what is the likelihood of states and other actors cooperating with sustained trust and transparency in the future?'. Instead, the relevant questions are (a) 'are there well-understood political, social and economic mechanisms that could bring about termination shock when the proposals for governing deployment are implemented?', and (b) 'are the conditions for the functioning of these mechanisms accumulating?'. My argument in this paper relates to (a). I have claimed that once we abandon Panglossian assumptions about levels of trust that can be sustained between states, the motives of private corporations, and the integrity of political leaders, we have good reasons to believe that Parker's and Irvine's governance proposals would function as mechanisms that would exacerbate the risk of the termination shock. The natural next step is to argue what would need to be changed in the global political, social and economic order so to make Parker's and Irvine's optimism warranted. Those who care about the governance of SRM to avoid termination shock should focus on this question before dreaming about a relatively simple world in which we need no longer worry about the worst outcomes of deployment.

Conclusion

Getting serious about minimizing the ethically unacceptable dangers of termination shock requires much more than the superficial proposals made by Parker and Irvine. Parker and Irvine claim that '[i]t is not justifiable to draw insights about the risk of termination shock by reporting the magnitude of the worst possible impacts, while leaving aside consideration of the likelihood of events that could cause or prevent termination' (Parker & Irvine, 2018). Read in one way, this is exactly right. In deciding how to act now to govern SRM we must take account of the likelihood of events that could cause or prevent termination. My objections to Parker and Irvine have been that they are sanguine to the point of naivete about the prospects for friendly conditions under which their proposals could reduce the risk of termination shock. If conditions of sustained and trusting cooperation between states and corporations do not obtain, their proposals will in fact make termination more likely, given the incentives they would create for powerful and self-interested actors.

Although we have ample evidence from history that states and corporations often yield to the temptation to exploit new technologies – and the governance of them – for their own ends (very often to benefit their most wealthy and powerful members), this is not set in stone forever. The reality is that we do not know how states and other actors will evolve in the future. But we are not warranted to assume the best of these actors, given this uncertainty and given the high stakes. The governance we need now for SRM must not proceed with these assumptions. Instead, it should offer immediate proposals for building out governance now in ways that can be ramped up to overhaul – perhaps, replace – our present ethically inadequate institutions.[12] It is not justifiable to draw insights about the risk of termination shock by assuming the best of states and corporations charged with managing deployment while leaving aside consideration of how deeply uncertain it is that these actors will act in the best interests of humanity as a whole.

Notes

1. SRM methods that have not generated serious interest from research scientists include placing mirrors in space, painting rooftops white, and genetically engineering crops to increase the reflectivity of their surfaces.
2. See The Marine Cloud Brightening Project: www.mcbproject.org. Accessed 5 February 2019. Thomas Ackerman leads on this project.
3. The experiment is called SCoPEx: https://projects.iq.harvard.edu/keutschgroup/scopex. Accessed 5 February 2019. David Keith used to lead on this project, which is now led by Frank Keutsch.
4. See https://keith.seas.harvard.edu/home. Accessed 5 February 2019.
5. The SRMGI is an NGO-driven project that aims to expand the global conversation around the governance of SRM research. See http://www.srmgi.org/. Accessed 5 February 2019.
6. Parker and Irvine accept that a *global* catastrophic event fit to disrupt deployment would not be addressed by their 'relatively easy and cheap' measures. They reflect that 'global domestic product would … need to drop by over 90% before maintaining SRM would cost more than 1% of the collective post-catastrophe GDP of the world's top 20 economies' and thus that 'if spare deployment capacity were maintained, or could be brought online quickly, a catastrophe would have to be on a scale unprecedented in modern history to force termination shock' (p. 6). This strikes me as a non sequitur for at least two reasons. First, there are good reasons to worry that in the Anthropocene we could face a number of catastrophes unprecedented in human, let alone modern, history (see Bostrom & Ćirković, 2008). Second, the fact (if it is one) termination shock caused by a global catastrophe would have to reduce GDP by over 90% for redeployment to cost more than 1% of GDP is just one factor in assessing how likely states would be to redeploy. Looking at the lamentable failure to date of states to take measures to tackle climate change that they can easily afford tells against Parker and Irvine's assumption that the affordability of redeployment significantly increases the likelihood of redeployment. That states can afford to do the right thing is tangential to whether they are likely to do the right thing.
7. One oddity of Parker's and Irvine's classification of the possible driver events for termination shock is that they fail to register climate surprises as a driver that does not fit their schema. Disruption of deployment as a result of, for example, the speedy shutdown of the THC is something humanity could not control (and

so looks like an external driver) and yet is caused by human choices (like an elective driver).

8. For criticism of the claim that cheapness per se is an advantage consider Stephen Gardiner's comment: '[s]aying that SSI is cheap because it does not cost much to spray particles into the stratosphere is a little like saying "brain surgery is cheap" because it would not cost much for me to buy a scalpel and start cutting into your head. While true in one sense, the point is largely irrelevant. What we care about in the brain surgery case is much wider, such as what the consequences of my digging will be, whether I am qualified, whether I have the right to do it, and what will happen if things go wrong. The same is true of climate engineering. To say that SSI is "cheap" is to ignore the most relevant "costs"' (Gardiner, 2019, p. 31).
9. There are parallels here with Stephen Gardiner's argument SSI deployment could start a geoengineering arms race: states with a legitimate interest in self-defence could undertake their own SRM deployments if they are threatened by the unilateral deployment of a different state, or if they believe that a different form of geoengineering is better for them than the one to which they are reacting. This dynamic could lead to an SSI arms race (Gardiner, 2013).
10. Shue's account of the conditions for precautionary action is influenced by (Gardiner, 2006). Other excellent recent treatments of precautionary approaches are (Hartzell-Nichols, 2017; Steel, 2014).
11. Shue identifies an additional third feature not listed here, as follows: '(3) *non-excessive costs*: the costs of prevention are not excessive (a) in light of the magnitude of the possible losses and (b) even considering other important demands on our resources'. (2010, p. 148). I shall take it to be straightforward that prevention costs are non-excessive in both the case of warming above 2C and the case of governance stimulating SRM research. In the climate case, the costs are those created by mitigation. In the SRM case, the costs are those created by not doing the research. Given the early-days state of research into SRM, these costs are almost entirely opportunity costs. Some SRM researchers make the case that this research could enable us to learn important things that would also benefit mitigation efforts, e.g. with respect to clouds. Granting this, if massive loss and threshold likelihood are satisfied in the case of governance stimulating SRM research, these opportunity costs are far from excessive, especially given that we may be able to learn these important things about clouds by other means.
12. There are a number of proposals already as to how to do this. For example, see Chhetri et al., 2018; Gardiner, 2014; González-Ricoy & Gosseries, 2016.

Acknowledgments

I am very grateful to Steve Gardiner and an anonymous reviewer for their constructive comments on this paper.

Disclosure statement

No potential conflict of interest was reported by the author.

ORCID

Catriona McKinnon http://orcid.org/0000-0003-0434-1083

References

Baum, S. D., Maher, T. M., & Haqq-Mistra, J. (2013). Double catastrophe: Intermittent stratospheric geoengineering induced by societal collapse. *Environment, Systems and Decisions*, *33*(1), 168–180.

Bostrom, N., & Ćirković, M. M. (2008). *Global catastrophic risks*. Oxford: Oxford University Press.

Callies, D. E. (2019). Institutional Legitimacy and Geoengineering Governance. *Ethics, Policy & Environment*, 21(3) 324–340.

Caney, S. (2005). *Justice Beyond Borders*. Oxford: Oxford University Press.

Chhetri, N., Chong, D., Conca, K., Falk, R., Gillespie, A., Gupta, A., … Nicholson, S. (2018). *Governing solar radiation management*. Washington, D.C.: Forum for Climate Engineering Assessment.

Cohen, G. A. (2009). *Why not socialism?* Princeton, N.J.: Princeton University Press.

Crutzen, P. J. (2006). Albedo enhancement by stratospheric sulfur injections: A contribution to resolve a policy dilemma? *Climatic Change*, *77*(3–4), 211–219.

Estlund, D. (2011). Human nature and the limits (If Any) of political philosophy. *Philosophy and Public Affairs*, *39*(3), 207–237.

Gardiner, S. M. (2006). A Core Precautionary Principle*. *Journal of Political Philosophy*, *14*(1), 33–60.

Gardiner, S. M. (2010). Is 'arming the future' with geoengineering really the lesser evil? Some doubts about the ethics of intentionally manipulating the climate system. In S. M. Gardiner, S. Caney, D. Jamieson, & H. Shue (Eds.), *Climate Ethics: Essential readings* (pp. 284–314). Oxford: Oxford University Press.

Gardiner, S. M. (2013). The desperation argument for geoengineering. *PS: Political Science and Politics*, *46*(1), 28–33.

Gardiner, S. M. (2014). A call for a global constitutional convention focused on future generations. *Ethics & International Affairs*, *28*(3), 299–315.

Gardiner, S. M. (2019). Climate engineering. In D. Edmonds (Ed.), *Ethics and the contemporary world* (pp. 29–43). London: Routledge.

Gardiner, S. M., & Fragnière, A. (2018). The tollgate principles for the governance of geoengineering: Moving beyond the Oxford principles to an ethically more robust approach. *Ethics, Policy & Environment, 21*(2), 143–174.

Goes, M., Tuana, N., & Keller, K. (2011). The economics (or lack thereof) of aerosol geoengineering. *Climatic Change, 109*(3–4), 719–744.

González-Ricoy, I., & Gosseries, A. (eds.). (2016). *Institutions For Future Generations*. Oxford: Oxford University Press.

Hartzell-Nichols, L. (2017). *A climate of risk precautionary principles, catastrophes, and climate change*. London: Routledge.

Heyward, C. (2015). Is there anything new under the sun? In A. Maltais & C. Mckinnon (Eds.), *The Ethics of Climate Governance* (pp. 133–154). London: Rowman and Littlefield International.

Hourdequin, M. (2019). Climate change, climate engineering, and the 'Global Poor': What Does justice require? *Ethics, Policy & Environment*, 21(3) 270–288.

Jones, A., Haywood, J. M., Alterskjaer, K., Boucher, O., Cole, J. N. S., Curry, C. L., ... Yoon, J.-H. (2013). The impact of abrupt suspension of solar radiation management (termination effect) in experiment G2 of the Geoengineering Model Intercomparison Project (GeoMIP). *Journal of Geophysical Research: Atmospheres, 118*(17), 9743–9752.

Keith, D. W., & MacMartin, D. G. (2015). A temporary, moderate and responsive scenario for solar geoengineering. *Nature Climate Change, 5*(3), 201–206.

Keller, D. P., Feng, E. Y., & Oschlies, A. (2014). Potential climate engineering effectiveness and side effects during a high carbon dioxide-emission scenario. *Nature Communications, 5*, 1–11.

Kintisch, E. (2010). *Hack the planet : Science's best hope– Or worst nightmare– For averting climate catastrophe*. London: John Wiley & Sons.

MacMartin, D. G., Caldeira, K., & Keith, D. W. (2014). Solar geoengineering to limit the rate of temperature change. *Philosophical Transactions of the Royal Society A, 372*, 20140134.

McKinnon, C. (2018). Sleepwalking into lock-in? Avoiding wrongs to future people in the governance of solar radiation management research. *Environmental.*

Morrow, D. R. (2014). Ethical aspects of the mitigation obstruction argument against climate engineering research. *Philosophical Transactions of the Royal Society A: Mathematical, Physical and Engineering Sciences, 372*(2031), 20140062–20140062.

Parker, A., & Irvine, P. J. (2018). The risk of termination shock from solar geoengineering. *Earth's Future*, 6, 456–457.

Preston, C. J. (2012). *Engineering the climate: The ethics of solar radiation management*. Lanham: Lexington Books.

Preston, C. J. (2016). *Climate justice and geoengineering : Ethics and policy in the atmospheric Anthropocene*. London: Rowman & Littlefield International, Ltd.

Reynolds, J. L., Parker, A., & Irvine, P. (2016). Five solar geoengineering tropes that have outstayed their welcome. *Earth's Future, 4*(12), 562–568.

The Royal Society. (2009). *Geoengineering the climate: Science, governance and uncertainty*. London: Royal Society. No. 10/09.

Shue, H. (2010). Deadly delays, saving opportunities: Creating a more dangerous world? In S. M. Gardiner, S. Caney, D. Jamieson, & H. Shue (Eds.), *Climate ethics: Essential readings* (pp. 146–162). Oxford: Oxford University Press.

Shue, H. (2015). Uncertainty as the Reason for Action: Last Opportunity and Future Climate Disaster. *Global Justice: Theory Practice Rhetoric*, *8*(2), 86–103.

Smith, P. T. (2014). Redirecting threats, the doctrine of doing and allowing, and the special wrongness of solar radiation management. *Ethics, Policy & Environment*, *17* (2), 143–146.

Steel, D. (2014). *Philosophy and the precautionary principle*. Cambridge: Cambridge University Press.

Svoboda, T., Keller, K., Goes, M., & Tuana, N. (2011). Sulfate aerosol geoengineering: The question of justice. *Public Affairs Quarterly*, *25*(3), 157–180.

Whyte, K. P. (2019). Indigeneity in geoengineering discourses: Some considerations. *Ethics, Policy & Environment*, 21(3) 289–307.

Zhang, Z., Moore, J. C., Huisingh, D., & Zhao, Y. (2015). Review of geoengineering approaches to mitigating climate change. *Journal of Cleaner Production*, *103*, 898–907.

Democratic authority to geoengineer

Holly Lawford-Smith

ABSTRACT
Does any existing single actor have, or could any existing single actor come to have, the authority to geoengineer? In this paper, I will focus on Solar Radiation Management strategies (leaving at least some Carbon Dioxide Removal strategies on the table). I'll argue that global democratic authorization is possible in principle, and could be obtained on the basis of large-scale representative sampling. I present experimental findings from the Australian context showing that democratic authorization would not be granted, and conclude that if we can expect this result to generalize, then the deployment of SRM by a single actor is impermissible.

Unilateral deployment

It is widely recognized that geoengineering comes with serious risks, including for land use, food security, biodiversity, pollution, and political stability. But there's another kind of risk, namely that of unilateral deployment: an actor or group of actors deploying geoengineering strategies, with effects that will be felt by other actors and in an unpredictable way (Virgoe, 2009; Lane & Bickel, 2009; Blackstock & Long, 2010; cf. Horton, 2011).

In a recent issue of *Science*, Janos Pasztor (Executive Director of the Carnegie Climate Geoengineering Governance Initiative), Cynthia Scharf (Senior Strategy Director for the Carnegie Climate Geoengineering Governance Initiative), and Kai-Uwe Schmidt (Professor of Mathematics at the University of Paderborn) state that:

> 'The greatest near-term risk, however, may be the unilateral deployment of SRM by one country, a small group of countries, or a wealthy individual'. (Pasztor et al., 2017)

Pasztor continues on this theme in another piece:

> 'We need to acknowledge that the aggregate environmental and socioeconomic risks of solar radiation management would probably be small in

> comparison with the benefits of reducing global temperatures. But those benefits and harms would be unequally spread among regions of the world, and between current and future generations. In the absence of multilateral agreements, there's no way of controlling who might execute such a geoengineering plan. It's possible that a small group of countries, or a single country, or a large company, or even a wealthy individual might take unilateral action on geoengineering. Others might subsequently engage in their own climate engineering strategies to counter such action'. (Pasztor, 2017)

For example, imagine that Australia decided to dedicate part of the 2020–2021 budget to Stratospheric Aerosol Engineering (SAI), dispersing sulphide gases (sulphuric acid, hydrogen sulphide, sulphur dioxide) into the stratosphere above Australia using a fleet of aircraft, and by the end of the financial year have in fact deployed SAI as planned. Or perhaps Bill Gates, or Coca-Cola, might decide to do this. Would that be permissible? Does any existing actor have, or could any existing actor come to have, the authority to geoengineer – in particular, to deploy SRM? (The same considerations will likely apply to some, but not all, Carbon Dioxide Removal (CDR) strategies).

I'll argue that any permissible deployment of SRM must have global democratic authorization, and present evidence that suggests such authorization is unlikely. In Section II, I'll ask whether disagreements over the permissibility of geoengineering are reasonable or unreasonable. I'll suggest that if they are reasonable, they should be settled democratically. In Section III, I'll ask who the 'demos' is when it comes to decisions about geoengineering, and argue that it's all individuals. Then I'll work through some of the challenges to taking decisions from this group. In Section IV, I defend the idea of representative sampling as an approximation of democracy. Finally, in Section V, I'll present evidence from the Australian context suggesting that geoengineering would not be democratically authorized.

Disagreements over deployment

When is the deployment of SRM by a single actor permissible? One way to approach this question is to ask whether disagreements between people over deployment are reasonable or unreasonable.[1] If they are unreasonable, there could be an obvious answer as to what we should do – clearly we *shouldn't* deploy SRM; clearly we *should* deploy SRM – and an explanation of why some groups refuse to concede that answer (excessive recklessness, technophilia, biased weighting of benefits against risks; excessive risk-aversion; technophobia; religious values in tension with human intervention upon nature).

If disagreement in this area is unreasonable, it is more plausible that that's because we obviously *shouldn't* geoengineer, rather than that we obviously should. There is a lot of uncertainty around the various solar geoengineering proposals, and their possible effects. The Carnegie Council Geoengineering Governance Initiative (C2G2) state on their FAQ page that in 2017 they established three priorities for geoengineering governance. One of these was to put solar geoengineering deployment on hold until 'the risks and benefits are better known', and 'the governance frameworks necessary for deployment are agreed' (Carnegie Climate Geoengineering Governance Initiative [C2G2], 2018). One of the Oxford Geoengineering Programme's five principles for the governance of geoengineering is 'Governance before deployment', namely that 'any decisions with respect to deployment should only be taken with robust governance structures already in place, using existing rules and institutions wherever possible' (Oxford Geoengineering Programme, 2009).

If that's right, then we can end the paper here: no single actor could have the authority to geoengineer, at least not with SRM. The enthusiasm of anyone who supports unilateral deployment of SRM can be explained away as unreasonable. But for the sake of argument, let's assume that disagreements over deployment are a matter of *reasonable* disagreement. Different types of SRM come with different costs, benefits, and risks. The threat of unmitigated climate change is great, and in a straight choice between facing that threat and deploying SRM to avoid it a reasonable person might choose the latter (see critical discussion of this apparently tragic choice in Gardiner, 2010). Perhaps a reasonable person could even prefer the uncertainty and risk associated with a strategy that protected her current lifestyle to a GHG emissions reduction plan that negatively affected it (even if only to a small degree).

When an issue is a matter of reasonable disagreement, it ought to be settled democratically. This is not to say that we can go straight to a vote; there might need to be a period of debate and discussion, and experts might need to weigh in to rectify common misunderstandings. When the issue is geoengineering, this will clearly be necessary, given the complexity of the interventions and the risks involved. There is also ample room for debate over the democratic procedure itself, including who should be enfranchised (children, future generations, and convicted criminals all raise challenges); which voting procedure to use (e.g. first past the post, preferential voting); and how to draw regional boundaries (e.g. the US's electoral college vs. popular vote). But these are debates that can be settled consistent with the general claim being true, namely that when the people disagree over a policy matter and their disagreement is reasonable, settling the disagreement should be done democratically.

On the charitable assumption that disagreements over using SRM are reasonable, we should settle them democratically. But that means we have to settle one of the procedural questions just raised: who is the 'demos' when it comes to decisions about geoengineering?

The 'demos' for geoengineering decision-making

If every country was a well-functioning democracy (and there were no stateless people), then we could take the demos when it comes to decisions about geoengineering to be the aggregate of these countries (to be *countries* rather than *individuals*, because countries could be relied on to represent the interests of their people). But not every country is democratic; not every democracy is well functioning (for example, in terms of the United States see discussion in Stanley, 2016); and there were an estimated 10 million stateless people in 2017 (Institute on Statelessness and Inclusion [ISI], 2018). Thus, when it comes to decisions about geoengineering, the demos cannot be aggregated out of existing democracies.

This means there's a 'constituting the demos' problem (alternately called the 'democratic boundary problem' and the 'problem of inclusion'). This is the problem of figuring out *what* the demos is, before taking a vote from it. One plausible solution to the constituting the demos problem is the all-affected principle (Arrhenius, 2005; Goodin, 2007). This principle includes everyone affected by a decision in the vote on the decision. So, for example, given that the deployment of SRM will affect everyone, everywhere – because its costs and benefits fall entirely unpredictably on some – the demos for geoengineering decision-making is everyone, everywhere. For discussion of the gendered aspects of geoengineering, one of which is that women will be disproportionately affected by it, see Buck, Gammon, & Preston, 2014. If we know that some groups will be affected more than others, we have reason to work especially hard to make sure they have a voice in decision-making).

Another solution that has been offered to the constituting the demos problem is the all-coerced principle. This principle includes everyone coerced by a decision in the vote on a decision (Abizadeh, 2008, pp. 39–42). Arash Abizadeh understands coercion as interfering with autonomy in three ways: by destroying or hindering the development of the mental capacities required for a person to formulate and pursue personal projects; by eliminating options such that a person is left with an inadequate range of valuable options; and by subjecting the person to another's will and thus violating her independence (*ibid*, 39–40).

The costs and risks of deploying SRM fall probabilistically on everyone, everywhere, and interfere with autonomy both actually (for those people the costs and risks in fact fall on) and prospectively (for everyone is dominated by the possibility that they will be subjected to these costs and risks in a way that is impossible to predict). A single actor's decision to deploy SRM violates the independence of those not involved in the decision; the negative effects of the decision constrain people's options, in some cases to the point that they no longer have an adequate range of options (for example, the severe effects on temperature if there is abrupt cessation of a SRM project; see discussion in

Currie & Lawford-Smith, 2017); and the negative effects of the decision may destroy or hinder mental capacities in those who are physically and psychologically harmed by deployment (e.g. those who suffer from the effects of rapid warming after there is abrupt cessation; those who are the victims when things go wrong in the attempt to deploy SRM).

These two principles *converge* on the answer to what the demos is when it comes to decisions about geoengineering: everyone, everywhere. This also fits with the second of the Oxford Principles, which is 'Public participation in geoengineering decision-making'. They say that 'wherever possible, those conducting geoengineering research should be required to notify, consult, and ideally obtain the prior informed consent of, those affected by the research activities' (Oxford Geoengineering Programme, 2009). If they say this about *research* then they will certainly want to say it about *deployment*. (For a critique of the Oxford Principles, see Gardiner & Fragniere, 2018).

'Everyone, everywhere' is an imperfect solution to the boundary problem, on both theories, because what it means in practice is *people who are currently alive*. Future people will be affected, and may have their autonomy interfered with as a result of SRM deployment, but they don't exist yet, so we can't give them a vote. Old people may *not* be affected, or have their autonomy interfered with as a result of SRM deployment, given the lag between deployment and environmental impacts, and yet will get a vote. People alive today may use their vote to burden the future with having to perpetually maintain the SRM put in place, in order to avoid the negative effects of abrupt cessation (see discussion in Wong, 2014). There's not a lot we can do to ensure that people vote ethically (especially considering there is unlikely to be a shared conception of what it means to do so).

Various people have made proposals for enfranchising the future (see, e.g. Gonzalez-Ricoy & Gosseries, 2017, Chs., pp. 7–24). Juliana Bidadanure, for example, suggests youth quotas in parliament as a proxy for representing future generations, and defends the suggestion on the grounds that more intergenerational diversity is likely to mean more innovative long-term problem-solving (Bidadanure, 2017). Simon Niemeyer & Julia Jennstal suggest deliberative mini-publics considering issues relating to intergenerational equity (Niemeyer & Jennstal, 2017). Such proposals could be introduced alongside standard voting, and once introduced might also increase the likelihood of current people voting in a way that is more sensitive to future people's likely interests.

Some might want to argue that *only* the future should be enfranchised on this issue, given that they will be disproportionately affected, so such proposals should replace a global democratic vote. But we should be cautious about such arguments. Average life expectancy in 2015 was between 49.3 (Sierra Leone, male) and 86.8 years (Japan, female) (World Health Organization [WHO], 2016). If some actor were to unilaterally deploy, a female person born in Japan on the day of deployment would have some 86.6 years of living with the effects of deployment. This might involve her generation paying the costs required to maintain SRM interventions, or suffering the effects of abrupt cessation if the intervention is not maintained. Other people born elsewhere will have less time living with the implications, but the amount of time may still be significant.

Furthermore, while there is reason to think that young people are not particularly good proxies for future generations in their own right (Bidadanure, 2017), among current generations there are people who are already suffering the effects of large-scale technological interventions (e.g. the people who were living in Chernobyl in 1986), and who may on the basis of that experience be better placed to represent the future than those lacking in such experience but tasked with imagining future people's perspective and interests. Enfranchising the future is no substitute for enfranchising the current global population, even if it can usefully supplement it (cf. Gardiner & Fragniere, 2018).

Unfortunately, there is no institutional framework in place for global democracy. Global democracy cannot be understood as existing through an institution like the United Nations. Not all member countries are democratic, and not all democratic member countries are well functioning enough that their representatives speak for their people. There is no framework in place capable of taking a global democratic decision. Does this mean there can be no democratic authorization of geoengineering?

We could take a hard line on democratic authorization, saying that deployment would only be permissible with global democratic authorization, and because global democratic authorization cannot be secured, deployment is impermissible. This would be like saying that it is impermissible to appropriate another person's property without her consent, and because her consent cannot be secured, appropriating her property is impermissible. (See Gardiner, 2013 for the concern that the 'consent' we might be able to secure for geoengineering is not sufficient to authorization, either because it can't be obtained from everyone who will be affected, because it is obtained under desperate conditions, or because it is to something consent alone cannot justify).

But we could also be a little more flexible, and acknowledge that it might be possible to *approximate* global democracy, in a way sufficient to authorizing the deployment of SRM. When really important political issues arise, democratic countries will sometimes have a referendum, or something like it (e.g. the referendum on changing the flag in New Zealand in 2015–6, the postal ballot on equal marriage in Australia in 2017). But sometimes the government or independent researchers just want to know what the people think about a particular issue, and in that case commission polls or surveys. These can give a good approximation of a democratic outcome. If the approximation is good enough, we should accept it as a substitute for full global democracy. There are things that can make the approximation better or worse, as I will explain below. But if we do the things that make it better, there's no reason not to make use of it.

It is possible to object in one of two ways, at this point. The first is to say that nothing short of a full global referendum is good enough. That would mean rejecting the claim that an approximation would do, perhaps by arguing that democracy has intrinsic value (insofar as it communicates equal respect for all voices and opinions). It would likely mean scuppering any chance of global democratic authorization, given the feasibility issues facing a full global referendum. Some countries, North Korea predictably among them, will simply refuse to conduct referenda. The second is to say that even if a full global democratic referendum is not required, a representative sample alone is not sufficient. That's like going straight to a vote without having a period of deliberation. A better way to approximate an *informed* vote would be to do deliberative polling. (It would certainly be interesting to know whether the results of deliberative polling would be significantly different to the results of representative sampling on this issue. Perhaps that is something that can be tested in future work.) Supposing neither of these objections go through, the path is clear for approximation.

Representative sampling as an approximation of democracy

Representative sampling is a method from the social sciences that surveys a representative sample of a target population in order to draw conclusions about what the target population think about some topic (see, e.g. Banerjee & Chaudhury, 2010). If we take a truly random sample, we can expect it to account for demographic differences in a roughly proportional way. For the sample to be truly random, the individuals must be selected entirely by chance, and every member of the target population must have an equal chance of being selected.

For example, imagine the Australian Capital Territory (ACT) Government wanted to find out whether ACT residents supported the widening of the bike lane down Northbourne Avenue. The Government would need to include *all* ACT residents as candidate respondents, and use a method of selecting respondents that gave them all an equal chance of being selected. Given that there are roughly equal numbers of men and women in the population, we can expect there to be roughly equal numbers of men and women in the sample. Likewise, for other demographic features. Whenever some in the target population do *not* have an equal chance of being selected, there's a risk of selection bias. Studies advertised to particular communities are at particular risk of selection bias, because we can expect only members of the community who feel strongly about the issue to respond.

Two variables are important in calculating the size of a representative sample relative to the size of the target population. They are (i) margin of error, and (ii) confidence level. Let's take Australia as the running example. The population of Australia is roughly 25 million people (as of 4 August 2019). If our margin of error is 5%, then supposing we get the result that 60% of the Australians approve of SRM deployment, that means we can be confident that between 55% and 65% of the Australians would have approved, had we polled the whole target population instead (similarly if the margin of error was 10%, we could be confident that between 50% and 70% of the Australians would have approved, and so on). Another way to put this is to say, if we selected 100 different random samples and ran the survey with them, as many as 5 out of 60 groups might deliver a result that *differed* from the result our actual group delivered, and as many as 5 of the 40 groups remaining might deliver a result that *agreed* with the result our actual group delivered. If our confidence level is 95%, then we can be extremely confident that the population lying within the confidence interval would have approved SRM deployment (if it is 90% we can be very confident, and so on).

Another variable that matters is response rate. If there were no reason to expect differences between people to track demographic lines, then we could simply figure out the likely response rate (say, about 20% of the people invited to respond to a survey will do so) and multiply that until we had assurance of the sample size we needed according to the variables discussed already. If we need 200 responses, then we invite 1000 people to respond. When there are no incentives offered for completion, response rates can be very low. The higher the response rate, the smaller the sample we need (e.g. if we think 50% of the people will respond, then we only need to invite twice as many people to respond as we need for the response to be representative).

A problem with simply increasing the size of the pool of people invited to respond kicks in when there *is* a reason to expect differences between people to track demographic lines. As James Jones puts it, '[s]ince nonresponders (i.e. people who were selected to be in the sample but did not complete the survey) often differ in meaningful ways from those who do respond, samples that have large proportions of nonresponses are unlikely to be representative of the population' (Jones, 1996, p. 49). The affects the extent to which we can expect the survey's results to be truly representative of the target population, and thus to generalize. One way to get around this is to incentivize those who have been selected, to respond (or use a company that does so).

The deployment of SRM is obviously a very important issue. It's important enough to have a full referendum on (consider that Australia, New Zealand, and the UK have recently had referenda or referenda-like things on gay marriage, changing the flag, and leaving the European Union, respectively). Procedural justice may require that on important issues, we actually give every person a vote, rather than use a method that will tell us with high reliability what the result would have been if we *had* given everyone a vote. There may also be other reasons to want people to vote: to 'bring the public along' on a policy reform, to increase the public sense of civic responsibility, to build community, or because there is expressive or symbolic value in voting. But absent those reasons, we need not have a full referendum, as long as we can get a near-perfect approximation of the democratic result. But that means making sure the margin of error is as small as possible, our confidence level is as high as possible, and our response rate is as close to perfect as we can get. After all, if we want a 51% majority supporting the deployment of SRM for it to be authorised, and if our result is that exactly 51% do authorize it, then a 5% margin of error means we can't be confident that we've actually gotten more than 51% of the vote. How expensive would it be to get this kind of result?

This is actually a fairly cheap option, when it comes to global sampling. For one thing, the sample size doesn't change once the target population exceeds a certain number (about 130,000) – a number which the global population is far in excess of. To get a representative sample of the global population, with a 5% margin of error and a 95% confidence interval, you only need to survey 383 people (Qualtrics, 2018). If we wanted to up that to a 1% margin of error and a 99% confidence interval, that's still only 14,702 people (Qualtrics, 2018).

The challenge for a global survey comes at the first hurdle: ensuring a truly random sample. Every member of the population must have an equal chance of being selected, and respondents must be selected by chance. This is not so difficult when we're talking about the ACT Government surveying its residents about bike lanes, but it's practically impossible when we're talking about *every person in the world*. There are people who we simply won't be able to access. We won't even be able to get a complete register of all people in order to select from them at random, and even if we could, we won't be able to get to some of them. Some people live in remote areas (research done in the field is significantly more expensive than research done online), some people live in war-torn areas (so it may be too dangerous to try to reach them), some people live in countries illegally, some people are stateless (an estimated 10 million) or living in refugee camps, some people are trafficked, or otherwise controlled by others in a way that meant it would be impossible to ensure a free response from them. Countries that are opposed to geoengineering (or have political reasons to oppose it because of their relationship with the country proposing to deploy it) may suppress their citizens' responses.

The response rate problem also kicks in here. There are demographic differences between the people of the world that can very much be expected to matter for the likelihood of approving geoengineering. Even if we could somehow get a reliable database of all people from which to select our sample at random, we'd have to bank on these 14,702 people all returning their surveys. Whoever doesn't respond will have a certain complex demographic profile, so they can't be replaced at random – they'd have to be replaced with someone of a similar profile. But it can be very difficult to establish who would count as a satisfactory replacement in a way that wouldn't affect the randomness of the original selection. The upshot: in principle, a global representative sample can be taken, in practice, it cannot.

When we can't ensure a truly random sample, we can change our survey method, and move to a stratified sample instead. Stratified sampling involves dividing the target population into groups that are mutually exclusive (no individual is a member of more than one group) and exhaustive (every individual is included in a group), then taking samples from each sub-group in proportion to the group's size in the target population. We divide the target population into subgroups expected to differ in their responses to the survey question. That way, we partly mitigate the response rate problem by *ensuring* the representation of people from particular demographics.

For example, suppose the ACT government has good reason to expect older residents to be opposed to the widening of the bike lane (because they tend to drive, and find bikes on the roads annoying) and younger residents to be in favour of it (because they tend to bike themselves, and be in favour of policy measures that are good for the environment). The target population can be divided into two groups, 'older' and 'younger', which are mutually exclusive and exhaustive, and the sample can be divided in a way that represents their proportions in the target population (suppose there are more older people than young, because fertility rates have dropped).

Whether a stratified sample will be able to address the problems with a random sample depends on which sub-groups within the global population we can expect to disagree over the permissibility of geoengineering. Might we expect disagreement to track country of birth (or citizenship, or residence)? Might we expect disagreement to track rich/poor, north/south, urban/rural, male/female, age group, occupation, religion, race, and culture (and maybe more)? (For some interview-based data on vulnerable populations' perspective on climate engineering see Carr & Preston, 2017). There is a longer discussion to be had over exactly the right social groups to sample. Those just mentioned are not mutually exclusive and exhaustive (one person will belong to multiple groups), so stratified sampling wouldn't work to track them. The only alternative then would be to run *multiple* surveys, and aggregate the results afterwards. This also throws up complications when it comes to the method of aggregation.

For simplicity's sake, I'll assume that the stratified sampling method can be made to work as a reasonable approximation of a truly random global sample. If the stratification is according to country (alone) then the subgroups will be mutually exclusive and exhaustive. We can set aside problems with dual nationals, stateless people, and refugees, because people from their origin country subgroups will be represented in the sample, even if *they* themselves will not be.

I don't have the resources to run a stratified survey of 14,702 people. But I'm curious enough about the results of such a survey to run something more limited, namely a single country survey. (This may indicative of how countries with a similar profile would vote; and gives us one piece of the puzzle when it comes to a stratified sample). In the next section, I report on a survey of Australians, asking about whether they would vote against, or in support of, the Australian Government deploying SRM.

Results from Australia

The population of Australia is roughly 25 million people. The number of people enrolled to vote is 16,176,487 (as of September 2018). If we were to use a 1% margin of error and 99% confidence level, the required sample size would be 16,560.[2] Qualtrics quoted AUD $112,000 for a survey of 16,000 people (*p.c.* October 2018). Alternatively, if we were to use a 5% margin of error and 95% confidence level, the required sample size would be 385. Qualtrics quoted AUD $2,100 for a survey of 300 people (*p.c.* October 2018). In comparison, the Australian postal ballot on marriage equality cost $80.5 million, and the New Zealand referendum on changing the flag cost $21.8 million (Australian Bureau of Statistics, 2018; Trevott, 2016).

$2,100 is obviously affordable to an individual researcher; $112,000 is affordable to a national government. If the main feasibility worry is price, then representative sampling is feasible. This cannot be used as an objection to establishing democratic authorization for geoengineering, then.

I had Qualtrics survey a representative sample of Australians in January 2019. I used a 95% confidence level and a 5% margin of error. Three hundred and thirty-five responses were collected. All individuals surveyed were eligible to vote in Australian national elections. In terms of demographics, 137 (40.90%) were male and 198 (59.10%) were female.[3] Twenty-two (6.57%) were 0–19 years old, 176 (52.54%) were 20–39 years old, 79 (23.58%) were 40–59 years old, 58 (17.31%) were 60–79 years old, and none were over 80. Eleven (3.28%) were Aboriginal or Torres Strait Islander, 223 (66.67%) were White Australian, 39 (11.64%) were Asian or Indian, 13 (3.88%) selected 'Other: person of colour' and 49 (14.63%) selected 'Other: white'. Sixty-five (19.40%) recorded their political affiliation as 'Left', 53 (15.82%) as 'Right', 92 (27.46) as 'Centre', and 125 (37.31%) as 'None'. The highest level of educational achievement for 5 (1.49%) was a PhD, for 66 (19.70%) was a postgraduate degree, for 95 (28.36%) was an undergraduate degree, for 137 (40.90%) was high school, and for 32 (9.55%) was 'other', e.g. TAFE, diploma, certificate, or trade qualification.

Participants were asked to read a brief statement about geoengineering, describing both carbon dioxide removal and solar radiation management (see Appendix). After confirming that they had read and understood this information, they were taken to a new page, which gave them two scenarios under which SRM might be deployed, and asked them to select one of three options: 'I would vote in support of the government using SRM', 'I would vote against the government using SRM', or 'I would refuse to vote/I would spoil the ballot'. In Scenario 1, participants were asked to suppose the Australian Government were conducting a referendum on geoengineering under present conditions. In Scenario 2, participants were asked to suppose the Australian Government were conducting a referendum on geoengineering under future conditions in which Australia and other countries had failed to make adequate cuts to GHGs, and temperatures had increased to a point that they were causing serious problems in parts of Australia (Scenario 2 follows the discussion in MacMartin, Caldeira, & Keith, 2014; McMartin et al., 2018). (See Appendix for exact wording).

The results were as follows. Under Scenario 1 (present conditions), 113 (33.73%) said they would vote in support of the government using SRM, 190 (56.72%) said they would vote against the government using SRM, and 32 (9.55%) said they would refuse to vote/spoil the ballot. Under Scenario 2, 112 (33.43%) said they would vote in support of the government using SRM, 187 (55.82%) said they would vote against the government using SRM, and 36 (10.75%) said they would refuse to vote/spoil the ballot.

Two things are immediately clear: the majority are against SRM in both scenarios, and at the group level, the difference between the two scenarios doesn't change the outcome in terms of authorization. Allowing for the margin of error, between 51.72% and 61.72% of the Australian voter population can be expected to be against SRM under Scenario 1. Under scenario 2, it's between 50.82% and 60.82%. For both, it's still a majority at the lower boundary.

One might have suspected that participants would be more open to SRM under the second scenario than the first. Of course, it could turn out that many people gave a different verdict between the two scenarios, and the totals were similar by sheer coincidence. And indeed, that is the case: approximately 60 participants gave different answers between the two scenarios, although it was more common to move between 'I would refuse to vote/I would spoil the ballot' and one of the other two answers than to move from voting in support to voting against, or *vice versa*.

Final thoughts

Given the structure of the climate challenge, the unilateral deployment of SRM would involve an actor – a country, a company, a super-rich individual – paying a cost alone, and both obtaining benefits that accrue to itself/themselves and others, and risking negative outcomes for itself/themselves and others. An actor who decides to do this could have any of a number of different motivations. They might be motivated by beneficence (*the*

Australian Government will pay the cost, and all countries will receive the benefit); by paternalism (*the Australian Government can see that deploying SRM is in all countries' best interests, and is willing to pay the cost to secure it*); by self-interest (*the Australian Government doesn't want climate change to negatively affect its citizens, so it will take the easiest route to avoiding it – not caring much if this has good or bad effects on other countries' citizens*); or in some other way.

Depending which of these motivate the decision, different moral considerations will be relevant. We might ask whether the actor had the right values, whether they were responsive to evidence, whether their action imposed acceptable or unacceptable risks on others, whether the risks were of a kind that could be traded off against benefits, and if they were, what the result of that tradeoff is. This complexity is not represented in the survey questions as they were posed, and might be expected to affect respondents' answers. It would certainly be interesting to see future studies explore what difference these alternative motivations might make to the prospects of democratic authorization.

It's not clear whether Australia's result can be expected to generalize. One might think that it's a good indication of how wealthy, developed nations with a strong ability to adapt to the effects of climate change and recover from the effects of natural disasters might vote. One might also be able to generalize on the basis of attitudes about structurally similar issues, so, for example, one could try to get more evidence for this by looking at whether other countries have voted in similar ways, or expressed similar opinions in national attitude surveys, when it comes to other high-tech large-scale interventions that have a 'hubris' aspect to them, like genetic modification, or stem cell engineering (on hubris in the context of geoengineering see Wong, 2015).

It is reasonable to expect that low-lying countries, and countries with less wealth and ability to adapt to and recover from the effects of climate change would be against SRM, at least in Scenario 1, because of a strong interest in seeing GHG levels reduced directly. But this interest may not carry over to Scenario 2. All of that is just to say that there's an open question and much further discussion to be had about what, if anything, we can read off the Australian result for the rest of the world. But *if* we think the result generalizes, and if those it generalizes to make up a majority of the global population, then we have reason to think that SRM would not be democratically authorized.

Of course, countries are welcome to pay the cost of finding out, so that we can be absolutely sure. The cost was not prohibitive. But the Australian result is at least suggestive: the deployment of SRM would *not* be democratically authorized, so the deployment of SRM – both under present conditions and future conditions where the temperature is higher due to lack of climate action – is impermissible.

Notes

1. There are other ways to approach the question; for the argument that governance ought to precede research into geoengineering see Blomfield (2015); for the argument that there is a high risk that geoengineering will involve moral corruption, and even 'moral schizophrenia', see Gardiner (2011), Ch. 10; for the argument that research into SRM should be constrained by a principle of non-domination see Taylor Smith (2019).
2. I used the Qualtrics online sample size calculator to get this number: www.qualtrics.com/blog/calculating-sample-size/. Another calculator gave a number that was 11 people higher: www.checkmarket.com/sample-size-calculator/.
3. It was specified that this was a question about biological sex, not gender identity.

Acknowledgments

I'm grateful to Jasper Hedges, Katie Steele, and Stephanie Collins; the audience at the University of Reading workshop 'Geoengineering, Justice and Legitimacy', September 2018, and the editors of this special issue for helpful comments and discussion.

Disclosure statement

No potential conflict of interest was reported by the author.

References

Abizadeh, A. (2008). Democratic theory and border coercion: No right to unilaterally control your own borders. *Political Theory*, *36*(1), 37–65.

Arrhenius, G. (2005). The boundary problem in democratic theory. In F. Tersman (Ed.), *Democracy unbound: Basic explorations I* (pp. 14–29). Stockholm: Filosofiska Institutionen, Stockholms Universitet.

Australian Bureau of Statistics. (2018, January 30). *Report on the conduct of the Australian marriage law postal survey* (p. 56). Australian Bureau of Statistics.

Banerjee, A., & Chaudhury, S. (2010). Statistics without tears: Populations and samples. *Industrial Psychiatry Journal*, *19*(1), 60–65.

Bidadanure, J. (2017). Youth quotas, diversity, and long-termism. In I. Gonzalez-Ricoy & A. Gosseries (Eds.), *Institutions for future generations*. Oxford: Oxford University Press. 266–280.

Blackstock, J. J., & Long, J. C. S. (2010). The politics of geoengineering. *Science, 327* (5965), 527.

Blomfield, M. (2015). Geoengineering in a climate of uncertainty. In J. Moss (Ed.), *Climate change and justice*. Cambridge: Cambridge University Press. 39–58.

Buck, H. J., Gammon, A., & Preston, C. (2014). Gender and Geoengineering. *Hypatia, 29* (3), 651–669.

Carnegie Climate Geoengineering Governance Initiative (C2G2). (2018, September 12). FAQs: What does C2G2 see as the top priorities for geoengineering governance? *c2g2.net*. Retrieved from https://www.c2g2.net//faqs/

Carr, W., & Preston, C. (2017). Skewed vulnerabilities and moral corruption in global perspectives on climate engineering. *Environmental Values, 26*, 757–777.

Currie, A., & Lawford-Smith, H. (2017). Accelerating the carbon cycle: The ethics of enhanced weathering. *Biology Letters, 13*(4), 20160859.

Gardiner, S. (2010). Is 'arming the future' with geoengineering really the lesser evil? Some doubts about intentionally manipulating the climate system. In S. Gardiner, D. Jamieson, S. Caney, & H. Shue (Eds.), *Climate ethics: Essential readings*. Oxford: Oxford University Press. 284–312.

Gardiner, S. (2011). *A perfect moral storm: The ethical tragedy of climate change*. Oxford: Oxford University Press .

Gardiner, S. (2013). The desperation argument for geoengineering. *Political Science and Politics, 46*(1), 28–33.

Gardiner, S., & Fragniere, A. (2018). The tollgate principles for the governance of geoengineering: Moving beyond the oxford principles to an ethically more robust approach. *Ethics, Policy & Environment, 12*(1), 143–174.

Gonzalez-Ricoy, I., & Gosseries, A. (Eds.). (2017). *Institutions for future generations*. Oxford: Oxford University Press.

Goodin, R. (2007). Enfranchising all affected interests, and its alternatives. *Philosophy and Public Affairs, 35*(1), 40–68.

Horton, J. (2011). Geoengineering and the myth of unilateralism: pressures and prospects for international cooperation. *Stanford Journal of Law, Science, and Policy, IV*, 56–69.

Institute on Statelessness and Inclusion. (2018, June). Statelessness in numbers: 2018. *institutesi.org*. Retrieved from https://www.institutesi.org/ISI_statistics_analysis_2018.pdf

Jones, J. (1996). The effects of non-response on statistical inference. *Journal of Health & Social Policy, 8*(1), 49–62.

Lane, L., & Bickel, E. (2009). Solar radiation management and rethinking the goals of COP-15. *Copenhagen Consensus on Climate: Advice for Policymakers* (Copenhagen Consensus Centre), *15*, 16–19.a.

MacMartin, D., Caldeira, & Keith, D. (2014). Solar geoengineering to limit the rate of temperature change. *Philosophical Transactions of the Royal Society A, 372*(2031), 20140134.

MacMartin, D., Ricke, K., & Keith, D. (2018). Solar geoengineering as part of an overall strategy for meeting the 1.5C Paris target. *Philosophical Transactions of the Royal Society A, 376*(2119), 20160454.

Niemeyer, S., & Jennstal, J. (2017). Youth quotas, diversity, and long-termism. In I. Gonzalez-Ricoy & A. Gosseries (Eds.), *Institutions for future generations*. Oxford: Oxford University Press. 247–265.

Oxford Geoengineering Programme. (2009). The principles. *geoengineering.ox.ac.uk*. Retrieved from http://www.geoengineering.ox.ac.uk/oxford-principles/principles/

Pasztor, J. (2017, April 25). Rules for geoengineering the planet. *MIT Technology Review*. Retrieved from https://www.technologyreview.com/s/604184/rules-for-geoengineering-the-planet/

Pasztor, J, Scharf, C, & Schmidt, K.-U. (2017). *How to Govern Geoengineering? Science, 357*(6348), 231. doi:10.1126/science.aan6794

Qualtrics. (2018, January). Sample size calculator. Retrieved from https://www.qualtrics.com/blog/calculating-sample-size/

Smee, B. (2018, May 3). Shift to renewables would save Australians $20bn a year – report. *The Guardian*. Retrieved from https://www.theguardian.com/environment/2018/may/03/shift-to-renewables-would-save-australians-20bn-a-year-report

Smith, T. (2019). Patrick. 'legitimacy and non-domination in solar radiation management research'. *Ethics, Policy & Environment*. [early view].*21*(3), 341–361.

Stanley, J. (2016). *How propaganda works*. New Haven: Princeton.

Trevett, C. (2016, June 15). Cost of flag referendum $4 million under budget. *The New Zealand Herald*. Retrieved from https://www.nzherald.co.nz/nz/news/article.cfm?c_id=1&objectid=11657032

Virgoe, J. (2009). International governance of a possible geoengineering intervention to combat climate change. *Climatic Change*, *95*(103), 115–116.

Wong, P. H. (2015). Confucian environmental ethics, climate engineering, and the 'playing god' argument. *Zygon*, *50*(1), 28–41.

Wong, P.-H. (2014). Maintenance required: The ethics of geoengineering and post-implementation scenarios. *Ethics, Policy & Environment*, *17*(2), 186–191.

World Health Organization. (2016, May 19). Life expectancy increased by 5 years since 2000, but health inequalities persist.

Appendix

Q7. Background material

On the next page, you will be asked how you would vote in a referendum on geoengineering, in two different scenarios. Please read this brief description of geoengineering and solar radiation management, so that you are able to make an informed decision about the questions on the next page.

Geoengineering

'Geoengineering' refers to the process of engineering the climate system. There are two main types of geoengineering, 'Carbon Dioxide Removal' (or 'CDR'), and 'Solar Radiation Management' (or 'SRM'). The first aims to remove carbon dioxide from the atmosphere, the most common proposal being to capture it at source and then bury it underground (this is 'carbon capture and storage'). There are many different CDR strategies. Even afforestation (planting forests) and reforestation (replanting forests) counts as CDR, because forests are carbon sinks. CDR strategies are generally considered less controversial than SRM strategies, even though some forms of CDR involve similarly large-scale and unpredictable interventions on the climate system. For example, 'enhanced rock weathering', a form of CDR, involves distributing finely ground rock particles over huge swathes of land in order to simulate natural carbon absorption processes at scale.

Solar Radiation Management

'Solar Radiation Management' refers to the process of attempting to reduce global or local temperature, by intervening on the way sunlight hits the surface of the earth. Like CDR, there are many different SRM strategies. Common proposals include brightening clouds so that they are denser and whiter and thus reflect more sunlight, or spraying reflective particles such as sulphites into the atmosphere using balloons or ships. There are also more low-tech proposals, include painting roofs white, covering deserts in huge sheets of white material, or sending white trash out into the ocean (to create a reflective version of the 'great ocean garbage patch'). The benefit of SRM is reduced temperature; the risks of SRM include unanticipated ecological consequences, unequal distribution of burdens/benefits across states and over time, conflict between states over deployment, consequences of ending SRM abruptly (such as a rapid increase in temperature), becoming locked into the use of SRM technology, and risks associated with reducing temperature but not greenhouse gas concentration (including ozone depletion, biodiversity loss, and ocean acidification).

If there's anything you want to know more about in relation to geoengineering in general, or SRM in particular, feel free to do some research online and come back to complete the survey.

Q8. Scenario 1:

Suppose the Australian Government is conducting a referendum on geoengineering, and you are required to vote. In the first scenario, things are exactly as they are now. Australia could choose to cut its greenhouse gas emissions (GHGs) directly, or to assist other countries in doing so, and this would have a positive effect on reducing global GHGs and thus global temperature. But instead it proposes to deploy a Solar Radiation Management (SRM) strategy in order to directly bring about cooling that would be beneficial to Australia, but which risks negative consequences including conflict with other states, rapid temperature increase upon cessation, and continued ocean acidification and biodiversity loss. If you were asked to vote in this referendum, how would you vote?

I would vote in support of the government using SRM.
I would vote against the government using SRM.
I would refuse to vote/I would spoil the ballot.

Q10. Scenario 2:

Suppose the Australian Government is conducting a referendum on geoengineering, and you are required to vote. In the second scenario, things are quite different to how they are now. Many years have passed, in which Australia and other countries have failed to make adequate cuts to greenhouse gas emissions (GHGs). Temperatures have increased in Australia and are causing serious problems in some parts of the country. Australia could choose to cut its GHGs directly, or to assist other countries in doing so, and this would have a positive effect on reducing global GHGs and thus global temperature. But instead, it proposes to deploy a Solar Radiation Management (SRM) strategy in order to directly bring about cooling that would be beneficial to Australia, but which risks negative consequences including conflict with other states, rapid further temperature increase upon cessation, and continued ocean acidification and biodiversity loss. If you were asked to vote in this referendum, how would you vote?

I would vote in support of the government using SRM.
I would vote against the government using SRM.
I would refuse to vote/I would spoil the ballot.

A mission-driven research program on solar geoengineering could promote justice and legitimacy

David R. Morrow

ABSTRACT
Over the past decade or so, several commentators have called for mission-driven research programs on solar geoengineering, also known as solar radiation management (SRM) or climate engineering. Building on the largely epistemic reasons offered by earlier commentators, this paper argues that a well-designed mission-driven research program that aims to evaluate solar geoengineering could promote justice and legitimacy, among other valuable ends. Specifically, an international, mission-driven research program that aims to produce knowledge to enable well-informed decision-making about solar geoengineering could (1) provide a more effective way to identify and answer the questions that policymakers would need to answer; and (2) provide a venue for more efficient, effective, just, and legitimate governance of solar geoengineering research; while (3) reducing the tendency for solar geoengineering research to exacerbate international domination. Thus, despite some risks and limitations, a well-designed mission-driven research program offers one way to improve the governance of solar geoengineering research relative to the 'investigator-driven' status quo.

Introduction

In 1942, Franklin D. Roosevelt instructed the US Army to undertake an ambitious project to build a nuclear weapon. Almost twenty years later, John F. Kennedy ordered NASA to put a man on the moon within a decade. Both undertakings – the Manhattan Project and the Apollo program – remain standard examples of mission-driven research programs. Many other mission-driven research programs aim, as those programs did, to build and deploy some kind of technological innovation, such as a supersonic airliner or a fusion power plant. Others, like the Human Genome Project, set out to answer a relatively well-defined scientific question. What binds these diverse projects together under the heading of 'mission-driven' research programs is that each explicitly aimed at achieving a specific, well-defined objective.

While many commentators have called for additional research into geoengineering over the last decade, recent suggestions have pushed the discussion toward *mission-driven* research programs (Keith, 2017; Long, 2017; MacMartin and Kravitz, 2019). Calls for research have differed in their preferred terminology, with some using 'solar radiation management' (SRM), while others use 'sunlight reflection methods,' 'albedo modification,' 'climate engineering,' or 'climate intervention.' More importantly, the various calls for research have differed in their proposed or implied goal: some would aim to develop and perhaps deploy solar geoengineering technology, while most would aim to evaluate solar geoengineering as a response to climate change. Building on the largely epistemic reasons offered by earlier commentators, this paper argues that a well-designed mission-driven research program that aims to evaluate solar geoengineering could promote justice and legitimacy, among other valuable ends.

To argue for this conclusion, the paper begins by examining and identifying various problems with the status quo in solar geoengineering research, which is characterized by an 'investigator-driven' research framework. It then argues that a mission-driven research program that meets certain conditions offers one way to overcome or mitigate those problems, including important problems related to justice and legitimacy. It draws on those arguments to identify key features to build into any mission-driven research program on solar geoengineering. Finally, it reviews some concerns and limitations of a mission-driven research program.

What is wrong with the investigator-driven status quo

Since Paul Crutzen broke the taboo on climate engineering research (2006), the academic literature on solar geoengineering has mushroomed (Belter & Seidel, 2013; Oldham et al., 2014). To date, though, research efforts have resembled the Wild West: a relatively small number of actors have carried out their own projects, individually or in small groups, without much oversight or support from governments, laying the groundwork for actions that could have serious consequences for vast numbers of people. In this 'investigator-driven' framework, researchers or research teams largely decide for themselves, in a decentralized way, what to study and how to study it based on their own intellectual interests and sense of what is important. Unlike in a mission-driven research program, an investigator-driven framework lacks a formal mechanism for centralized decision-making, planning, and coordination that can specify an end-goal and direct the collective research enterprise toward it.[1] Because the distinction between investigator-driven and mission-driven research depends heavily on the existence of a mechanism for centralized coordination, the concepts apply primarily to broad, socially distributed research efforts, rather than to the efforts of individual researchers or labs. (While it is possible for a small research team to pursue mission-driven research on their own, missions like those of the Manhattan Project, the Apollo program, the Human Genome Project, or a hypothetical solar geoengineering research program are too broad for a single research team. Thus, despite the possibility of borderline cases, the resulting vagueness is unimportant for present purposes.)

Despite having produced significant advances in the research community's ability to model solar geoengineering (Kravitz, n.d.; Kravitz et al., 2011; Tilmes et al., 2018), understand the physical science behind it (Irvine, Kravitz, Lawrence, & Muri, 2016), and understand its social, political, legal, and ethical implications (Flegal, Hubert, Morrow, & Moreno-Cruz, 2019), this investigator-driven approach to solar geoengineering research suffers from several important problems that could be mitigated by a well-designed mission-driven research program. First, it might not produce the knowledge societies need to make wise decisions about whether and how to use solar geoengineering. Second, in the absence of strong international governance of solar geoengineering research, the investigator-driven framework makes it difficult to govern research in an efficient, effective, just, and legitimate way. Third, it creates conditions that exacerbate the unjust domination of developing countries by developed countries.

An ineffective way to learn what societies need to know

An investigator-driven framework for solar geoengineering research is unlikely to deliver the knowledge societies would need in order to make wise, well-informed decisions about further research and potential deployment of solar geoengineering. In contrast to a mission-driven research program, an investigator-driven framework is less likely to deliver systematic, comprehensive assessments of solar geoengineering technologies (Long, 2017; MacMartin & Kravitz, 2019). That is, an investigator-driven is more likely to overlook or ignore important questions (MacMartin & Kravitz, 2019) and relevant risks (Keith, 2017). There are several reasons for this. Large-scale engineering projects involve too many interlocking parts to be handled effectively through uncoordinated, investigator-driven research (Long, 2017). Investigator-driven research may not prioritize research questions correctly (MacMartin & Kravitz, 2019). Relatedly, the status quo incentivizes researchers to pursue questions that illuminate both solar geoengineering and the climate system in general. While such 'multiple benefit research' is valuable and has some advantages (National Research Council, 2015), there are likely to be other important research questions that do not produce much benefit for climate science more generally, and focusing too heavily on multiple benefit research risks ignoring those questions (Keith, 2017). Furthermore, policymakers and publics may have questions about solar geoengineering that scientists would not identify or would not find sufficiently important or interesting to investigate on their own (Long, 2017). Finally, despite some recent progress in building research capacity throughout the Global South (Rahman, Artaxo, Asrat, & Parker, 2018) and significant research efforts in China (Cao, Gao, & Zhao, 2015) and India (Bala & Gupta, 2018), the solar geoengineering research community remains predominately Western. Investigator-driven research is therefore likely to overlook important questions of local and regional interest in developing countries.

An inefficient, ineffective, low-legitimacy, and burdensome game of 'governance whack-a-mole'

Societies have a legitimate interest in guiding solar geoengineering research. One source of society's authority over solar geoengineering research comes from the so-called 'right to science,' more formally known as 'the right to enjoy the benefits of scientific progress and its applications' (Reynolds, 2019, pp. 107–109). Although this right finds expression in various human rights agreements, including the Universal Declaration on Human Rights and the legally binding International Covenant on Economic, Social, and Cultural Rights, its actual content had long remained underdeveloped (Hubert, in press; Shaver, 2015; Wyndham & Vitullo, 2018). More recently, various institutions and legal scholars have taken up the task of clarifying the content of the right to science (Hubert, in press, Muller 2010; Shaver, 2015). These efforts have interpreted the right to science as having several parts. The most salient aspect of the right to science, for the purpose of justifying society's authority over solar geoengineering research, is 'the rights of ... stakeholders and the lay public to contribute to the definition and shaping of the scientific enterprise through democratic participation' (Hubert, in press, p. 9). A key UN statement interpreting the right to science also 'calls upon states to take measures to protect all sectors of society from the harmful effects and misuse of science and technology' (Hubert, in press, p. 12), which could be interpreted to give societies some authority over research and development into potentially harmful technologies.

In the absence of strong international governance of solar geoengineering research, however, the investigator-driven research framework makes it very difficult for societies to exercise that authority in an effective, efficient, just, and legitimate way.

To see why, consider two bespoke governance institutions created for specific solar geoengineering research projects: the Stratospheric Particle Injection for Climate Engineering (SPICE) project and the Stratospheric Controlled Perturbation Experiment (SCoPEx). The SPICE project aimed to study various aspects of stratospheric aerosol injection as a method of solar geoengineering. For one part of the study, researchers had planned to pump about a bathtub's worth of water through a kilometer-long hose suspended from a balloon, for the express purpose of investigating whether hose-tethered balloons might someday deliver sulfur to the stratosphere. The experiment involved no direct environmental risk; it raised concerns because and only because it represented a step toward the capacity for solar geoengineering. Because the project focused on such a controversial and ethically fraught emerging technology, the project planners and funder devised a 'stage-gate' process in which a panel would review each phase of the project before the researchers proceeded to the next one. This review required the researchers to 'reflect on the wider risks, uncertainties, and impacts surrounding the test and the geoengineering technique to which it could lead' (Macnaghten & Owen 2011, p. 293). Amid considerable public controversy, the review panel ultimately scuttled the outdoor balloon experiment, in part because of a potential conflict of interest on the part of one of the researchers (Schäfer et al., 2015, pp. 67–68).

More recently, researchers at Harvard have been planning the Stratospheric Controlled Perturbation Experiment (SCoPEx), in which a balloon would carry a set of scientific instruments into the stratosphere. The instruments would release a few kilograms of small particles and then monitor their behavior in the stratosphere to learn more about the physics and chemistry relevant to stratospheric aerosol injection (Keutsch Research Group, 'SCoPEx,' n.d.). Like the SPICE balloon project, SCoPEx would have no impact on the climate and would create no direct environmental risk. Recognizing that its explicit link to solar geoengineering raised concerns, however, the researchers created an independent advisory committee to issue governance recommendations to the Harvard administration (Keutsch Research Group, 'SCoPEx Governance,' n.d.). Designing and establishing this advisory committee has required considerable effort on the researchers' part.

While these governance efforts were and are justified, given the circumstances, and while the researchers and their funders appear to have made sincere and good faith efforts to do the best they can, these bespoke governance institutions fall short in several ways.

The most important problem arises from the fact that while the fundamental question confronting both efforts is whether scientists should be conducting (certain kinds of) research on solar geoengineering at all, governance efforts like this are not an appropriate mechanism for doing so. These experiments matter not because of their direct environmental impact, but because they might represent the 'camel's nose under the tent,' opening the way for further research and development of solar geoengineering. But if the right to science means that 'stakeholders and the lay public [have a right] to contribute to the definition and shaping of the scientific enterprise through democratic participation,' having elite panels of experts from one corner of the globe makes such weighty decisions hardly constitutes 'democratic participation,' even if those experts do their best to incorporate others' perspectives. Thus, these institutions are inadequate to protect the right to science. Second, because they depend on small panels of non-democratically chosen, unrepresentative experts, these institutions exhibit low levels of political legitimacy. A governance institution has legitimacy to the extent that it has the (normative or perceived) authority to make decisions about a particular topic (Callies, 2018; Jinnah, Nicholson, & Flegal, 2018). Third, such institutions are relatively ineffective at directing the solar geoengineering research enterprise as a whole because their direct authority extends only to particular research projects. While small advisory committees' decisions could reverberate more broadly through the diffusion of governance models and ideas (Jinnah et al., 2018) and by setting precedents, a single committee's decisions about solar geoengineering research could not (and arguably should not) constrain researchers beyond the specific research project governed by that committee. Because of these problems, such institutions are highly inefficient. They require significant amounts of time from a wide range of actors but produce limited social benefits. Developing new institutions for each research project amounts to a game of 'governance whack-a-mole,' with separate institutions attempting to govern each new research project as it arises.

One final problem with these arrangements involves another aspect of the right to science: the right to scientific freedom. In analyzing the relationship between the right to science and international environmental law, Anna-Maria Hubert explains the right to scientific freedom as a right to freedom from political or other interference in the selection and execution of scientific research projects and the dissemination of scientific findings (Hubert, in press). Other interpreters of the right to science reach similar conclusions about the right's implications for scientific freedom (Muller, 2010; Shaver, 2015). While this is a strictly negative right, meaning that it merely protects against interference rather than providing a positive right to research support, the right to scientific freedom does imply that social efforts to 'define and shape the scientific enterprise' ought to be balanced against researchers' freedom. Given that establishing and working with bespoke governance institutions involves a significant burden on researchers for relatively little social benefit, it is worth considering alternative governance arrangements that do a better job of balancing these competing aspects of the right to science.

It is worth reiterating that individual research projects only need these bespoke governance institutions because researchers are working under such a weak governance regime. Governance of solar geoengineering research currently exists in various forms.[2] At the international level, various legal instruments and decisions by international institutions have some implications for solar geoengineering research (Talberg, Christoff, Thomas, & Karoly, 2018; Reynolds, 2019), but many of these apply only to large-scale outdoor research or apply only in virtue of some other feature of certain kinds of research, such as potential impact on stratospheric ozone. At the national and supranational level, some existing laws and regulations in some countries might be relevant, at least for large-scale outdoor research (Bracmort & Lattanzio, 2013; Bodle et al., 2014; Schäfer et al., 2015). These institutions, however, are rather ineffective in governing near-term research because they apply mainly to individual large-scale outdoor experiments, which are likely to be many years away, and are therefore ineffective in governing near-term research or in directing the overall research enterprise.[3] More informally, authoritative reports on solar geoengineering, mainly written by academics, play an important role in *de facto* governance by steering research (Gupta & Möller, 2019). Arguably, efforts to articulate principles for solar geoengineering research, including the Oxford Principles (Rayner et al., 2013) and the Tollgate Principles (Gardiner & Fragnière, 2018) counts as a form of *de facto* governance, too, as do efforts to devise a voluntary code of conduct for solar geoengineering researchers (Hubert, 2017). This sort of governance exerts only limited power without uptake from formal institutions, however, and at any rate, the legitimacy of such efforts suffers from the same problems as the bespoke institutions created for SPICE and SCoPEx. While the remainder of this paper focuses on the role that a mission-driven research program on solar geoengineering could play in addressing problems with the current governance landscape, developing other forms of robust governance could have similar benefits. What the world needs is a legitimate international body with real power over solar geoengineering research that can serve as a locus of public deliberation and democratic participation in defining and shaping solar geoengineering research.

Exacerbating international domination

A third problem with the status quo is that research into solar geoengineering is currently exacerbating the relation of domination between technologically advanced countries and the rest of the world.[4] One party dominates another, in the relevant sense of domination, when the first party has superior power over another and the capability to wield that power arbitrarily (Smith 2018, p. 352). To say that someone can wield power arbitrarily is to say that they face no genuine external constraints on their use of their power that 'systematically and reliably lead [the dominating party] to consider, take seriously, and respect [the dominated party's] interests' when using that power (Smith 2012, p. 48).

To illustrate the concept and highlight the moral importance of domination, consider two cases in which one party wields superior power over another. Police officers generally wield superior power over ordinary citizens. In a just state, however, the rule of law effectively constrains officers' exercise of power, so that officers may only use that power in certain ways and must respect citizens' rights and interests in doing so. Thus, despite wielding superior power over ordinary citizens, a police officer in a just state does not dominate ordinary citizens (Smith 2012, p. 49). Slaveholders in the antebellum United States, by contrast, faced no such constraints. Thus, slaveholders dominated the people they held in slavery, regardless of how well or badly they treated them. Patrick Taylor Smith asks us to imagine a slaveholder, Emma, who has held Harriet in slavery for Harriet's entire life. Imagine, Smith says, that Harriet anticipates and fulfils Emma's needs 'without command or complaint' and that, in return, Emma ensures that Harriet lives comfortably. Even under such circumstances, the relationship between Emma and Harriet is problematic precisely because Emma dominates Harriet: Harriet enjoys the material comfort and relative freedom that she does entirely at Emma's discretion (Smith 2012, pp. 47–48). As these examples illustrate, domination is a morally problematic relationship because it involves arbitrary power over another, and it remains problematic even when the dominating party refrains from exercising its power or exercises it in ways that materially benefit the dominated party.[5]

It is worth emphasizing that domination does not depend on the intentions of the dominating party, but only on the power dynamics between the dominating and dominated party. In particular, the connection between solar geoengineering research and domination in no way depends on researchers' intentions. To say that such research exacerbates domination is emphatically not to say that researchers intend or desire to exacerbate domination. Rather, it is to say that, under current governance arrangements, some kinds of solar geoengineering research have the undesirable side effect of exacerbating international domination.

As Smith argues, some kinds of solar geoengineering research can give technologically advanced countries even greater power over other countries because it brings them (closer to having) the capacity to alter the climate to suit their own interests. This gives the technologically advanced countries asymmetric power over the environment and options that other countries face. For example, research into stratospheric aerosol injection – especially on tailoring deployment to meet specific objectives (MacMartin, Irvine, Kravitz, & Horton, 2019) – not only brings technologically advanced countries closer to being able to alter the global climate, but also grants them greater knowledge about the potential risks and benefits that could follow from such deployments (Smith, 2018).

Current suggestions and mechanisms for governing investigator-driven solar geoengineering research cannot do much to mitigate domination. Clearinghouses and codes of conduct that emphasize transparency can slightly reduce the power asymmetry that research produces, but simply sharing information about solar geoengineering would not shrink the enormous gap in technological capabilities between countries. Furthermore, because the existing governance arrangements are voluntary, they do not genuinely constrain technologically advanced countries' arbitrary exercise of power: their effectiveness depends entirely on the more powerful parties' voluntary forbearance (Smith, 2018, pp. 348–351). Various commentators have suggested principles for governing solar geoengineering and solar geoengineering research (Chhetri et al., 2018; Gardiner & Fragnière, 2018; Morrow, Kopp & Oppenheimer, 2009; Rayner et al., 2013), some of which could reduce domination by, for instance, giving vulnerable communities greater power in decision-making about research. In general, however, there is currently no institution to implement or enforce these principles, and individual researchers and research teams would be hard-pressed to live up to them on their own, and doing so would still depend on researchers' voluntary compliance.

How a mission-driven research program could mitigate these problems

A well-designed mission-driven research program offers one way to mitigate the aforementioned problems with the investigator-driven status quo. Thinking through these problems, therefore, provides a way to think through some crucial questions about how to structure such a research program.

Identifying, prioritizing, and answering necessary questions

What distinguishes a mission-driven research program from an investigator-driven research agenda is that a mission-driven research program is organized around achieving a (relatively) well-defined objective. Achieving that objective requires identifying, prioritizing, and answering essential questions and solving essential problems. Thus, compared to an investigator-driven approach, a well-designed and well-executed mission-driven research program is more likely to provide the information and capacities that societies need in order to evaluate or deploy solar geoengineering.

Focusing on this desideratum highlights the central design choice for a mission-driven research program on solar geoengineering: what is the program's objective?

One obvious possibility is that the program aims to develop the capacity to deploy solar geoengineering to achieve some well-defined effect, such as halving additional warming each year (Keith, 2013) or stabilizing temperatures while keeping certain key variables within acceptable limits (Tilmes et al., 2018). Such a 'deployment-oriented' mission-driven research program would closely resemble classic mission-driven research programs, such as the Manhattan Project, the Apollo program, and the Supersonic Transport Aircraft Committee that spearheaded the development of the Concorde, with two important qualifications. First, such a program would aim to develop the capacity for solar geoengineering, but not necessarily deploy it. (Commentators differ on the plausibility of developing the technology without deploying it. I will return to this issue below in discussing concerns about a 'slippery slope' toward deployment.) Second, developing a genuine capacity to deploy geoengineering would require developing institutions to make decisions about deployment and to manage any deployment, and thus the research program would have a more robust legal, political, and social scientific side to them than programs like the Manhattan Project or Apollo program did.

A second possibility is that the program aims to produce enough knowledge to enable well-informed decisions about deploying solar geoengineering. Such an 'information-oriented' mission-driven research program (as opposed to a 'deployment-oriented' one) would more closely resemble the Human Genome Project, which aimed to answer a relatively well-defined scientific question, rather than the more outcome-oriented programs like the Manhattan Project. The central question in designing such a program would be what information it needs to obtain. We can divide answers to that question into two kinds. A narrower information-oriented program would focus exclusively on natural science and engineering questions about the possible uses and impacts of solar geoengineering. MacMartin and Kravitz (2019), for instance, sketch a mission-driven research program that would focus on three clusters of questions: (1) Which objectives could feasibly be met through solar geoengineering, and how could societies deploy solar geoengineering to meet those objectives? (2) What are the projected climate and other direct environmental impacts of deploying solar geoengineering? (3) How much uncertainty is there about the outcomes of deployment, and how could that uncertainty be reduced?

A broader information-oriented mission-driven research program would also encompass questions of social science, governance, ethics, politics, and law. Defining exactly what questions need to be answered by such a program would itself be part of the program's mission, but sample questions include: (1) How could solar geoengineering be governed, and what are the advantages and disadvantages of different governance regimes? (2) What are people's attitudes toward and concerns about solar geoengineering? (3) How would different uses of solar geoengineering be evaluated from different ethical perspectives?

Devising a comprehensive list of questions to investigate will require input from diverse communities. The need for diverse voices in setting the agenda and a diverse set of researchers for implementing it strongly suggests that any mission-driven research program, and especially an information-oriented one, will require significant and genuine international participation. Without it, the program is unlikely to deliver the information that the excluded countries and communities would need to have in order to make well-informed decisions about solar geoengineering. Thus, without international participation, an information-oriented program is unlikely to succeed in its mission.

Given their rather open-ended nature, such information-oriented research programs may strain the definition of a 'mission-driven research program.' What would justify the 'mission-driven research program' label for such a program is a commitment to continue the program until all essential questions had been identified and adequately answered. Contrast this with the closest analogue to date, which was a time-limited Special Priority Program of the German Research Foundation, initiated in 2012 with the explicit aim of *evaluating* solar geoengineering and carbon removal rather than working toward deployment (Oschlies & Klepper, 2017). While that program aimed to make progress toward the same overall goal of an information-oriented mission-driven research program into solar geoengineering, it was not set up to ensure the attainment of that goal. Setting definitions aside, a long-term, well-resourced, and well-coordinated research program is, relative to the status quo, more likely to provide societies with the information and capacities that they need to evaluate solar geoengineering.

Balancing social control with scientific freedom

Recall that the status quo, characterized by an investigator-driven research framework in the context of weak international governance, creates several challenges for effective, efficient, just, and legitimate governance of solar geoengineering research. The basic problem is that without a single locus for public deliberation and democratic participation, it is very hard for societies to shape the solar geoengineering research enterprise as a whole. Depending on its design, a mission-driven research program could address that problem by creating a single site for international public engagement, contestation, and governance.

To the extent that a research program exercises authority over all or most solar geoengineering research, it offers societies a means to control both individual experiments or research projects and the solar geoengineering research enterprise as a whole. A program could acquire such authority if most states and private entities funding solar geoengineering research channel their research funds for solar geoengineering through the program. It could further extend its authority by adopting a code of conduct for solar geoengineering research and encouraging other funders to require grantees to abide by that code of conduct (Chhetri et al., 2018; Hubert, 2017). These mechanisms would not bring all solar geoengineering research under the authority of the research program, as much computer modelling and social scientific research on solar geoengineering occurs without specific, project-based funding. By identifying research questions and setting research priorities, however, such a research program could still influence the research conducted outside the program's authority, thereby encouraging research to remain responsive to societal needs.

To the extent that a mission-driven research program shoulders the responsibility of creating appropriate governance institutions for individual research project, engaging with stakeholders on an international basis, compiling and distributing information about solar geoengineering research, directing the overall research enterprise, and so on, it improves legitimacy and reduces the inefficiencies of redundant bespoke institutions and the administrative burdens that researchers face by choosing to study solar geoengineering

Mitigating International Domination

If it were designed in the right way, a mission-driven research program could mitigate the tendency for solar geoengineering research to exacerbate international domination, which is a form of injustice. Once again, the choice between a deployment-oriented program and an information-oriented program proves relevant: other things being equal, a deployment-oriented program would exacerbate domination more than an information-oriented one because it would, by developing an actual capacity for deployment, do more to increase technologically advanced countries' capacities to control the climate and because it would likely leave less technologically advanced countries with less information about the impacts of solar geoengineering on their territories and citizens. But another set of design choices also plays a crucial role in the program's effect on international domination – namely, the extent and ways in which countries vulnerable to domination participate in the program.

To understand how the participation of countries vulnerable to domination (for short, 'vulnerable countries') relates to domination, consider three strategies for preventing domination: equalizing power, making power accountable, and limiting power. To equalize power (with respect to some particular type of power) means to ensure that each party has equal power (in that particular respect), so that if one party has power over another, the second party has equal power over the first. To make power accountable (with respect to some particular type of power) is to impose adequate constraints on the use of that power, so that no party can wield that power arbitrarily. To limit power (of a certain kind) is to limit an actor's capability to wield the relevant kind of power at all. Smith illustrates the first two strategies with reference to nuclear weapons: Some early proposals for the governance of nuclear weapons involved placing the power to govern them in the United Nations' hands, which would restrain arbitrary use. Rather than adopt these approaches to hold power accountable, the United States and the Soviet Union opted instead for a power-equalizing approach (with respect to one another, at least) in which both the Americans and the Soviets developed the power to annihilate the other with nuclear weapons, each striving to ensure that neither had superior power over the other (Smith, 2018, p. 352). Nonproliferation treaties and disarmament treaties exemplify a power-limiting strategy aimed at reducing various actors' capability to use nuclear weapons at all.

How could a research governance institution help hold technologically advanced states accountable with respect to the power derived from solar geoengineering capabilities? In exploring this question, Smith imagines a powerful international institution with broad international participation and comprehensive control over solar geoengineering research. This institution would include both legislative and judicial functions, giving it the power to make decisions about solar geoengineering research, adjudicate disagreements, and demand that states penalize those who violate the institution's rules about research. To support such an institution, Smith says, is to say that 'the kind of robust regime that many imagine for [solar geoengineering] *deployment* should also be developed for [solar geoengineering] *research*' (Smith, 2018, p. 354, original emphasis). As Smith acknowledges, however, it is politically infeasible to create such a powerful international institution (Smith, 2018, pp. 354–355).

A mission-driven research program, however, could still exercise considerable power over solar geoengineering research, thereby helping to hold technologically advanced countries accountable. The problem, however, is that even an institution as powerful as the one that Smith imagines could do only so much to hold states accountable for their solar geoengineering-related power. The relevant power, after all, is the power to *deploy* solar geoengineering arbitrarily, not to research it; the research matters to domination because it could empower actors to deploy solar geoengineering. The main way that a research governance institution could help hold power accountable, then, is by ensuring that researchers study the impacts of different deployment strategies on vulnerable countries – something that would be more likely to happen in an information-oriented program with meaningful participation by vulnerable countries. This would strengthen vulnerable countries' ability to convince powerful entities, such as the United Nations, to intervene if some actor planned to deploy solar geoengineering in a way that threatened the vulnerable countries' interests. While this provides only a limited check on technologically advanced countries' power with respect to solar geoengineering, it provides another reason in favor of an information-oriented mission-driven research program in which vulnerable countries had sufficient power to ensure that researchers studied the impacts that most concerned those countries.

What about equalizing power with respect to solar geoengineering? Two distinct options present themselves. The first would be to use a research governance institution to distribute the capacity to deploy solar geoengineering broadly throughout the international community. As Smith notes, this would require much more than simply sharing information: it would require actively developing the scientific, engineering, and technological capacities of vulnerable countries so that they had a genuine capability to deploy solar geoengineering (Smith, 2018, p. 355). It is worth noting that asymmetries in climate preferences might make it impossible for the broad distribution of deployment capacity to completely equalize different countries' power with respect to solar geoengineering: because of the 'free-driver problem' facing solar geoengineering (Weitzman, 2015), whichever actor preferred the strongest cooling effect would have a greater ability to impose its preferred climate on others. (To see why, imagine that the US preferred to reduce incoming solar radiation by 1 Watt per meter squared, whereas Indonesia preferred to reduce incoming solar radiation by 2 Watts per meter squared. If the US deployed solar geoengineering at its preferred intensity, Indonesia could deploy additional solar geoengineering to reach its preferred intensity, but not vice versa.) At any rate, Smith's historical analogue for this approach – nuclear proliferation during the Cold War – points to an obvious source of concern about it: while giving every state the capacity to deploy solar geoengineering would put states on a more equal footing, it also increases the risk of reckless or contested deployment and conflict. A second option – which is arguably more feasible than broadly distributing the capacity to deploy solar geoengineering – is to promote research on counter-geoengineering techniques. Counter-geoengineering refers to 'action taken through technical means to counter the change in radiative forcing caused by [solar geoengineering]' (Parker, Horton, & Keith, 2018, p. 1059). The most obvious way to do this would be to emit powerful, short-lived greenhouse gases to increase radiative forcing, but it might also be possible to physically neutralize cooling agents such as stratospheric aerosols (Parker et al., 2018, pp. 1060–1061). The widespread availability of counter-geoengineering methods would have significant but ambiguous strategic implications (Heyen, Horton, & Moreno-Cruz, 2019), and promoting it could, therefore, reduce the power asymmetry that research into solar geoengineering would exacerbate. It is well within the power of a mission-driven research program to ensure active research into widely available methods of counter-geoengineering.

A third option for reducing domination, which Smith does not discuss, is to limit the dominating party's power. How might a mission-driven research program limit actors' power to deploy solar geoengineering in an arbitrary way? An information-oriented program could attempt this by refusing to fund research that would bring the world too close to developing the capability to deploy solar geoengineering. This might take the form of prohibitions on certain kinds of engineering-related research or on experiments that exceed some threshold, defined, perhaps, in terms of duration, geographical extent, and impact on radiative forcing. A sufficiently advanced, well-resourced actor could, however, develop the technology outside the confines of the program, which limits the efficacy of this approach. A subtler approach would be to steer research away from specific solar geoengineering technologies that seem to offer greater possibilities for domination, such as those that might enable more fine-grained control over regional climates. To see how this would reduce domination, consider a simplified case involving just two countries, technologically advanced Wakanda and highly vulnerable Motunui, and two approaches to solar geoengineering, stratospheric aerosol injection (SAI) and marine cloud brightening (MCB), each of which could be deployed in two different ways, which we will call SAI_1, SAI_2, MCB_1, and MCB_2. Suppose that research reveals that SAI_1 would be best for both Wakanda and Motunui whereas SAI_2 would be very bad for both countries; but that MCB_1 would be very good for Wakanda but very bad for Motunui, whereas MCB_2 would be reasonably good for both countries. Further developing MCB would leave Motunui at Wakanda's mercy; if Wakanda chooses to deploy MCB, Motunui's fate depends on Wakanda's willingness to take Motunui's interests into account. Developing SAI instead of MCB, however, would mean that even if Wakanda acts only in its own interests, it would behave as if taking Motunui's interests into account when deciding whether and how to deploy solar geoengineering. While Wakanda would still have superior power over Motunui, Wakanda's self-interest would act as a check on its exercise of that power. Putting this the other way around, the research program could try to prioritize solar geoengineering technologies whose optimal deployment for technologically advanced countries most closely resembles the optimal deployment for vulnerable countries. Judgments about which solar geoengineering technologies those are would likely evolve over time as scientists learn more about the distributional and long-term effects of different technologies, which suggests that this 'steering' toward particular technologies would need to be an ongoing process.

Implications for the design of a mission-driven research program

Reviewing the ways in which a mission-driven research program could address the problems with the investigator-driven status quo points to several lessons for designing the program:

(1) An *information-oriented mission-driven research program* is preferable to a deployment-oriented one. That is, the mission of the research program should be to identify and produce the knowledge needed to support wise decisions about solar geoengineering deployment, rather than developing the capability to deploy solar geoengineering.

(2) The program should, to the extent possible, be *able to coordinate and exert appropriate influence, directly or indirectly, over the majority of solar geoengineering research around the world*. This might be accomplished

by placing the program in an international institution and channelling most or all national funding for solar geoengineering research through it, as well as promulgating standards and a code of conduct that research outside its auspices could follow.

(3) The program should serve as a *focal point for societal deliberation and public engagement on solar geoengineering*. This will both increase the efficiency, effectiveness, and legitimacy of those engagements and liberate individual research teams from the burden of conducting such engagements and devising governance structures on their own.

(4) The governing body of the program should have *strong international representation*, including representation from highly vulnerable countries or blocs (e.g. the African Union and the Alliance of Small Islands States), and representatives from those countries should have genuine power to ensure that the program's activities reflect their needs and interests.

(5) The program should seek to *identify and prioritize approaches to solar geoengineering that offer the least opportunity for international domination*. It is worth acknowledging, however, that appropriate design choices could at best lessen the prospects for international domination. Preventing domination altogether would require robust international governance of solar geoengineering deployment, which is beyond the scope of any research governance institution.

Concerns and limitations

Creating a mission-driven research program – even a well-designed one – will not resolve all of the concerns about solar geoengineering research, much less concerns that would accompany any deployment. This section reviews three major concerns that apply specifically to a mission-driven research program.

One concern is that creating such a high-profile organization to research solar geoengineering exacerbates the risk of mitigation obstruction – also known as a moral hazard effect, risk compensation, or mitigation deterrence. The basic concern here is that if publics or policymakers draw the (mistaken) conclusion that solar geoengineering gives them an excuse not to cut emissions, they will cut emissions less aggressively than they would have in the absence of a solar geoengineering research program. Researchers have debated the existence, strength, and ethical implications of this hypothesized effect (Baatz, 2016; Callies, 2019a; Fairbrother, 2016; Hale, 2012; Lin, 2013; Lockley & Coffman, 2016; Markusson, McLaren, & Tyfield, 2018; McLaren, 2016, Merk, Pönitzsch, and Rehdanz, 2016, 2018; Morrow, 2014; Ott, 2018; Quaas, Quaas, Rickels, & Boucher, 2017; Raimi, Maki, Dana, & Vandenbergh, 2019; Reynolds 2015, 2019). While commentators generally agree that governance of solar geoengineering research should aim to minimize any mitigation obstruction, identifying governance mechanisms to do this remains challenging (Baatz, 2016; Jinnah et al., 2019), especially in light of the tendency toward what Stephen Gardiner calls 'moral corruption' (Baatz, 2016; Gardiner, 2011). Given that a mission-driven research program on solar geoengineering would have no power over decisions related to mitigation, there is only so much that could be done in designing such a program to minimize mitigation obstruction. Establishing an information-oriented program, rather than a deployment-oriented one, is likely to help, since it would reduce the perception that the world is on its way to deploying solar geoengineering. The program could also help ensure that researchers' study designs, messaging, and public engagements highlight the limitations of solar geoengineering and the necessity of emission reductions (Morrow, 2014), though this would not, by itself, be sufficient to counteract the powerful social and political forces that could drive mitigation obstruction (Baatz, 2016).

A second concern is that establishing a mission-driven research program would set the world on a slippery slope toward deployment. Like the worry about mitigation deterrence, fear of a slippery slope appears as a frequent theme in discussions of solar geoengineering. Detailed analyses of the argument remain sparse, with different commentators disagreeing about how much weight to attach to slippery slope concerns (Callies, 2019b; McKinnon 2018). To the extent that such concern is legitimate, a mission-driven research program threatens to exacerbate the risk by accelerating research into solar geoengineering. Here, too, an information-oriented program enjoys some advantage over a deployment-oriented one. In an early articulation of the slippery slope argument, Dale Jamieson argues that societies rarely invest in developing a new technology and then decline to use it (1996). Although Callies (2019b) contests this claim, those who share Jamieson's fear will find it provides another reason to focus a research program on evaluating solar geoengineering rather than developing it. A cooperative international research program would also address one of the potential mechanisms that Catriona McKinnon (2018) identifies, which is an 'arms race' in which different countries rush toward deployment capacity because they see other countries developing their own capacity. More generally, the slippery slope argument provides an argument for building 'stage-gates' into the research program, as was done for the SPICE project, so that moving from one phase of research to the next requires a deliberate, well-considered decision, and decisionmakers have a genuine opportunity to slow, redirect, or shut down the research program if necessary (McKinnon, 2018). As Jinnah and colleagues note, a legitimate research governance regime must be able to conclude that certain kinds of research are illegitimate or undesirable (Jinnah et al., 2018).

A third concern is that concentrating solar geoengineering research in a single research program may make research more vulnerable to capture by special interests, which might seek to direct research in certain ways (Long &

Scott 2013; Wolff, in press). This risk could be reduced by ensuring diversity in the governance of the research program – both in terms of international representation and diversity in backgrounds, as Jinnah et al. (2019, p. 375) advise for state-level commissions: it would be much harder for special interests to capture the program if the individuals in charge of the program represent a broad range of interests and perspectives.

One important limitation of a mission-driven research program is its inability to compel any state or researcher to abide by its restrictions or recommendations. While states could push researchers to comply with it by channelling all research funding through it, the international system has no means to compel states to do so. States might have some scientific and strategic incentives to cooperate, however, provided that other states do as well. If some state or private actor, however, decided that the research program's rules were too stringent or that it was failing to pursue an important line of research, the research program itself would have very little power to prevent those actors from pursuing the research on their own; this imposes important limits on how much power the research program could exercise over technologically advanced states that disagreed with its judgments about research. Despite the benefits of a mission-driven research program, then, governing solar geoengineering research may eventually require more powerful institutions.

Conclusions

Creating an international, information-oriented, mission-driven research program on solar geoengineering could provide several benefits over the current, investigator-driven framework. First, it would provide a more effective way to identify and answer the questions that policymakers would need to answer to make wise, responsible decisions about solar geoengineering. Second, it would improve the efficiency, effectiveness, legitimacy, and justice of research governance. Third, it would reduce the tendency for solar geoengineering research to exacerbate international domination. Thus, even though a mission-driven research program comes with risks and limitations, it offers one way to improve the governance of solar geoengineering research relative to the status quo. The most important ethical concerns about solar geoengineering, however, apply not to research but to deployment. These include concerns about procedural justice surrounding any possible deployment and about distributive and restorative justice during deployment. These are largely beyond the reach of a mission-driven research program and would need to be addressed by some other means.

Notes

1. There is some formal coordination of research efforts by the scientists themselves, through efforts such as GeoMIP and the Geoengineering Modeling

Research Consortium, and some research decisions have been guided by funders, as in the German Research Foundation's Special Priority Program, but these are deviations from the investigator-driven norm.

2. Governance includes not just regulation and hard law, but also 'softer' or less formal activities that shape behavior (Bevir, 2012; Rosenau, 1995). Jesse Reynolds defines it as 'the goal-oriented, sustained, focused, and explicit use of authority to influence behavior,' citing 'unwritten norms, nonbinding principles and rules, laws, administrative regulations, market instruments, procedures, institutions, funding, and international law' as examples of forms of governance (Reynolds, 2019, p. 7). Aarti Gupta and Ina Möller add that activities that 'have governance effects' can count as *'de facto* governance' even without the explicit, goal-oriented use of authority (Gupta & Möller, 2019, p. 481).
3. For steps by which existing institutions could strengthen their governance of near-term research, see Chhetri et al. (2018).
4. I will focus here on Patrick Taylor Smith's argument that solar geoengineering research can exacerbate international domination. A number of philosophers, including Smith, have also argued that solar geoengineering deployment could or would involve international and intergenerational domination. See Gardiner (2013); Hourdequin (2019); Nolt (2011); Smith (2012). For other philosophical work on solar geoengineering research that relates to issues of domination, see also Blomfield (2015); Whyte (2012).
5. Smith is drawing on a broader philosophical conversation about the nature and wrongness of domination. For an overview of that conversation, see McCammon (2018). For more detailed examinations of domination, see Lovett (2010); McCammon (2015); Pettit (2005, 2012); Rigstad (2011).

Acknowledgments

Thanks to Zach Dove for a helpful discussion of these ideas, and to Catriona McKinnon and Steve Gardiner for helpful critical comments. The positions taken in this paper represent the author's own views, and not those of the Forum for Climate Engineering Assessment.

Disclosure statement

No potential conflict of interest was reported by the author.

References

Baatz, C. (2016). Can we have it both ways? On potential trade-offs between mitigation and solar radiation management. *Environmental Values, 25*(1), 29–49.

Bala, G., & Gupta, A. (2018). Solar geoengineering research in India. *Bulletin of the American Meteorological Society, 100*(1), 23–28.

Belter, C. W., & Seidel, D. J. (2013). A bibliometric analysis of climate engineering research. *Wiley Interdisciplinary Reviews: Climate Change, 4*(5), 417–427.

Bevir, M. (2012). *Governance: A very short introduction* (1st ed.). Oxford: Oxford University Press.

Blomfield, M. (2015). Geoengineering in a climate of uncertainty. In J. Moss (Ed.), *Climate change and justice* (pp. 39–58). Cambridge: Cambridge University Press.

Bodle, R. S., Oberthür, S., Donat, L., Homann, G., Sina, S., & Tedsen, E. (2014). *Options and proposals for the international governance of geoengineering.* Berlin: Ecologic Institute.

Bracmort, K., & Lattanzio, R. K. (2013). *Geoengineering: Governance and technology policy* (pp. R41371). Washington, DC: Congressional Research Service.

Callies, D. E. (2018). Institutional legitimacy and geoengineering governance. *Ethics, Policy & Environment, 21*(3), 324–340.

Callies, D. E. (2019a). *Climate engineering: A normative perspective.* Lanham: Lexington Books.

Callies, D. E. (2019b). The slippery slope argument against geoengineering research. *Journal of Applied Philosophy, 36*(4), 675–687.

Cao, L., Gao, C.-C., & Zhao, L.-Y. (2015). Geoengineering: Basic science and ongoing research efforts in China. *Advances in climate change research, 6*(3), 188–196.

Chhetri, N., Chong, D., Conca, K., Falk, R., Gillespie, A., Gupta, A., ... Nicholson, S. (2018). *Governing solar radiation management.* Washington, DC: Forum for Climate Engineering Assessment, American University.

Crutzen, P. J. (2006). Albedo enhancement by stratospheric sulfur injections: a contribution to resolve a policy dilemma? *Climatic Change, 77*(3), 211.

Fairbrother, M. (2016). Geoengineering, moral hazard, and trust in climate science: Evidence from a survey experiment in Britain. *Climatic Change, 139*(3), 477–489.

Flegal, J. A., Hubert, A.-M., Morrow, D. R., & Moreno-Cruz, J. B. (2019). Solar geoengineering: Scientific, legal, ethical, and economic frameworks. *Annual Review of Environment and Resources, 44*(1), 399–423.

Gardiner, S. M. (2011). *A perfect moral storm: The ethical tragedy of climate change.* New York: Oxford University Press.

Gardiner, S. M. (2013). The desperation argument for geoengineering. *PS: Political Science & Politics, 46*(1), 28–33.

Gardiner, S. M., & Fragnière, A. (2018). The tollgate principles for the governance of geoengineering: moving beyond the oxford principles to an ethically more robust approach. *Ethics, Policy & Environment, 21*(2), 143–174.

Gupta, A., & Möller, I. (2019). De facto governance: How authoritative assessments construct climate engineering as an object of governance. *Environmental Politics, 28*(3), 480–501.

Hale, B. (2012). The world that would have been: moral hazard arguments against geoengineering. In C. J. Preston (Ed.), *Engineering the climate: The ethics of solar radiation management*(pp. 113–131). Lanham, MD: Rowman & Littlefield.

Heyen, D., Horton, J., & Moreno-Cruz, J. (2019). Strategic implications of counter-geoengineering: Clash or cooperation? *Journal of Environmental Economics and Management, 95*, 153–177.

Hourdequin, M. (2019). Geoengineering justice: The role of recognition. *Science, Technology, & Human Values, 44*(3), 448–477.

Hubert, A.-M. (2017). *Code of conduct for responsible geoengineering research*. Calgary: Geoengineering Research Governance Project.

Hubert, A.-M. (in press). *The human right to science and its relationship to international environmental law*. European Journal of International Law.

Irvine, P. J., Kravitz, B., Lawrence, M. G., & Muri, H. (2016). An overview of the Earth system science of solar geoengineering. *Wiley Interdisciplinary Reviews: Climate Change, 7*(6), 815–833.

Jamieson, D. (1996). Ethics and intentional climate change. *Climatic Change, 33*(3), 323–336.

Jinnah, S., Nicholson, S., & Flegal, J. (2018). Toward legitimate governance of solar geoengineering research: A role for sub-state actors. *Ethics, Policy & Environment, 21* (3), 362–381.

Jinnah, S., Nicholson, S., Morrow, D. R., Dove, Z., Wapner, P., Valdivia, W., ... Chhetri, N. (2019). Governing climate engineering: A proposal for immediate governance of solar radiation management. *Sustainability, 11*(14), 3954.

Keith, D. W. (2013). *A case for climate engineering*. Cambridge, MA: The MIT Press.

Keith, D. W. (2017). Toward a responsible solar geoengineering research program. *Issues in Science and Technology, 33*(3).

Keutsch Research Group. (n.d.). *SCoPEx* [online]. Retrieved from https://projects.iq.harvard.edu/keutschgroup/scopex

Keutsch Research Group. (n.d.). *SCoPEx governance*[online]. Retrieved from https://projects.iq.harvard.edu/keutschgroup/scopex-governance

Kravitz, B. (n.d.). The Geoengineering model intercomparison project (GeoMIP) [online]. *GeoMIP*. Retrieved from http://climate.envsci.rutgers.edu/GeoMIP/index.html

Kravitz, B., Robock, A., Boucher, O., Schmidt, H., Taylor, K. E., Stenchikov, G., & Schulz, M. (2011). The Geoengineering model intercomparison project (GeoMIP). *Atmospheric Science Letters, 12*(2), 162–167.

Lin, A. C. (2013). Does geoengineering present a moral hazard. *Ecology Law Quarterly, 40*, i–712.

Lockley, A., & Coffman, D. (2016). Distinguishing morale hazard from moral hazard in geoengineering. *Environmental Law Review, 18*(3), 194–204.

Long, J. C. S. (2017). Coordinated action against climate change: A new world symphony. *Issues in Science and Technology; Washington, 33*(3), 78–82.

Long, J. C. S., & Scott, D. (2013). Vested interests and geoengineering research. *Issues in Science and Technology, 29*(3), 45–52.

Lovett, F. (2010). *A general theory of domination and justice*. Oxford: Oxford Univ. Press.

MacMartin, D. G., Irvine, P. J., Kravitz, B., & Horton, J. B. (2019). Technical characteristics of a solar geoengineering deployment and implications for governance. *Climate Policy 19*(10), 1325–1339.

MacMartin, D. G., & Kravitz, B. (2019). Mission-driven research for stratospheric aerosol geoengineering. *Proceedings of the National Academy of Sciences 116*(4), 1089–1094.

Macnaghten, P., & Owen, R. (2011). Good governance for geoengineering. *Nature, 479* (7373), 293.

Markusson, N., McLaren, D., & Tyfield, D. (2018). Towards a cultural political economy of mitigation deterrence by negative emissions technologies (NETs). *Global Sustainability, 1*, 1–9.

McCammon, C. (2015). Domination: A rethinking. *Ethics, 125*(4), 1028–1052.

McCammon, C. (2018). Domination. In E. N. Zalta (Ed.), *The stanford encyclopedia of philosophy*[online]. Metaphysics Research Lab, Stanford University.Retrieved from https://plato.stanford.edu/archives/win2018/entries/domination/

McKinnon, C. (2018). Sleepwalking into lock-in? Avoiding wrongs to future people in the governance of solar radiation management research. *Environmental Politics*, 28 (3), 441–459.

McLaren, D. (2016). Mitigation deterrence and the "moral hazard" of solar radiation management. *Earth's Future*, *4*(12), 596–602.

Merk, C., Pönitzsch, G., & Rehdanz, K. (2016). Knowledge about aerosol injection does not reduce individual mitigation efforts. *Environmental Research Letters*, *11*(5), 054009.

Merk, C., Pönitzsch, G., & Rehdanz, K. (2018). Do climate engineering experts display moral-hazard behaviour? *Climate Policy 19*(2), 231–243.

Morrow, D. R. (2014). Ethical aspects of the mitigation obstruction argument against climate engineering research. *Philosophical Transactions of the Royal Society A: Mathematical, Physical and Engineering Sciences*, *372*(2031), 20140062.

Morrow, D. R, Kopp, R. E, & Oppenheimer, M. (2009). Toward ethical norms and institutions for climate engineering research. *Environmental Research Letters*, *4*(4), 045106. doi:10.1088/1748-9326/4/4/045106

Muller, A. (2010). Remarks on the venice statement on the right to enjoy the benefits of scientific progress and its applications (article 15 (1)(b)ICESCR). *Human Rights Law Review*, *10*(4), 765–784.

National Research Council (U.S.). (2015). *Climate intervention: Reflecting sunlight to cool earth*. Washington, DC: The National Academy Press.

Nolt, J. (2011). Greenhouse gas emission and the domination of posterity. In D. G. Arnold (Ed.), *The ethics of global climate change* (pp. 60–76). Cambridge: Cambridge University Press.

Oldham, P., Szerszynski, B., Stilgoe, J., Brown, C., Eacott, B., & Yuille, A. (2014). Mapping the landscape of climate engineering. *Philosophical Transactions of the Royal Society A: Mathematical, Physical and Engineering Sciences*, *372*(2031), 20140065.

Oschlies, A., & Klepper, G. (2017). Research for assessment, not deployment, of Climate Engineering: The german research foundation's priority program spp 1689: Research for assessment, not deployment. *Earth's Future*, *5*(1), 128–134.

Ott, K. K. (2018). On the political economy of solar radiation management. *Frontiers in Environmental Science*, *6*, 43.

Parker, A., Horton, J. B., & Keith, D. W. (2018). Stopping Solar Geoengineering through technical means: A preliminary assessment of counter-Geoengineering. *Earth's Future*, *6*(8), 1058–1065.

Pettit, P. (2005). The domination complaint. *Nomos*, *46*, 87–117.

Pettit, P. (2012). *On the people's terms: A republican theory and model of democracy*. Cambridge: Cambridge University Press.

Quaas, M. F., Quaas, J., Rickels, W., & Boucher, O. (2017). Are there reasons against open-ended research into solar radiation management? A model of intergenerational decision-making under uncertainty. *Journal of Environmental Economics and Management*, *84*, 1–17.

Rahman, A. A., Artaxo, P., Asrat, A., & Parker, A. (2018). Developing countries must lead on solar geoengineering research. *Nature*, *556*(7699), 22.

Raimi, K. T., Maki, A., Dana, D., & Vandenbergh, M. P. (2019). Framing of geoengineering affects support for climate change mitigation. *Environmental Communication*, *13*(3), 300–319.

Rayner, S., Heyward, C., Kruger, T., Pidgeon, N., Redgwell, C., & Savulescu, J. (2013). The Oxford principles. *Climatic Change, 121*(3), 499–512.

Reynolds, J. (2015). A critical examination of the climate engineering moral hazard and risk compensation concern. *The Anthropocene Review, 2*(2), 174–191.

Reynolds, J. L. (2019). *The governance of solar geoengineering: Managing climate change in the Anthropocene*. Cambridge, UK: Cambridge University Press.

Rigstad, M. (2011). Republicanism and geopolitical domination. *Journal of Political Power, 4*(2), 279–300.

Rosenau, J. N. (1995). Governance in the twenty-first century. *Global Governance, 1*, 13–44.

Schäfer, S., Lawrence, M., Stelzer, H., Born, W., Low, S., Aaheim, A., Vaughan, N. (2015). *The European Transdisciplinary Assessment of Climate Engineering (EuTRACE): Removing greenhouse gases from the atmosphere and reflecting sunlight away from earth.*

Shaver, L. (2015). The right to science: Ensuring that everyone benefits from scientific and technological progress. *European Journal of Human Rights, 2015*(4), 411–430.

Smith, P. T. (2012). Domination and the ethics of solar radiation management. In C. J. Preston (Ed.), *Engineering the climate: The ethics of solar radiation management* (pp. 43–61). Lanham, MD: Lexington Books.

Smith, P. T. (2018). Legitimacy and non-domination in solar radiation management research. *Ethics, Policy & Environment, 21*(3), 341–361.

Talberg, A., Christoff, P., Thomas, S., & Karoly, D. (2018). Geoengineering governance-by-default: An earth system governance perspective. *International Environmental Agreements: Politics, Law and Economics, 18*(2), 229–253.

Tilmes, S., Richter, J. H., Kravitz, B., MacMartin, D. G., Mills, M. J., Simpson, I. R., Ghosh, S. (2018). CESM1(WACCM) stratospheric aerosol geoengineering large ensemble project. *Bulletin of the American Meteorological Society, 99*(11), 2361–2371.

Weitzman, M. L. (2015). A voting architecture for the governance of free-driver externalities, with application to geoengineering. *The Scandinavian Journal of Economics, 117*(4), 1049–1068.

Whyte, K. P. (2012). Now this! Indigenous sovereignty, political obliviousness and governance models for SRM research. *Ethics, Policy & Environment, 15*(2), 172–187.

Wolff, J. (in press). Fighting risk with risk: Solar radiation management, regulatory drift, and minimal justice. *Critical Review of International Social and Political Philosophy*.

Wyndham, J. M., & Vitullo, M. W. (2018). Define the human right to science. *Science, 362*(6418), 975.

OPEN ACCESS

Geoengineering the climate and ethical challenges: what we can learn from moral emotions and art

Sabine Roeser, Behnam Taebi and Neelke Doorn

ABSTRACT

Climate change is an urgent problem, requiring ways and approaches to address it. Possible solutions are mitigation, adaptation and deployment of geoengineering. In this article we argue that geoengineering gives rise to ethical challenges of its own. Reflecting on these ethical challenges requires approaches that go beyond conventional, quantitative methods of risk assessment. Quantitative methods leave out important ethical considerations such as justice, fairness, autonomy and legitimacy. We argue that emotions and art can play an important role in ethical deliberation about geoengineering. Emotions can point out what morally matters. We also examine the role that works of art can play. Recently, artists have become involved with risky technologies. We argue that such artworks can contribute to emotional-moral reflection and public deliberation on geoengineering, by making abstract problems more concrete, letting us broaden narrow personal perspectives, exploring new scenarios, and challenging our imagination.

Introduction

Climate change – if not averted adequately and in time – could cause serious disruptions in society including issues associated with global warming and sea-level rise. It has been argued that geoengineering could potentially help alleviate at least some of these disruptions (Keith, Parson, & Morgan, 2010; Tuana, 2019). Geoengineering is the 'deliberate large-scale manipulation of the planetary environment to counteract anthropogenic climate change' (The Royal Society, 2009, p. 1). One can distinguish two categories of climate alterations caused by absorbing CO_2 out of the air – also referred to as Carbon Dioxide Removal (CDR) – and climate alterations caused by partially reflecting sunlight back into space – also called Solar Radiation Management (SRM) (The Royal Society, 2009). While it was already known in the previous century that large volcanic eruptions could have an impact on the regional and global climate, it was only after the seminal paper of the Nobel laureate Paul Crutzen (2006) that the idea of SRM started to receive increasing attention in the policy world (Blackstock & Low, 2019a). A specific form of SRM concerns the idea of injecting aerosols, i.e. tiny particles, often sulphate, into the stratosphere in order to partially reflect sunlight; this is also referred to as Stratospheric Aerosol Injection (SAI). SAI is considered to be an effective and affordable method that could, in principle, be deployed within years. However, critics have pointed out that SRM and, more specifically, SAI could also cause droughts, ozone depletion and impact on agriculture, leading to potentially profound societal disruptions.

Hence, geoengineering could also be disruptive of its own, due to its potentially large-scale risks. This means that decisions about geoengineering involve a trade-off between different risks (Huttunen, Skytén, & Hildén, 2015; Linnér & Wibeck, 2015). One major challenge for such a trade-off is that geoengineering risks could be temporally dispersed in unequal ways. In other words, the risks imposed on future people by any decision in the present to pursue geoengineering could be greater than the risks imposed on the people making this decision. Another major challenge relates to the geographical distribution of risks in any time slice. For example, geoengineering could be used to promote the interests of the world's most advantaged and powerful people to the detriment of the global poor.

These and other ethical challenges raised by geoengineering must be addressed. In this paper, we will provide a new perspective concerning ethical deliberation about geoengineering, by focusing on emotions and art. We will argue that works of art can help people to reflect on, partially unfamiliar, ethical questions that might be raised by geoengineering deployment in the future. This is because art can prompt moral emotions, and as we argue, these are key to moral reflection. We will propose emotions and art as hitherto overlooked but potentially helpful ingredients for ethical deliberation. A commonly heard objection to involving emotions in the public discourse is that it is supposed to make the discourse vulnerable to populistic thinking. We will discuss a framework of moral emotions and art that avoids this concern, and we will argue that emotions and art can actually help us to critically reflect on the desirability of potentially disruptive technologies that involve ethical challenges, such as SAI. We propose that art that engages with new technologies can enable us to reflect on the social and ethical implications of these technologies, also involving (moral) emotions. We argue that this offers a promising avenue for thinking about the ethical challenges posed by geoengineering

We will proceed as follows. In section 2 we will discuss how the deployment of geoengineering could give rise to unanticipated risks. That is, partially unanticipated negative effects of geoengineering could take place, and negative effects could happen at a yet undefined moment at a yet undefined place in the future, as a result of which ethical problems could emerge. In Section 3 we will then explore in more detail how art and emotions can contribute to such ethical deliberation.

Ethical challenges of geoengineering

Decision making on SAI and other types of geoengineering also involves a decision on the acceptability of risks. Conventional approaches to making such decisions are based on a quantitative notion of risk, most often expressed in terms of the likelihood of unwanted consequences of a technology or activity and their severity (Hansson, 2009). Risk assessors then often use expected utility and risk cost benefit analysis in order to assess and compare risks. While the risk management literature shows an increasing attention to the ethical aspects of risk management (Doorn, 2015), it is still based on this quantitative notion of risk, which does not cover the full range of ethical aspects of risk (Shrader-Frechette, 1991). First of all, the quantitative approach already makes normative assumptions as to what counts as an unwanted effect. Mostly this is in terms of annual fatalities and economic damage. However, arguably one should also consider severe illness, effects for the environment, and other impacts on people's well-being, such as privacy and ways of life, and the distribution of such effects (Asveld & Roeser, 2009; Roeser et al. 2012). Furthermore, while quantitative notions of risk are able to distinguish between different levels of severity, they cannot differentiate between consequences of fundamentally different nature, for example between consequences that are reversible and those that are irreversible (Doorn, 2018). Especially in the context of climate change and the implications of SRM and SAI, the to all intents and purposes irreversible impacts may be a weighty consideration. SAI therefore prompts a need for methods or approaches for risk assessment that go beyond mere quantitative assessment.

Indeed, there is literature reflecting on the ethical desirability of geoengineering, discussing possible positive and negative implications of SRM and, specifically SAI (Burns & Strauss, 2013; Gardiner, 2010; Keith et al., 2010; The Royal Society, 2009; Tuana, 2019). Some scholars argue that SRM's potential is simply too large to be neglected. Baard and Wikman-Svahn (2016) have considered potential obligations for developed nations to provide for SRM options, in case that their other obligations to mitigate or adapt to climate change fail. Horton and Keith (2016, p. 80) are resolute in their conclusion: that is, since climate change risks will disproportionately affect the poor – in developing countries but also in industrialized countries – we have a consequentialist 'moral obligation to conduct research on solar engineering' because its benefits are for all. This argument has been criticized by Hourdequin (2018, p. 270) for disregarding the fundamental question of justice that needs to first be answered before we can recommend its application for the benefit of some people, especially concerning the world's poor.[1] Other scholars argue that the potential risks of SRM are too large to justify its application. Robock (2008), for instance, lists '20 reasons why geoengineering might be a bad idea', mostly focusing on risks such as regional droughts, ozone depletion and impact on agriculture due to fewer sunlight. Kortetmäki and Oksanen (2016) discuss the food production risks that would result from these effects. Gardiner (2010) argues that geoengineering could create political inertia in achieving the actual solutions to climate change, namely mitigation and reduction of emission gases. Because of the risks, the deployment of geoengineering has often been considered by many as a last resort, only to be used if 'the political will needed to effectively mitigate climate change might not emerge in time to avoid serious, potentially catastrophic damage to future populations around the world' (Blackstock & Low, 2019a, p. 2). In other words, mitigation (i.e. reducing emission gases) is the gold standard, but if we do not act in time, we could (soon) reach a tipping point with the accumulated emission gases in the atmosphere after which mitigation efforts – along with adaptation – may no longer suffice to limit the consequences of climate change. As the argument goes, beyond this point mitigation would not help any more to avoid the further exacerbation of climate change. We will then need – in addition to adaptation – 'technological fixes' such as SAI. Hence, while capable of avoiding potential disruptions caused by climate change, SAI and other forms of geoengineering could cause certain societal disruptions due to their potentially large-scale risks that will be dispersed both spatially and temporally. This means that decisions about geoengineering involve a trade-off with risks of insufficient mitigation (Blackstock & Low, 2019b, p. 41; Linnér and Wibeck (2015); Huttunen et al., 2015). What morally exacerbates this problem is what Gardiner (2011, p. 143) calls 'the tyranny of the contemporary'; that is, when a 'fix' helps to resolve the worst impacts in the short or medium-term while it worsens those impacts in the longer term, all of which could enable the application of 'parochial geoengineering' that only provides benefits to the present generations (and immediately following ones) while disregarding the interest of long-term future generations (Gardiner, 2013, p. 522). Fragnière and Gardiner (2016, p. 15) therefore argue that SAI should not be considered as 'Plan B'.

There are other important aspects that further complicate addressing the ethical challenges of SAI. We here highlight two. First, aerosols have a limited life-time, while the intended positive impacts may require that aerosols would be injected continuously. More importantly, if the aerosol injection would stop in the future the temperature could again increase to the temperatures before the injections started, or even to higher temperatures. Stopping could therefore exacerbate previous effects. It has been argued that when SAI is applied for realizing temperature change, termination of the deployment could 'produce warming rates up to five times greater than the maximum rates under the business-as-usual CO_2 scenarios' (Irvine, Sriver, & Keller, 2012, p. 97). Can it be justified from a perspective of intergenerational justice to apply SAI, thereby potentially imposing perpetual responsibilities on future generations? How should we deal with these questions throughout the period of deployment of SAI and, more specifically, the obligations that each generation would then impose on the subsequent generations? Preston (2016) argues that for seriously considering geoengineering, there must be a 'cessation requirement' in place from the outset that stipulates how easy it could be stopped after it has been taken into use.[2] It should be noted that some advocates of SAI argue that if we commit to aggressive mitigation while deploying SAI, then deployment could eventually be scaled down and even stopped, without previous impacts kicking in again. There will therefore be no perpetual responsibilities imposed on future generations. However, given that large-scale and aggressive mitigation has proven to be hard to achieve so far, it is doubtful whether this will work in combination with geoengineering. To the contrary, policy makers, industry and society could be even be less committed to aggressive mitigation given the promise of the technological fix provided by the deployment of geoengineering.

Second, the long-term risks of geoengineering are mostly unknown, because there have not yet been large-scale experiments with SAI and other forms of geoengineering. Some scholars argue that the effect of geoengineering experimenting cannot be tested without full-scale application, but that this 'can only be tested by injection into an existing aerosol cloud, which cannot be confined to one location' (Robock, Bunzl, Kravitz, & Stenchikov, 2010, p. 530). As a result of this, the consequences are not only dispersed temporally, but will also be dispersed spatially. In other words, a problem such as drought could happen somewhere, sometime in the future, but it is difficult to anticipate when and where it will happen and how severe it will be. It is commonly accepted that there will be regional disparities when SAI is applied; these disparities are also sometimes proposed to be effectively used to create regional climate change, such as cooling (Irvine, Ridgwell, & Lunt, 2010). Of course, this example assumes that we are familiar with the nature of such risks, which is sometimes not the case with new technological innovations, leading to problems of ignorance and unanticipated risks. What morally exacerbates this problem is that we cannot quickly stop negative potential impacts in the future, because – as mentioned above – stopping the deployment of aerosols could lead to rapid warming, to levels substantially more than before the SAI deployment (Robock, Marquardt, Kravitz, & Stenchikov, 2009).

In this section we discussed how ethical challenges could be relevant for reflecting on the desirability of SAI as an option for geoengineering the climate. It is in this regard important to realize that this reflection on the desirability of SAI and other forms of geoengineering does not require a one-off engagement. Instead, it requires continuous ethical reflection from early stages of development of these options throughout the process of development, implementation as well as (long-term) deployment.

While risk ethics has developed as a relatively new area within moral philosophy to provide tools to deal with the ethical considerations of risks and uncertain outcomes (Hayenhjelm & Wolff, 2012), SAI poses challenges that go beyond the current state of risk ethics. Social scientists and philosophers studying risks have argued that decision-making about risks is not a purely scientific and quantitative issue, rather it requires ethical considerations, such as due attention for distributive issues, autonomy, availability of alternatives (cf. Krimsky & Golding, 1992; Asveld & Roeser, 2009; Hansson, 2013; Doorn, 2015; this is also starting to be acknowledged by risk managers Aven & Renn, 2009). However, these ethical considerations are less helpful in case of technologies, and associated risks, that are unfamiliar to us and at the same time potentially of such a disruptive nature as SAI. Such technologies require methods to deliberate on and make sense of the disruptive nature itself, rather than guidelines that look, for example, primarily at the distributive aspects of the technology or the question whether there are less risky alternatives available. What is needed therefore, are approaches that spark our moral imagination and that help us to deliberate about the risks of these potentially disruptive technologies, before they will be applied but also continuously while they may be introduced into society. We will discuss such approaches in the next section.

Emotions and art as resources for moral reflection

As we have argued in the previous section, decision making about geoengineering requires ethical reflection. It is plausible – although we shall not argue the case here – that only public deliberation about the ethical questions raised by SAI and other forms of geoengineering will be adequate to make progress towards answers (cf. e.g. Roeser et al. 2012; Roeser, 2018a). As well as raising ethical questions, new technologies also often give rise to emotional responses (Slovic, 2010). This is seen by many scholars as a reason why deliberation about new technologies is difficult, as they see emotions as an obstacle to rational deliberation (Loewenstein, Weber, Hsee, & Welch, 2001; Sunstein, 2005). Furthermore, as we have seen in the previous sections, given that the scientific information about risky technologies involves uncertainty, ethical considerations concerning risky technologies can themselves be uncertain. This is especially the case with new technologies such as geoengineering and SAI that may involve unexpected technological developments and their effects on nature and society. We will argue that art and emotions can play an important role in ethical deliberation about such hard to predict developments.

Emotions and values are typically considered as matters on which people differ and which lead to problems and conflicts. However, as indicated in the previous sections, values are inherent to decision making about risky technologies such as geoengineering. We will argue that emotions can serve as important indicators of what people value, and emotional reflection and deliberation can be facilitated by works of art. We will discuss this step by step in what follows.

Values

There are many core values which people can find important, for example sustainability, wellbeing, justice and autonomy. People may disagree as to how to prioritize these different values (Perlaviciute & Steg, 2015). However, in polarized debates, this may be portrayed or perceived as a fundamental clash of values, where one group only seems to care about one value and another group only about another value (e.g. Dignum et al., 2016). This is frequently also mirrored in the media, leading to further polarization and people withdrawing into their own virtual or real-life 'bubbles', which serves as a centrifugal force, taking people further apart. However, public deliberation could also be construed in a very different way: by trying to take as the basis for deliberation the values that people agree about. People could be encouraged to first find common ground instead of emphasizing differences of viewpoints from the outset. Starting from this common ground can provide for better understanding of where people's viewpoints diverge, and for which reasons. This relates to the role of the imagination and of empathy, leading us to our next point.

Emotions

In the academic literature on risk, emotions are frequently portrayed as opposed to rationality and as a threat to decision-making (Sunstein, 2005, 2010, cf. Kahneman, 2011 on Dual Process Theory). However, emotion research emphasizes that emotions can have cognitive aspects, which means that they can be of vital importance for practical, ethical and political decision making (e.g. Frijda, 1986; Hall, 2005; Kingston, 2011; Nussbaum, 2001, 2013). Moral emotions can be an important source of moral wisdom (Roeser, 2011).[3] Research by the neuropsychologist Antonio Damasio (1994) on amygdala patients as well as research on sociopaths (cf. Nichols, 2004 for a review) shows that without emotions, we would not be able to make particular moral judgments and behave socially. In other words, emotions do their work unnoticed, all the time. They serve as the social glue that lets us communicate, understand and relate with each other. Emotions can be a gateway to a better understanding of each other's perspectives, and particularly, how we value things and why. For example, by listening to the story of people who are upset about the negative side-effects of a technology on their life, others can sympathize with their experiences and understand why they matter, which can lead to different ideas on how a technology could be developed or implemented. Emotions could then be seen as a starting point for deliberation about moral values in decision making about risky technologies (Roeser, 2018a; Roeser & Pesch, 2016).

This can also provide for an important perspective in the context of climate change and the possible role of SAI and other forms of geoengineering. Moral emotions can help us to fully grasp the moral implications of climate change for people who are geographically or temporarily far away, which can in turn provide us with motivation to change our behavior and make personal sacrifices in order to mitigate climate change, for example, by changing our lifestyle (Roeser, 2012). Effects of climate change are continuous or chronic, which can make it easy to ignore them, and people may lack motivation to change their behavior (McKinnon, 2011). Climate engineering can play a role in mitigating and adapting to climate change, but it gives rise to additional ethical questions as it introduces potential burdens for society. However, mitigating as well as adapting to climate change requires more awareness of the problems than is currently the case, and a willingness to make personal sacrifices by, for example, changing one's lifestyle. Of course there are also powerful political forces at stake, but next to that, mitigating climate change requires a moral appeal to individual human beings to reflect on and adapt their behavior. Emotions can further help us deliberate about normative aspects of geoengineering. Because the normative aspects of such new technologies are partially uncertain, we cannot fall back on predefined moral norms; rather, we have to engage in ongoing reflection, also involving introspection into our own values and caring about implications for other people. Emotions can help us in this reflection, as they serve as signals as to what we and others value. Making emotions explicit can bring latent concerns to the fore and encourage people to investigate ethical implications that are not yet clearly developed.

However, explicating such latent concerns and undefined values can be challenging. Furthermore, emotions can also be biased and misleading, for example by being grounded in self-interest. Moral emotions can play a role in overcoming such biases. For example, shame, guilt and feelings of responsibility can let us critically asses our initial emotions and broaden our outlook to also include the perspective of others (Roeser, 2010). However, this can be difficult, as moral emotions can themselves be misleading, due to being grounded in stereotypes, triggered by irrelevant influences, and also because people's emotions and moral views are deeply ingrained in their personality as well as in their culture and surroundings (Greene, 2013; Haidt, 2012; Kahan, 2012). This means that we need approaches that further facilitate emotional-moral reflection. Art might provide for such a perspective, by creating space to explore and reflect on the moral ambiguities, paradoxes and complex moral questions involved in technological developments such as SAI and other forms of geoengineering.

Art

Various philosophers have developed accounts concerning the importance of how art can contribute to moral and political reflection, also involving emotions (e.g. Carroll 2001; Nussbaum, 2001; Gaut, 2007; Kingston, 2011, p. 209; Kompridis, 2014). Art typically engages our imagination and reflection and gives rise to emotional responses, all of which can help to reflect on and understand different perspectives and scenarios. Presumably, this could also be the case concerning art that engages with new technological developments (e.g. Roeser, Alfano, & Nevejan, 2018; Roeser & Steinert, 2019). Indeed, we owe paradigmatic points of reference in moral reflection on technologies to artists and writers who developed visions on technological developments long before they were a reality. Think of novels such as *Frankenstein, Brave New World* and *1984*. In foreshadowing possible developments, negative as well as positive, works of art and literature can serve as a guide on where to go, as well as a warning sign on where not to go, or which implications to prevent, by developing more responsible technologies. In this way technology-focused art can contribute to ethical reflection on technological developments, also and specifically when these are hard to predict, by exploring possible scenarios in a more tangible way.

Over the last decades, more and more artists and writers have developed works that engage with technological developments; this is what we would like to call 'techno-art' (cf. Reichle, 2009; Wilson, 2010; Myers, 2015 for extensive overviews). Bioartists experiment with and reflect on biotechnology. For example, Adam Zaretsky plays with the possibilities of genetic modification, by creating zebrafish with two heads, thereby challenging legal and ethical boundaries. The Culture and Art project has created a 'victimless leather' from tissue engineering. Anna Dimitriu makes artworks from bacteria. There are other artists who experiment with AI, robotics, and nuclear energy, to mention just a few controversial areas of technology with which artists engage. These artworks focus on risks and potential benefits for society, also involving emotional responses of the audience. Works of techno-art can shed important light on complex ethical questions related to technological innovations. This different focus of techno-art means that current philosophical theories on the relation between art and morality do not suffice in studying these kinds of artistic developments and their relevance for emotional-moral reflection. This is largely uncharted territory that has so far not been explored by many philosophers (for some exceptions see Zwijnenberg, 2014; Roeser et al., 2018; Roeser & Steinert, 2019).

Technological risks give rise to ethical challenges that require a reexamination of conventional ethical theories as these are not adequately equipped to deal with risk and uncertainty, by typically assuming full knowledge of consequences (Hansson, 2012, also see Gardiner, 2011). Similarly, in order to understand the role of technology-engaged art in public debates about risks, this requires new aesthetic theories. Existing philosophical approaches to the relationship between art and morality do not focus on artworks that engage with science and technology. There are empirical studies on the contribution of images and narratives on emotions, awareness and behavior change related to climate change (Leiserowitz, 2006; Spence & Pidgeon, 2010), and foresight scenarios to explore the impacts of SRM (Low, 2017). Works of visual art and literature could play a crucial role in such contexts (Mehnert, 2016; Mobley, Vagias, & DeWard, 2010). We distinguish visual techno-art from literary techno-art, i.e. visual art works versus works of literature which engage with science and technology. Visual techno-artists often do not use traditional materials and techniques such as painting, photography and sculpture. Rather, they use scientific and technological techniques, such as biotechnology, robotics or new media, to develop artworks. This is less the case with literary techno-art, but in both cases, the artist or author engages with scientific or technological developments. Furthermore, these artists and authors engage with different topics than other artists, frequently concerning the implications of a technological development for society, and these also inspire different emotional responses.

This relates to a currently hotly debated topic in epistemology and cognitive science concerning the role of external features to aid our thinking, cognition and knowledge (cf. Clark & Chalmers, 1998 on the extended mind; Giere, 2002; Palermos & Pritchard, 2013 on socially extended knowledge). However, not only practical devices such as maps and notebooks can play the role of extended cognition and knowledge, but artworks can do so too (cf. Krueger & Szanto, 2016 on the role of music and emotions for extended knowledge). Techno-art can also be seen as a form of 'socially extended knowledge' (Roeser, 2018b). This idea needs further elaboration and can draw on as well as contribute to the debate on extended cognition and knowledge. For example, artists often think out of the box and can help us take our imagination further than the more strictly regimented steps in which scientific researchers and engineers tend to proceed. Furthermore, they provide for much more concrete images and narratives than the abstract argumentation of philosophers, thereby appealing to people's imagination, sympathy and understanding, which can provide for different and complementary ethical insights than purely cognitive and analytical reasoning.

For example, techno-art can present society with visions that give rise to emotional engagement with technology, emphasizing positive prospects as well as risks and ambiguities. Techno-art can explore the boundaries of emotionally laden moral notions such as dignity, suspicion, and trust. Techno-art can explore moral dimensions of technologies in a very visible or tangible way that can lead to a more direct experience and more concrete, context-specific ethical insight than abstract reasoning. In this way, techno-art can make a constructive contribution to the public debate as well as to the academic ethical debate on technological risks, by providing additional insights and perspectives that might get overlooked in a purely theoretical academic or public debate. In that way, it can make ethical deliberation more accessible for a broader range of stakeholders. In what follows we will discuss examples of techno-art in the context of climate change and climate engineering, and how these can contribute to emotional-moral reflection.

There are novelists who write about climate change, such as Cormac McCarthy (*The Road*) and Lauren Groff (*Florida*). They provide for powerful, dystopian narratives that show the ultimate implications of our choices. Such narratives appeal to people's imagination and sympathy, which can provide for additional motivation to adapt one's behavior. Emotions that are inspired by such artworks can provide for more powerful ethical insights than abstract reasoning, as well as for more motivational force (Roeser, 2012). Recently, leading novelist Amitav Ghosh (2016) has argued that more writers should engage with climate change as it is one of the most pressing problems of our times, and writers can uniquely contribute to bringing these largely abstract and long-term developments closer to people's awareness by creating narratives that appeal to our imagination.

Next to climate novelists, there are climate artists working with visual art forms and installations, such as David Buckland and Boo Chapple. Boo Chapple has created an interactive project that plays with the suggestion from geoengineers to shield the earth under a white layer to reflect sunlight away from the earth as a way to combat climate change. This is an idea that resembles SAI, but it would be even more invasive. Chapple asked people to wear reflecting white hats and to deliberate on the impact of such technologies and whether they are desirable. In this way she appeals to people's imagination and reflective emotional capacities, inspiring ethical deliberation that is fueled by concrete experiences.

In 2018, there was a widely discussed exhibit at the Stedelijk Museum Amsterdam called 'Coded Nature' by the artistic duo Studio Drift. Their works reflect on our relation with technology and nature. At this exhibit, a film with the name 'Drifters' was shown. In this poetic film, concrete blocks rise up from a lake in a hilly landscape and rise seemingly weightless, becoming more and more like a flock of birds and finally collapsing into a monolithic whole. The concrete blocks are paradoxical: feathery and at the same time heavy, coming from nature but also strange and ultimately dominating. In a very subtle way, this film touches on our emotions and therefore allows us to reflect on our relationship with nature on the one hand and technology on the other. Like the concrete blocks in the film, we originate from nature, but we also change it. We are natural beings on the one hand and cultural beings on the other, and without people there would be no technology and no concrete blocks that are part of our current landscape but also threaten this. The blocks of concrete can be seen to symbolically relate to geoengineering: they protect nature while also dominating it, and once this technology is in place there may be no way back. We will then irrevocably be locked[4] in such a system, just like the blocks of concrete that eventually cover the sky in the video 'Drifters'.

These examples illustrate that artworks can make a powerful contribution to ethical deliberation throughout the process of development, implementation and (long-term) application of geoengineering such as SAI. Art can help to make climate change more salient and probe people to take actions, and to let people critically reflect on the possible role of for example geoengineering. The vast challenges posed by climate change require our best possible efforts to reflect on ethical implications on technological developments that are hard to foresee at this moment. Artists can help in deliberation, by providing works that can spur critical reflection on which values may be furthered or threatened, by triggering our imagination and moral emotions. This can provide for an important new resource for existing approaches to participatory technology assessment (cf. Van Asselt & Rijkens-Klomp, 2002 for an overview of such approaches). In such approaches, scenarios are sometimes developed by policy makers, communication experts or social scientists and used for reflection (e.g. Boenink, Swierstra, & Stemerding, 2010). However, arguably, artists and writers can provide for more challenging and intriguing scenarios and images, as they are experts in creating images and narratives that profoundly challenge our imagination and trigger reflection. In conventional technology assessment, the focus is on scientific information and to the extent that it includes ethical reflection, that is based on rational argumentation. Focusing on art, values and emotions can provide for much more profound reflection, understanding and insight (Roeser & Pesch, 2016). This can make ethical challenges explicit, which would be more difficult in abstract, rational reflection. Artworks and narratives can make scenarios and unclear and ambiguous normative implications of geoengineering more tangible and easier to imagine, thereby stimulating critical reflection, based on imagination, compassion, sympathy, introspection and understanding.

Of course it is important to note that techno-art is not a foolproof solution for ethical deliberation on geoengineering. Like all other forms of insight and deliberation, it can be biased, mislead or even intentionally used for manipulation. In other words, techno-art is not a 'silver bullet' to ethical deliberation about geoengineering, and in general, there are no 'silver bullets' to such complex issues. However, given the profound challenges we are facing it is important to draw on all resources that we have, and techno-art can provide for such a possible additional resource which has until now not been sufficiently recognized. If techno-art will be included in deliberation on geoengineering, it will be important to build in checks and balances, for example, involving different artists who provide complementary perspectives (cf.

Roeser & Steinert, 2019 for further discussion of this). In this way, techno-art can broaden people's horizons and challenge their imagination, thereby contributing to critical ethical reflection.

Conclusion

Climate change could cause disruptive effects for society, requiring new, innovative strategies on how to combat or adapt to climate change. Geoengineering, for example in the form of SAI is a *technological* strategy to address some of the challenges of climate change, but it would also create environmental and societal disruptions of its own, due to its potentially large-scale risks. This means that decisions about geoengineering require a trade-off between different types of risk. What makes such a trade-off problematic is that these risks are spatially and temporally dispersed. All these issues pose not only technological challenges but also ethical challenges, requiring explicit ethical deliberation. Also, unanticipated technological risks could in the future give rise to unforeseen ethical challenges and, by that, make earlier applied moral norms for assessing the desirability of geoengineering less relevant. Similarly, our understanding of moral norms and what we consider 'good' in society may change over time.

We need, therefore, a continuous ethical 'monitoring' in which the performance of SAI and other types of geoengineering is continuously assessed from a technological as well as from a normative-ethical perspective, and based on this, adjusted and adapted in an iterative process. This also requires new deliberative strategies. We have argued that emotions and art can play crucial roles in this, as they can be important gateways to values and critical reflection. 'Techno-art', especially art that engages with climate change and geoengineering can trigger our reflection and imagination concerning future scenarios, bringing these closer to home and thereby bridging the problematic gap between our current actions and their remote, yet profound impacts.

Techno-art can make a powerful contribution to important debates facing contemporary society, by providing for a new, not yet explored avenue of public deliberation and emotional-moral reflection about technological risks. Organizations that can use such an approach are (inter)national governments, policy advisory boards, technological research organizations and NGOs representing citizens' interests, in order to facilitate public dialogue. Researchers who develop new technologies can be inspired by techno-art to derive insights into emotional-moral considerations that can contribute to more responsible innovations.

This approach is hitherto largely unexplored, while it could contribute to making progress in decision making concerning one of the most complicated challenges that have ever faced humanity. If we do not act now, it could be too late; yet, imprudent and hasty action to promote technologies such as SAI and other types of geoengineering could lock us in situations where there is no way back. Art, emotions and values can help us to reflect on complexity and uncertainty, providing us with wisdom in the light of the ethical challenges presented by climate change and geoengineering.

Notes

1. See also (Hourdequin, 2016).
2. Preston argues that between the two types of geoengineering, CDR will probably be the easiest to stop, while SRM (including SAI) will be much more difficult to stop. Also see McKinnon (2019) on the ethical problems with the risk of getting 'locked in' SRM.
3. With the notion 'moral emotions', we refer to tokens of emotions that can be relevant for moral insight and reflection. Hence, next to paradigmatic moral emotions such as guilt and shame, also fear can be a moral emotion when it draws attention to morally relevant issues (cf. Roeser, 2011).
4. Cf. McKinnon (2019) on lock-in of SRM.

Acknowledgments

We would like to thank Catriona McKinnon, Steven Gardiner and an anonymous reviewer for extremely helpful feedback on earlier versions of this paper.

Disclosure statement

No potential conflict of interest was reported by the authors.

ORCID

Neelke Doorn http://orcid.org/0000-0002-1090-579X

References

Asveld, L., & Roeser, S. (Eds.). (2009). *The ethics of technological risk*. London: Earthscan.

Aven, T., & Renn, O. (2009). On risk defined as an event where the outcome is uncertain. *Journal of Risk Research*, *12*(1), 1–11.

Baard, P., & Wikman-Svahn, P. (2016). Do we have a residual obligation to engineer the climate as a matter of justice? In C. J. Preston (Ed.), *Climate justice and geoengineering* (pp. 49–62). London: Rowman & Littlefield.

Blackstock, J. J., & Low, S. (2019a). Geoengineering our climate: An emerging discourse. In J. J. Blackstock & S. Low (Eds.), *Geoengineering our climate? Ethics, politics and governance* (pp. 1–10). London: Routledge.

Blackstock, J. J., & Low, S. (Eds.). (2019b). *Geoengineering our climate? Ethics, politics and governance*. London: Routledge.

Boenink, M., Swierstra, T., & Stemerding, D. (2010). Anticipating the interaction between technology and morality: A scenario study of experimenting with humans in biotechnology. *Studies in ethics, law, and technology*, 4. Available at: http://www.bepress.com/selt/vol4/iss2/art4.

Burns, W. G. G., & Strauss, A. L. (Eds.). (2013). *Climate change geoengineering: Philosophical perspectives, legal issues, and governance frameworks*. New York: Cambridge University Press.

Carroll, N. (2001). *Beyond aesthetics: Philosophical essays*. Cambridge: Cambridge University Press.

Clark, A., & Chalmers, D. (1998). The extended mind. *Analysis*, *58*, 10–23.

Crutzen, P. J. (2006). Albedo enhancement by stratospheric sulfur injections: A contribution to resolve a policy dilemma? *Climatic Change*, *77*(3–4), 211.

Damasio, A. R. (1994). *Descartes' error: Emotion, reason and the human brain*. New York: G.P. Putnam.

Dignum, M, Correljé, A, Cuppen, E, Pesch, U, & Taebi, B. (2016). Contested technologies and design for values: the case of shale gas. *Science and Engineering Ethics*, *22*(4), 1171–1191. doi:10.1007/s11948-015-9685-6

Doorn, N. (2015). The blind spot in risk ethics: Managing natural hazards. *Risk Analysis, 35*(3), 354–360.

Doorn, N. (2018). Distributing risks: Allocation principles for distributing reversible and irreversible losses. *Ethics, Policy & Environment, 21*(1), 96–109.

Fragnière, A., & Gardiner, S. M. (2016). Why geoengineering is not 'Plan B'. In C. J. Preston (Ed.), *Climate justice and geoengineering* (pp. 15–32). London: Rowman & Littlefield.

Frijda, N. H. (1986). *The emotions*. Cambridge: Cambridge University Press.

Gardiner, S. (2011). *A perfect moral storm: The ethical tragedy of climate change*. New York: Oxford University Press.

Gardiner, S. M. (2010). Is 'arming the future' with geoengineering really the lesser evil? Some doubts about the ethics of intentionally manipulating the climate system. In S. M. Gardiner, S. Caney, D. Jamieson, & H. Shue (Eds.), *Climate ethics: Essential readings* (pp. 284–312). New York: Oxford University Press.

Gardiner, S. M. (2013). Why geoengineering is not a 'global public good', and why it is ethically misleading to frame it as one. *Climatic Change, 121*(3), 513–525.

Gaut, B. (2007). *Art, emotion and ethics*. Oxford: Oxford University Press.

Ghosh, A. (2016). *The great derangement: Climate change and the unthinkable*. Chicago: University of Chicago Press.

Giere, R. (2002). Scientific cognition as distributed cognition. In P. Carruthers, S. Stitch, & M. Siegal (Eds.), *Cognitive bases of science (pp. 285–299)*. Cambridge: Cambridge University Press.

Greene, J. (2013). *Moral tribes*. New York: Penguin.

Haidt, J. (2012). *The righteous mind: Why good people are divided by politics and religion*. New York: Vintage Books.

Hall, C. (2005). *The trouble with passion: Political theory beyond the reign of reason*. New York: Routledge.

Hansson, S. O. (2009). An agenda for the ethics of risk. In L. Asveld & S. Roeser (Eds.), *The ethics of technological risk*(pp.11–23). London: Earthscan.

Hansson, S. O. (2013). The ethics of risk. *Ethical analysis in an uncertain world*. Basinstoke: Palgrave Macmillan.

Hansson, S.O. (2012). A panorama of the philosophy of risk. In RoeserS., HillerbrandR., PetersonM. & SandinP. (Eds), *handbook of risk theory* (pp. pp. 27–54). Dordrecht: Springer.

Hayenhjelm, M., & Wolff, J. (2012). The moral problem of risk impositions: A survey of the literature. *European Journal of Philosophy, 20*(1), E26–E51.

Horton, J., & Keith, D. (2016). Solar geoengineering and obligations to the global poor. In C. J. Preston (Ed.), *Climate Justice and geoengineering* (pp. 79–92). London: Rowman & Littlefield.

Hourdequin, M. (2016). Justice, recognition and climate change. In C. J. Preston (Ed.), *Climate justice and geoengineering* (pp. 33–48). London: Rowman & Littlefield.

Hourdequin, M. (2018). Climate change, climate engineering, and the 'Global poor': What does justice require? *Ethics, Policy & Environment, 21*(3), 270–288.

Huttunen, S., Skytén, E., & Hildén, M. (2015). Emerging policy perspectives on geoengineering: An international comparison. *The Anthropocene Review, 2*(1), 14–32.

Irvine, P. J., Ridgwell, A., & Lunt, D. J. (2010). Assessing the regional disparities in geoengineering impacts. *Geophysical Research Letters, 37*, 18.

Irvine, P. J., Sriver, R. L., & Keller, K. (2012). Tension between reducing sea-level rise and global warming through solar-radiation management. *Nature Climate Change, 2*(2), 97–100.

Kahan, D. (2012). Cultural cognition as a conception of the cultural theory of risk. In S. Roeser, R. Hillerbrand, M. Peterson, & P. Sandin (Eds.), *Handbook of risk theory* (pp. 725–759). Dordrecht: Springer.

Kahneman, D. (2011). *Thinking fast and slow*. New York: Farrar, Straus and Giroux.

Keith, D. W., Parson, E., & Morgan, M. G. (2010). Research on global sun block needed now. *Nature, 463*, 426.

Kingston, R. (2011). *Public passion: Rethinking the grounds for political justice*. Montreal: McGill-Queen's University Press.

Kompridis, N. (Eds.). (2014). *The aesthetic turn in political thought*. London: Bloomsbury Academic.

Kortetmäki, T., & Oksanen, M. (2016). Food systems and climate engeering: A plate full of risks. In PrestonC. J. (Ed.), Climate Justice and Geoengineering (pp. 109–120). London: Rowman & Littlefield.

Krimsky, S., & Golding, D. (1992). *Social theories of risk*. Westport: Praeger.

Krueger, J., & Szanto, T. (2016). Extended emotions. *Philosophy Compass, 11*, 863–878.

Leiserowitz, A. (2006). climate change risk perception and policy preferences: The role of affect, Imagery, and values. *Climatic Change, 77*, 45–72.

Linnér, B. O., & Wibeck, V. (2015). Dual high-stake emerging technologies: A review of the climate engineering research literature. *Wiley Interdisciplinary Reviews: Climate Change, 6*(2), 255–268.

Loewenstein, G. F., Weber, E. U., Hsee, C. K., & Welch, N. (2001). Risk as feelings. *Psychological Bulletin, 127*, 267–286.

Low, S. (2017). Engineering imaginaries: Anticipatory foresight for solar radiation management governance. *Science of the Total Environment, 580*, 90–104.

McKinnon, C. (2011). Climate change justice: Getting motivated in the last chance saloon. *Critical Review of International Social and Political Philosophy, 14*, 2.

McKinnon, C. (2019). Sleepwalking into lock-in? Avoiding wrongs to future people in the governance of solar radiation management research. *Environmental Politics, 28* (3), 441–459.

Mehnert, A. (2016). *Climate change fictions*. Basingstoke: Palgrave Macmillan.

Mobley, C., Vagias, W., & DeWard, S. (2010). Exploring additional determinants of environmentally responsible behavior: The influence of environmental literature and environmental attitudes. *Environment and Behavior, 42*, 420–447.

Myers, W. (2015). *Bio art: Altered realities*. London: Thames and Hudson.

Nichols, S. (2004). *Sentimental rules*. Oxford: Oxford University Press.

Nussbaum, M. C. (2001). *Upheavals of thought: The intelligence of emotions*. Cambridge: Cambridge University Press.

Nussbaum, M. C. (2013). *Political emotions: Why love matters for justice*. Cambridge, MA: Harvard University Press.

Palermos, S. O., & Pritchard, D. (2013). Extended knowledge and social epistemology. *Social Epistemology Review and Reply Collective, 2*(8), 105–120.

Perlaviciute, G., & Steg, L. (2015). The influence of values on evaluations of energy alternatives. *Renewable Energy, 77*, 259–267.

Preston, C. J. (2016). Climate engineering and the cessation requirement: The ethics of a life-cycle. *Environmental Values, 25*(1), 91–107.

Reichle, I. (2009). *Art in the age of technoscience: Genetic engineering, robotics, and artificial life in contemporary art*. Vienna: Springer.

Robock, A. (2008). 20 reasons why geoengineering may be a bad idea. *Bulletin of the Atomic Scientists, 64*(2), 14–18.

Robock, A., Bunzl, M., Kravitz, B., & Stenchikov, G. L. (2010). A test for geoengineering? *Science, 327*(5965), 530–531.

Robock, A., Marquardt, A., Kravitz, B., & Stenchikov, G. (2009). Benefits, risks, and costs of stratospheric geoengineering. *Geophysical Research Letters, 36*(19), L19703.

Roeser, S. (2010). Emotional reflection about. In R. S. Risks' (Eds.), *Emotions and risky technologies* (pp. 231–244). Dordrecht: Springer.

Roeser, S. (2011). *Moral emotions and intuitions*. Basingstoke: Palgrave Macmillan.

Roeser, S. (2012). Risk communication, public engagement, and climate change: A role for emotions. *Risk Analysis, 32*, 1033–1040.

Roeser, S. (2018a). *Risk, technology, and moral emotions*. Routledge. doi:10.4324/9781315627809

Roeser, S. (2018b). *Socially extended moral deliberation about risks: a role for emotions and art*. In J. A. Carter, A. Clark, J. Kallestrup, S. Orestis palermos, & D. Pritchard (Eds.), Socially extended epistemology. Oxford: Oxford University Press. doi:10.1093/oso/9780198801764.003.0009

Roeser, S., Alfano, V., & Nevejan, C. (2018). 'The role of art in emotional-moral reflection on risky and controversial technologies: The case of BNCI. *Ethical Theory and Moral Practice, 21*, 275–289.

Roeser, S., & Pesch, U. (2016). An emotional deliberation approach to risk. *Science, Technology and Human Values, 41*, 274–297.

Roeser, S., & Steinert, S. (2019). Passion for responsible technology-development: The philosophical foundations for embedding ethicists and artists in technology-projects. *Philosophy,* 85: 87–109.

Shrader-Frechette, K. S. (1991). *Risk and rationality*. Berkeley, CA: University of California Press.

Slovic, P. (2010). *The feeling of risk*. London: Earthscan.

Spence, A., & Pidgeon, N. F. (2010). Framing and communicating climate change: The effects of distance and outcome frame manipulations. *Global Environmental Change, 20*, 656–667.

Sunstein, C. R. (2005). *Laws of fear*. Cambridge: Cambridge University Press.

Sunstein, C.R. (2010), Moral heuristics and risk, in S. Roeser, (ed.) Emotions and Risky Technologies (pp. 3–16). Dordrecht: Springer

The Royal Society. (2009). *Geoengineering the climate. Science, governance and uncertainty*. London: Author.

Tuana, N. (2019). The ethical dimensions of geoengineering: Solar radiation management through sulphate particle injections. In J. J. Blackstock & S. Low (Eds.), *Geoengineering our climate? Ethics, politics and governance* (pp. 71–85). London: Routledge.

Van Asselt, M., & Rijkens-Klomp, N. (2002). A look in the mirror: Reflection on participation in integrated assessment from a methodological perspective. *Global Environmental Change, 12*, 167–184.

Wilson, S. (2010). *Art + science now: How scientific research and technological innovation are becoming key to 21st-century aesthetics*. London: Thames and Hudson.

Zwijnenberg, R. (2014). Biotechnology, human dignity and the importance of art. *Teoria: Revista di Filosofia*, 34, 131–148.

Index